W9-CDF-490

TRACK CHANGES

TRACK CHANGES

A LITERARY HISTORY OF WORD PROCESSING

MATTHEW G. KIRSCHENBAUM

THE BELKNAP PRESS OF
HARVARD UNIVERSITY PRESS
Cambridge, Massachusetts
London, England
2016

Printed in the United States of America

First printing

Library of Congress Cataloging-in-Publication Data

Names: Kirschenbaum, Matthew G., author.
Title: Track changes : a literary history of word processing / Matthew G. Kirschenbaum.
Description: Cambridge, Massachusetts : The Belknap Press of Harvard University Press, 2016. | Includes bibliographical references and index.
Identifiers: LCCN 2015041450 | ISBN 9780674417076 (alk. paper)
Subjects: LCSH: Word processing—History. | Writing—Technological innovations. | Creation (Literary, artistic, etc.)—Technological innovations.
Classification: LCC Z52.4 .K57 2016 | DDC 005.52—dc23 LC record available at http://lccn.loc.gov/2015041450

For my parents,
Arlene and Mel,
Who brought home an Apple

CONTENTS

"Well, salvage whatever you can,
threadbare mementos glimmering in recollection."

—Ecclesias (a word processor) in Henry Roth,
A Star Shines over Mt. Morris Park (1994)

PREFACE

Track Changes began, as many books do, with a question: What was the first novel written with a word processor? Being an English professor interested in the history of writing as well as computers, I thought it was the sort of thing I should know, but I didn't. It is a commonplace of literary history that Mark Twain's long, meandering memoir *Life on the Mississippi* (1883) was the first piece of *belles lettres* to be submitted to a publisher as a typescript: "I was the first person in the world to apply the type-machine to literature," he informs us in a sketch written some years after the fact and featured in an ad by the typewriter's manufacturer, Remington, otherwise best known for firearms.[1] In fact, the book was typed for him by an assistant. Within a decade, in a gesture that would soon become commonplace, Oscar Wilde entrusted the manuscript of *The Picture of Dorian Gray* to Miss Dickens's Typewriting Service in London.[2] Other equally famous early adopters can also be named: Friedrich Nietzsche, for example, and Henry James, who reportedly was so enamored of the sound of the keys of his secretary's typewriter that they became a kind of metronome—he found he couldn't dictate without their rhythmic counterpoint.[3] (James is typically thought to have begun dictating his prose after he suffered a repetitive strain injury writing by hand.) But when it came to computers and word processing, it appeared there were no widely recognized historical counterparts to Twain and his typist.

In the course of my research I was able to answer my question to my own satisfaction—that is what researchers do after all, and thus my candidates for who was first are duly presented in the pages that follow. But firsts are always relative—both to what came before and to what came after. In my case the answer largely depended on how I chose to define terms like "write" and "word processor." As historian Michael R. Williams explains in his discussion of the competing contenders for the mantle of who built the first computer, "if you add enough adjectives to a description you can always claim your own favorite."[4] Besides the inevitable proliferation of criteria, though, the question of who was first also usually implies a linear, causal view of historical progress, often circumscribed in the telling by a great man's genius and imagination (my gender specificity here is intentional) and the resulting technological marvels overtaking a passively accepting public. Word processing, however, permits us none of these simplifications, as even a casual survey of its history will show. Does word processing belong to the history of writing or the history of computing? The answer is not obvious. The conjoining of the word "word" with the word "processing" or "processor" has at various times been used to denote principles of office management, an actual person (typically female), a physical hardware device, or a piece of intangible software. It has also encompassed not only the written production of text with a keyboard but also verbal dictation, shorthand note taking, and the duplication, mailing, and filing of documents.

There is another kind of first associated with any technology as pervasive as word processing, at least for those of us of a certain age: Each of us remembers our own first time. I am just old enough to have lived through word processing's popular advent, but my own personal experience is one of solely vicarious contrasts. Going into my adolescence I had never used a typewriter for any purposeful task; but I knew that one of the hallmarks of the transition from grade school to middle school was that handwritten reports were no longer acceptable. I vividly remember the family typewriter's periodic appearances on our kitchen table whenever my older brother or sister had a school paper to write. It was a big, heavy, noisy thing. The volume of the television set in the adjoining family room would rise accordingly. I remember my siblings' exertions over the unforgiving apparatus and my own sense of mild dread, knowing my turn was almost at hand. (Such were the pallid anxieties of my coming of age in suburbia.) And then, miraculously it seemed, the home computer arrived, the product *Time* magazine gave its cover to in January 1983 as the "Machine of the Year." My family bought an Apple IIe, and with it a very simple word processing program called the Bank Street Writer. Accompanied by a screeching dot matrix printer, I used the Apple to write my school papers as well as reams of juvenile science fiction and fantasy. I quickly learned to insert and delete, move blocks of text, and find and replace words and phrases. (Features such as spell- and grammar-check and any kind of advanced formatting were beyond the program's capabilities.) The tearing of the spindled edges from the fanfold paper signaled that a job was complete. As for the typewriter, it was unceremoniously relegated to the back of the closet, where it was soon forgotten, and so technological progress and middle-class privilege had conspired to let me sidestep a rite of passage that only a few years earlier had seemed inevitable.

Writing *Track Changes* (mostly in Word, on a couple of small, lightweight laptops) has taken me back to a time in my youth that I recognize only in retrospect as the pivotal moment in the growth and widespread adoption of word processing for literature. Many of the writers I stayed up late reading as a teenager—best-selling authors like Stephen King, Frank Herbert, Anne McCaffrey, and Tom Clancy—were themselves experimenting with the technology, getting their own first computers at more or less the same time we got our Apple. (Clancy got an Apple IIe himself, in fact.) All of this explains one of the attractions of this project to me: a chance to explore some of the connections that tied me, a kid in my upstairs bedroom poking around with the Bank Street Writer or Apple BASIC, to the writers I idolized—who were also, it turns out, wrestling with the very same technology themselves. They were learning the same jargon and terminology that I was, scratching their heads over dot matrix and ink jet, and struggling to initialize a diskette or recover a file they had just inadvertently deleted. It was a leveling moment of sorts, and for me, that realization served to humanize these famous authors.

Of course those are not my only motives. Almost from their inception during and after the Second World War, modern digital computers "processed" letters just about as easily as they did numbers. This was in the nature of the computer as a general-purpose machine for manipulating and storing arbitrarily encoded symbols, as noted by the technology historian Thomas Haigh.[5] Nonetheless (or better, precisely because of this), the story of writing in the digital age is every bit as messy as the ink-stained rags that would have littered the floor of Gutenberg's print shop or the hot molten lead of the linotype machine. This book's title is derived from the popular feature used to record revisions to a document in Microsoft Word; and the fate of digital manuscripts—what an electronic document file can tell us about its own composition history—is a special scholarly and technical interest of mine. But here I am interested in tracking even broader kinds of changes: in particular, recovering how authors thought about their newly acquired computers and their relationship to their writing, and how audiences—the reading public—imagined the same. There was a time when the future of word processing in literary circles seemed like an open question, and we will revisit that time in this book. Today, however, computers and word processors are simply part of literary life. The literary history of word processing is in need of documenting, just as scholars have done for previous literary technologies, from charcoal-stained sticks to the "type-machine."[6]

Which is not to say that the place of word processing in literature is now settled, or that its past is easily divisible from its present. I am completing this book at a moment when, for the first time in over two decades, the word processing market is no longer dominated by a single enterprise-level package. Alternatives to Word, ranging from so-called artisanal software tools like Scrivener or WriteRoom to cloud-based apps and services, now offer more choices in our digital writing environments than at any time since the early 1980s. Screens are both expanding in size (through plasma technology) and contracting (the supercomputers in our pocket we still plaintively call "phones"). Voice recognition, which allows us (like Henry James) to write simply by speaking, is an everyday option. "Text" itself has become a verb.

Track Changes is not a stylistic study. It does not seek to tease out subtleties of how individual authors' writing styles may have been altered following their adoption of word processing. I do not engage in the kind of computational text analysis or text mining we nowadays associate with the digital humanities. Neither do I seek to appraise whether word processing has been "good" or "bad" for literature as such. My agenda is both more ambitious and more modest: I accept that computers and word processing are aspects of the literary, and thus I try to help reconstruct the way in which this came to be and its significance for how we think about the material act of writing. In this I take my general cues from what we term book history, a field glossed by Ezra Greenspan and Jonathan Rose as the "social, cultural, and economic history of authorship, publishing, printing, the book arts, copyright, censorship, bookselling and distribution, libraries, literacy, literary criticism, reading habits, and reader response."[7]

Many researchers and commentators have lately focused with much urgency on questions of reading in particular, and what words on the screen may mean for us—or literally do to us—as readers. There have been popular treatises and academic studies of reading digitally, there have been populist rants and jeremiads, and there have been scientific, humanistic, and industry accounts of the state of reading today.[8] But whereas e-books and e-reading devices did not become a major fixation for readers and publishers until after the turn of the millennium (despite years of promises and false starts), word processing, by contrast, has been a fixture of the literary since the start of the personal computer era. *Track Changes* thus seeks to narrate and describe in material and historical terms how computers, specifically word processing, became integral to literary authorship and literary writing.

The use of the indefinite article in my subtitle is deliberate and essential. This is unequivocally *a* literary history of word processing. By no means does my work accord equal attention to all aspects of Greenspan and Rose's admirably capacious definition above. Most apparently, this book is overwhelmingly biased toward Anglo-American authors writing prose fiction (to a lesser extent poetry) in English. This bias is a function of my limitations as a monolingual researcher. I do not wish to imply that Anglo-American authors were the inevitable early adopters or innovators, and if *Track Changes* spurs research into the literary history of word processing in other languages and writing systems, I will be enormously pleased, even if it shows the parochialism of my narrative to be all the more pronounced.[9]

The technological focus of this book is likewise predetermined. Mainframe line editors or text editors, computer typesetting, desktop publishing, hypertext and interactive fiction, networked writing environments (like email or bulletin boards), and Web publishing technologies are all addressed where appropriate, but they are not the main actors here. My focus is on word processors, by which I mean hardware and software for facilitating the composition, revision, and formatting of free-form prose as part of an individual author's daily workflow. The functional boundaries among the preceding technologies are of course porous and to a great extent arbitrary; some will wish that my attention had been more expansive in this regard. Nonetheless, to the extent a distinction can be maintained, I have sought to write a literary history of word processing, not of electronic writing and publishing at large.

And while I must be accountable to the contours of the particular history I have limned, in many cases the specific choice of focus is arbitrary—arbitrary in the sense that these were the stories that were recoverable to me in the course of my research. But every author who uses a word processor has a story to tell. For the ones I have included here, I have come to them through my own interviews and oral histories, through authors' published interviews and statements, through the findings of other scholars, biographers, critics, and fans, and through sleuthing in primary sources ranging from literary archives and special collections to back issues of *Writer's Digest, Byte* magazine, and *InfoWorld,* as well as Facebook pages, blogs, Twitter feeds, and other content from the online world. I ran Google Books and Lexis-Nexus searches, and I scoured the online archives of the *New York Review of Books* and the *Paris Review.* Still, my scope and methods are imperfect and idiosyncratic, and I have relied on serendipity as well as due diligence. There will be individuals and episodes I have missed.[10]

The story I tell has a particular center of gravity: the years 1964 to 1984. Although this book advances to (indeed, looks beyond) the present day, this twenty-year period is where much of the narrative unfolds. Why this particular span? 1964 was the year IBM released its Magnetic Tape Selectric Typewriter or MT/ST, the product that would, by the end of that same decade, become the core of their word processing product line. And of course 1984 has at least one large significance in the history of personal computing: it was the year the Apple Macintosh was released (it came to the attention of many people for the first time during a memorable sixty-second Super Bowl commercial, replete with Orwellian imagery and directed by Ridley Scott). The Macintosh introduced the public at large to the mouse, to windows, icons, bitmapped graphics, and to the whole point-and-click paradigm of working with a computer. But it had relatively little impact on literary word processing, at least at first—the program that shipped with it, MacWrite, was considered weak (it could only handle documents of up to around eight pages), and it was several years before it would have seemed an attractive choice for most authors. One exception: the novelist Mona Simpson, who got a Mac very soon after they became available. Then working as an intern at the *Paris Review,* she was writing her first novel, *Anywhere but Here* (1986). Her biological brother—with whom she had recently been reunited—was Steve Jobs.[11]

In the context of this book, however, 1984 is also notable for another reason. By then, according to data attributed to the Association of American Publishers, somewhere between 40 percent and 50 percent of all literary authors in America were using word processors.[12] Clearly a watershed had been reached, at least in the United States.[13] "Writers of all descriptions are stampeding to buy word processors," grumped Thomas Pynchon in the *New York Times Book Review* that year, the same year Gore Vidal declared that word processing was "erasing" literature.[14] In fact, the relevant time frame can be even more narrowly delimited, beginning in 1977. Zilog's Z-80 microprocessor, the workhorse of the first generation of home computer systems, had been released just the previous year. Electric Pencil, the first word processor written for a personal computer, had likewise become available at the very end of 1976. In 1977, the first generation of integrated systems (the TRS-80 Model I, Apple II, and Commodore PET) appeared. WordStar debuted in 1979. WordPerfect in 1980. The Osborne 1 and IBM PC in 1981. The Kaypro II in 1982. Microsoft Word in 1983. The pace of technological advance (driven not just by idealistic notions of "progress" but by rapidly coalescing market forces) was intense, creating radically foreshortened time frames.[15] In 1978 or 1979, writers using a word processor or a personal computer were part of the vanguard. By 1983 or 1984 they were merely part of the zeitgeist.[16] A longtime associate of the annual Bread Loaf Writers' Conference remembers that by the second half of the decade, "there was probably as much talk about word processing and hardware as there were complaints about the publishing industry and gossip."[17]

As its title makes plain, *Track Changes* lacks the usual middle distance between the subject matter of a book and the book's means of composition. I have therefore adopted a reverse chronological trajectory, generally working backward in time with my chapters. This maneuver, sometimes identified with what is called media archaeology, helps me avoid the temptation of lapsing into easy or self-fulfilling narratives of technological progress while simultaneously exposing the inherent strangeness of "word processing" through its increasingly remote—rather than increasingly familiar—incarnations. Where I can, I also allow my history to be related through the words of the persons who participated in it, preserving quotations from the authors themselves, critical commentators, and other contemporaries as artifacts of a discourse that, while still proximate to us in time, will sometimes sound peculiar to us in its particulars. This book thus seeks to honor the curatorial mandate of its epigraph. Above all, I have strived to offer an account of what was perceived to be at stake with word processing, and address the question of why the technology—which at first may seem little more than a welcome upgrade of the typewriter—proved so contentious. Individual people and systems, ideas and themes, even figures of speech and frames of reference, reoccur; however, I hope I have also managed to convey the rich variety of writers' experiences with computers, and the deep and abiding individuality of the act of writing—a quality no different in a digital medium than in any other.

TRACK CHANGES

INTRODUCTION

It Is Known

By any account George R. R. Martin's interview with late-night talk show host Conan O'Brien was revealing. The *Song of Ice and Fire* author confessed his occasional squeamishness about killing off major characters, disclosed that the very earliest drafts of the stories didn't have any dragons, and told of a missing persons case in Russia that was solved by a fan of HBO's *Game of Thrones* who knew from watching the series that carrion-eating crows might lead to the victim's location before it was too late. "They were sort of waiting for her to die so they could peck out her eyes," O'Brien deadpanned. "It's not exactly *Lassie*." But none of this is what stuck with viewers the day after the author's May 13, 2014, appearance on the show.

Prompted by a question about the prodigious length of his books, Martin also talked about his writing process—specifically his reliance on a DOS-era computer, with no Internet connection, on which he runs a program called WordStar. He called his antique computer his "secret weapon" and suggested that the lack of distraction (and isolation from the threat of computer viruses, which he apparently regards as more rapacious than any dragon's fire) accounted for his long-running productivity.

That was the segment that went viral on social media. The clip was posted to YouTube and from there embedded in innumerable tweets, Facebook feeds, and blogs.[1] Some commenters immediately, if indulgently, branded Martin a Luddite, while others opined it was no wonder it was taking him so long to finish the whole story (or less charitably, no wonder that it all seemed so interminable). But what was it about these seemingly obscure details that people found so compelling? Part of it was no doubt the unexpected blend of novelty and nostalgia: Many fans would be old enough to remember WordStar for themselves, and the intricacy of its interface seems somehow in keeping with Martin's quirky persona, part paternalistic grandfather and part *Doctor Who* character. WordStar thus becomes an accessory to his public image, like the black fisherman's cap he is frequently photographed wearing. But it is also clearly much more than that. Martin's passion for the program is unmistakable: "It does everything I want a word processing program to do and it doesn't do anything else," he declared. "I don't want any help . . . I hate some of these modern systems where you type a lowercase letter and it becomes a capital. I don't want a capital; if I'd wanted a capital I would have typed a capital, I know how to work the Shift key."

And thus, to echo a refrain from the epic high-fantasy series, it is known.[2] The surprise, of course, was not that Martin uses a computer, but that he uses *that* computer and *that* particular word processing program (WordStar 4.0 was released in 1987, and Windows had surpassed DOS as Microsoft's most popular operating system by the early 1990s). Martin's response brought home to viewers that we can become habituated to something like a piece of software just as we do a favorite pen or a particular weight of paper.

Indeed, writers' responses to writing instruments have always been equal parts intimate and uncompromising, and word processing is no different in this regard. Dennis Baron, who has discussed WordStar in the context of his own personal engagement with writing technologies, chooses to emphasize its rough edges, the clutter of its on-screen menus and the cumbersome nature of the keyboard commands necessary to perform tasks we now accomplish with just a click of the mouse. Though he acknowledges its significance in its moment, he is also happy to bid it good riddance.[3] But WordStar was no Rube Goldberg-like contraption or a half-baked bit of code: on the contrary, it was a triumph of both software engineering and what we would nowadays call user-centered design. WordStar dominated the home computer market in the first half of the 1980s before losing out to WordPerfect, itself to be eclipsed by Microsoft Word. Initially sold through mail order (it came on diskette in a plastic baggie), WordStar set the standard for word processing in the first generation of personal computers. It was the brainchild of a typical Silicon Valley marriage between a so-called suit and a beard. The program was marketed by a company named MicroPro, the creation of software pioneer Seymour Rubinstein. The programming, however, was done by Rob Barnaby, who single-handedly wrote an astonishing

137,000 lines of code in a six-month period in 1978. The initial market was not "Windows" or "Mac" (neither of which existed) but an operating system that predated even DOS called CP/M. Little known today, CP/M at the time powered a number of top-selling personal computers, including the Osborne and the Kaypro, both originally designed for the business executive market. Rubinstein would ensure that WordStar was available for both products, an arrangement that also made them attractive to many home users—including writers. By 1984 WordStar had captured nearly a quarter of the word processing market.[4] Authors who cut their teeth on it include figures as diverse as Michael Chabon, Arthur C. Clarke, Ralph Ellison, William F. Buckley Jr., Eve Kosofsky Sedgwick, and Anne Rice (who in turn equipped her vampire Lestat with WordStar when it came time for him to write his own eldritch memoirs).

WordStar today still enjoys something of a cult following among science fiction and fantasy novelists, including (besides Martin), for much of his career, Frederick Pohl as well as Robert J. Sawyer, who has written a long exposition explaining his attachment to the program.[5] Its longevity is all the more remarkable because WordStar was initially inspired by what we would today call a hack: Barnaby had previously taken the command-line editor already built into CP/M (which allowed a programmer to enter and revise code one line at a time) and made the screen display instead a full page of text. From there it was but a short step to the so-called WYSIWYG paradigm, which was the cornerstone of Rubinstein's marketing strategy: "What You See Is What You Get." What was on the screen was what would be on the page when you printed it, or such was the claim. WordStar included such early features as word-wrap, automatic hyphenation, full justification of text, and the graphical display of page breaks. All were seemingly modest enhancements, but all of them were carefully designed to reinforce the illusion of a seamless transition from the cramped monochrome cathode ray display to the crisp, clean surface of a freshly printed sheet of paper.[6]

But if WYSIWYG was the marketing mantra, this alone wasn't what made the program so attractive to writers. Here Sawyer's insights, later extended by Ralph Ellison scholar Adam Bradley, are vital. On the one hand, WordStar—paying close attention to the physical layout of the keys to minimize wasteful finger movements—was designed to maximize the efficiency of the touch typist.[7] Nonetheless, as Sawyer demonstrates, WordStar actually adheres to a model of composition derived from longhand. Part of this has to do with its keyboard commands: "WordStar's powerful suite of cursor commands lets me fly all over my manuscript, without ever getting lost," Sawyer writes.[8] Whereas the strike of the typewriter's keys forces the writer ever forward, character by character, line by line, WordStar's intricate layers of push-button inputs allowed far more freedom and flexibility. So extensive were the combinations and permutations that MicroPro even provided a cardboard stencil of crib notes that could be fitted conveniently around a keyboard.

More fundamentally WordStar's debt to longhand rests on what in software terms is called a "mode." Many readers will recognize modes from their own experience: When you switch back and forth between different "screens" or interfaces within the same application to accomplish different kinds of tasks, you are working within different modes. Though most of today's software makes mode transitions as invisible and seamless as possible, they are often still conspicuous on newer platforms and applications. When I read a Google doc on my iPad, for example, I find I have to manually switch from a View to an Edit mode in order to make changes to the text. Sawyer's insight is that a modal model replicates the typewriter (correction tape and other mechanical hacks notwithstanding, composition and editing remained two very distinct activities when working on the typewriter); whereas the modeless model to which most software aspires in fact resembles a longhand approach to document composition, with all the freedom and flexibility that it affords.

Barnaby and Rubinstein grasped this at the very outset of the home computer era. In effect, WordStar's genius was that it used the layout of the typewriter keyboard to replicate the freedom of longhand composition, combining the efficiency of typing with a hands-on, no-nonsense approach to really handling a manuscript—almost as if the writer was somehow ruffling the electronic pages, marking them up and blocking off passages, sorting them into piles and flagging them with bits of scrap paper or colored ribbon or bubble gum wrappers. "The distinction between the modes is no more distracting than the lifting of ball-point from paper to reposition one's pen," says Sawyer.[9] Adam Bradley, in his scholarship on Ralph Ellison's use of WordStar, similarly notes that "with the entire document as his workspace, he had unfettered flexibility."[10] It should be added that Bradley is not just speaking impressionistically here; in editing the posthumous manuscripts for publication (the unfinished second novel that was redacted as *Juneteenth* and published in its fuller form as *Three Days Before the Shooting . . .*), he has made a detailed study of all of Ellison's manuscript materials, both paper and digital, and written extensively about the patterns of attention and revision he is able to discern from them.

Word processing thus means much more than turning a computer into a souped-up typewriter. (There was much that typewriter users had to unlearn when confronting their first word processor, such as the need to zing the carriage return—a point also made by Baron.) Michael Heim, who wrote the first book to treat word processing not technically or vocationally but philosophically, saw the fundamental difference as one of symbols. Word processors, which is to say computers, reduce all of their input and output to symbolic tokens that are themselves numeric—mere indicators of discrete physical events like voltages and currents and magnetic polarities. The user sees only distant shadows of these phenomena, which are recast as phosphorescent flickers on her screen—shapes and letters, still and moving pictures, all of which we recognize and manipulate. The result of all those dense layers of mediation amplifying and echoing one another is an utterly unique experience, one that many of us can still remember from our own first encounter with a word processor. Writing in 1987, Heim casts it in terms of joy and emancipation: "After the first five months, many people who write on computers are enraptured," he says. "This is bliss. Here is true freedom."[11]

This is a profound claim. For the truth is that whether it's writer's cramp or the pencil point that breaks from too much pressure or the typewriter keys that jam, writing invariably runs into resistance. It is not as if Heim does not understand this. He quotes Emile-Auguste Chartier ("Alain"): "My pen is always trying to go through the paper; my writing is like wood sculpture."[12] Yet Heim sees word processing as a profound rupture with this history of material resistance, this push and pull between aching bodies and blank surfaces, instruments, and inscriptions. Others followed him in this regard. For example, Mark Poster influentially yoked electronic writing and electronic media to poststructuralist theory, and spoke in explicitly Derridean terms about computers "dematerializing" the "trace" of writing; he celebrated the "evanescent, instantly transformable" texts thus created.[13] Hannah Sullivan, a superb scholar who studies habits of revision in modernist authors like Joyce and Woolf, also accepts this emancipatory logic, noting that "nowadays the cost of revision has fallen almost to zero." Because every electronic text is always in a state of indefinite flux (at least theoretically), she says, revision, although technically easier to practice, is also (practically speaking) less pressing—you can always change the text tomorrow instead of today.[14]

These observations resonate because they comport with our own day-to-day experiences of writing on devices as diverse as tablets and laptops and phones with both traditional and touchscreen keyboards, or perhaps even of writing with voice recognition and hand gestures. But computers, word processors, and other digital devices are themselves only technologies—artifacts designed, machined, and built by human hands. Software programs are technological artifacts too—Rob Barnaby's 137,000 lines of assembly code that formed the basis for the original WordStar were a kind of virtual machine, humming along without a hitch only because the wrenches in the system were systematically removed one by one in the painstaking process of debugging. Where, then, can we locate the resistance—a concept, after all, fundamental to electrical engineering—and, *pace* Poster, where might we locate the *materiality* of word processing? If we can, then George R. R. Martin's preference for WordStar becomes no more idiosyncratic than, say, novelist Claire Messud's insistence on a particular brand of graph paper notebook.[15]

Why does a literary history of word processing matter, and what should it be besides a compendium of authors and dates and programs or machines? Is knowing that George R. R. Martin uses WordStar the same as knowing that he prefers chocolate ice cream to vanilla (something I just made up)? Or is it more like knowing that he is a student of England's War of the Roses (something that is true). It is not hard to see the influence of the latter in the tangled epics Martin weaves around knights, battles, and royal intrigue, and one could reasonably expect to derive some insight into his plots, characterization, and themes from a study of the conflict. In truth, however, neither of these analogies seems right to me. Certainly knowing that Martin uses WordStar means more than knowing his favorite ice cream. But knowing that Martin uses WordStar, even knowing something about its particular features and affordances as a word processing program, is at best a dim pretense for any real illumination of his fiction; Martin's books, genre fiction though they may be, mean many things to many different people, and they are not reducible to a single explanatory agent or element. Computational text analysis can show us much, but it would take a lot of convincing for me to believe that Martin's sentence structures (for example) are tied in any significant degree to the specifics of WordStar's keyboard commands. Adam Bradley, in his consideration of Ralph Ellison's use of

WordStar during the later phases of the long-stymied composition of his second novel, similarly rejects the idea that the computer was somehow responsible for Ellison's failure to complete it.[16] Any analysis that imagines a single technological artifact in a position of authority over something as complex and multifaceted as the production of a literary text is suspect in my view, and reflects an impoverished understanding of the writer's craft.[17] By the same token, however, knowing that Martin uses WordStar still seems important; it seems like something we probably *should* know, just as we know the precise make and model of the typewriter Twain once owned, and just as we know details of the design of the tiny table on which Jane Austen did her writing at Chawton Cottage (it was twelve-sided, made of walnut wood, and set on a tripod). We don't know exactly why it is important to know these things, but we know we would rather know them than not.

In Martin's case, this knowledge is perhaps easiest to appreciate when we consider his reputation for writing very long books, with divergent plotlines involving dozens of different characters, their paths (that is, their literal plots) now converging, now diverging. It is not as if *Game of Thrones* were some kind of kabbalistic commentary on WordStar's antiquated formalisms, which we are only now, with this newfound knowledge, clever enough to decode; nor is it that WordStar enables Martin's productivity at any simple instrumental level—many authors before him have written very long books involving multithreaded story arcs. But it surely does facilitate it in some very practical ways, as Martin's own remarks on *Conan* attest. The image of the writer's hands fused to the keyboard, never needing to lift themselves to manipulate a mouse as the sentences unfurl themselves beneath his flying fingers is a compelling one, comporting as it does with the kind of commitment and endurance such an oeuvre demands.[18]

For years the *Guardian* newspaper has run an occasional series called "Writers' Rooms," which was joined in 2011 by "Writers' Desktops," showcasing the virtual workspaces of individual authors' daily computer environments. Skeptics may ascribe such a feature to vulgar author worship, the pedant's or fan's (or voyeur's) insatiable, irrational craving for intimacy, ownership, and omniscience that A. S. Byatt once summarized with the title of her novel *Possession* (1990). But novelist Tom McCarthy, subject of the inaugural installment, disagrees. "Technology reveals us to ourselves as we always in fact were," he says. "Networked, distributed, laced with code."[19] Robert Pinsky, writing some years ago in *Wired* magazine, is more prosaic: "I find the computer utterly a human artifact. It reeks of us, as do our trombones, and cars, and scissors, and parades, and pizzas. Technology is exactly like humanity. It is our baby, and we are its."[20] Artifacts have politics, and details about particular systems and software could well be of interest to future scholars, much as bibliographers have reconstructed the minutiae of work arrangements in eighteenth-century printing houses or archaeologists make models and mock-ups of Sumerian tablets or labor historians interview workers at a newspaper.[21] Moreover, there is an eminently practical component to this kind of knowledge, particularly where it concerns a technology as complex as a personal computer. Efforts to preserve authors' electronic manuscripts—say, a unique version of a text delicately held in the symbolic array of magnetic polarities suspended atop a 5¼-inch plastic disk's iron oxide coating—will depend on our precise knowledge of the operating systems, software, hardware, and data structures then in use. We will return to this topic in Chapter 10, but documentation of an author's computing environment is invaluable to the mission of contemporary archives and special collections, where a hard drive is as likely to be the object of preservation as a Moleskine notebook or a sheaf of typescript tied with a threadbare cord.

If the requirements of future document preservationists were all that mattered, we could cede this ground to specialists and allow them to collect such data as they needed in the course of their dealings with authors, agents, and estates. But the exigencies of preservation do not, I believe, serve to capture our collective fascination with George R. R. Martin and WordStar, or for that matter Jane Austen and her writing table. To fully articulate what is at stake, we have to recognize that these are profoundly humanistic concerns. We can find conscious expressions of them going back at least to Vico's *verum factum* principle in the early eighteenth century.[22] More recently such concerns bring us to what the literary and textual scholar Jerome McGann has termed the scholar's art: "Not only are Sappho and Shakespeare primary, irreducible concerns for the scholar, so is any least part of our cultural inheritance that might call for attention," asserts McGann. "And to the scholarly mind, every smallest datum of that inheritance has a right to make its call. When the call is heard, the scholar is obliged to answer it accurately, meticulously, candidly, thoroughly."[23]

Such datum points can sometimes open surprising avenues for critical interventions. Evija Trofimova, for example, meditates on the mundane accoutrements often accompanying Paul Auster's scenes of writing, in both his fiction and his personal circumstances—cigarettes, (red) notebooks, and typewriters—finding in them entrée points to the key formal elements of the oeuvre, including "*mises en abyme,* narrative splits, and alternative storylines." "But all of this," she continues, "can only become visible if one dares to turn away, for a moment, from the centered intent of the human author and to look more closely at the work of 'things.'"[24] (Nor is she the first to sense this: Auster's typewriter has been repeatedly painted by the artist Sam Messer, and the book he and Auster subsequently collaborated on naturally becomes a primary text in Trofimova's analysis.)[25] Not all authors will reward such consideration equally—the choice of Auster was no more arbitrary for Trofimova than it was for Messer—but McGann (and Vico) serve to remind us that Auster is anomalous only in the degree to which this is so, not in the more elemental truths of what the scholar's art encompasses. *All* writers write with the aid of particular things, be they DOS-enabled computers, Olympia typewriters, or Rhodia graph paper—the literary world's pizzas and trombones.

George R. R. Martin has sold some thirty million books. His work has been translated into more than thirty different languages. But we can't know what goes on inside his head as he writes them; neither, for that matter, can George R. R. Martin. This is the universal reality of writing: the wrenching of thoughts and ideas through the sieve of language, one line, one sentence, one page at a time.[26] But close up against the chasm between composition and inspiration we find—always—material things and technologies. "There may be speech without tools or tangible materials," notes John Durham Peters, "but writing always requires both."[27] Even a piece of software like WordStar, for all of its seeming elusive ineffability, is tangible in this important sense: it is ineluctably there, pushing back against our infinitely wider array of human foibles and fumbling as we settle in before the screen, its blinking cursor the gateway to an exactingly scripted space of logic, action, and reaction. This is the essence of those machines (like computers) that are purpose-built to function as formal systems: "We know every step in the production of a computer microprocessor, and every part of every part," writes W. Brian Arthur, a scientist at the Santa Fe Institute. "We know exactly how the processor operates, and all the pathways of the electrons inside it."[28] The novelist Vikram Chandra, who for many years also worked as a computer programmer, explains his own delight at this same realization: "There were mysteries, things I didn't understand, but there were always answers. If I tried hard, there was always a logic to discover, an internal order and consistency . . . I could produce these harmonies, test them, see them work."[29]

Martin's intimate knowledge of WordStar's functions and keyboard patterns might be best characterized as tacit knowledge, the extraordinary combination of muscle memory and unarticulated experience that enables us to perform very complex tasks without conscious effort or consciously knowing how to do them. Tacit knowledge is necessary for the flow states many writers cite as characteristic of their most productive sessions, and they deeply resent anything that jolts them out of that zone.[30] (Thus Martin's comment about autocorrect, for example.) In the wake of the widespread adoption of word processing by writers in the 1980s, composition researchers such as Daniel Chandler and Christina Haas worked to articulate the specific components of tacit knowledge in computing in order to better understand the new realities of the writing process.[31] Haas found, for example, that screen size and other aspects of the graphical interface had a measurable impact on what she termed an author's "sense of the text," his or her ability to conceive of it as a whole, a gestalt—and that this in turn had implications for how the writer approached the task of revision.[32] Chandler, meanwhile, called such tacit knowledge "resonance," the way in which small, seemingly insignificant details of a particular technology ended up mattering a great deal to a writer, often for inexplicable or inarticulate reasons. An example might be the feel of a keyboard like IBM's now-defunct Model M, the way in which its keys, once depressed, spring back into their upright position with a distinct, authoritative "click." (So integral is this experience to some users' confidence and comfort at the keyboard that Model M devotees have been known to stockpile spares.)[33]

Friedrich Nietzsche famously declared that our writing tools shape our thoughts, a terse aphorism whose own presentation as such reflects the influence of the Malling-Hansen Writing Ball (an early typewriter whose keys were arranged in a half-sphere) that he began using in 1882 as his eyesight deteriorated.[34] Word processors, futuristic in design but also often awkward or arbitrary in their actual operation, provoked just such reflections among writers. "Do we think differently about what we are writing if we are writing it with a reed pen or a pen delicately whittled from the pinion of a goose, or a steel pen manufactured in exact and unalterable replication in Manchester," asked William Dickey, a poet who would experiment with Apple's HyperCard software late in his life. "Do we *feel* differently, is the stance and poise of our physical relationship to our work changed, and if it is, does that change also affect the nature and forms of our ideas?"[35] Dickey's self-questioning here anticipates the exigencies of what McGann would come to term the scholar's art, for the questions Dickey asks are unresolvable absent close and careful attention to the particulars of individual writers and their writing instruments.

It would be difficult to point to an author at a greater remove from George R. R. Martin and his vocation of epic high fantasy than Lucille Clifton, the African American poet, children's book author, essayist, and educator who was also a poet laureate of the state of Maryland. "The strength of her poetry, which seldom contained capital letters or standard punctuation, came from its simplicity," noted the *Washington Post* in her obituary.[36] Nevertheless, she and Martin shared this much at least: she relied on an obsolete word processor for much of her career, in her case a Philips/Magnavox VideoWRITER 250. An ungainly device not unlike a *Star Wars* droid in appearance, the VideoWRITER featured an oblong monitor thrust forward over a keyboard attached to it by a coiled cable (one early reviewer described it as resembling a microwave oven).[37] Its thermal transfer printer was built in as well, so a sheet of paper would have come curling out of the back of the unit. It was not a general purpose computer but rather a dedicated word processor, marketed to consumers who wanted to avoid the complexities (and expense) of the former in favor of a machine devoted solely to writing. It was therefore designed to be "user-friendly" (unlike WordStar, with its baroque chorded keyboard sequences, its keys came with labels like Spell, Print, and Help). First introduced in 1985, it retailed for $800.

Clifton appears to have begun using hers in 1988, at the very height of her career.[38] She relied on it, not to grind out millions of words of prose fiction, but to keyboard the sparse lines of her poetry for all of her books except her final collection, *Voices* (2008). Indeed, by her own account, she did not "write" on it at all but rather composed predominantly in her head and then simply used the screen to lay out the text. "By the time I get to where I'm sitting down to print out on a word processor, I've edited in my head," she says in an interview. "I work in my head a lot," she continued. "I can't work with a pen and paper either. I need to see the look of the thing as close to print as I can."[39]

A word processor is ideally suited to furnish hard copy on demand, but what accounts for her long-term attachment to this one in particular? Perhaps it was its promised ease of use, at least initially, followed by the comfort of a familiar routine. But Clifton was no technophobe; she owned at least one other computer late in her life and had become a prolific user of email. Notably, however, the VideoWRITER's screen, which displayed text in monochrome amber, had an unusual aspect ratio, the term by which its lateral dimension is measured. It was thus considerably wider than it was high, displaying 100 characters across as opposed to the more typical 80 or 40. Clifton, for whom the visual appearance of the poem as printed on the page was paramount, may have stuck with it because the screen had such ample room in the horizontal, so that she never had to worry about the rhythm or look of a line being disrupted by a forced break. In other words, it preserved—indeed, enabled—her sense of the text. She would maintain the VideoWRITER in good working order for two decades, and regardless of how much of her composition she did in her head, it clearly meant something more to her than just a contrivance for producing hard copy. This particular writing instrument—and its resonance—has left a subtle but insistent imprint on her work. Surely the scholar's art demands that we be as sensitized to the VideoWRITER's distinctive characteristics as any other aspect of Clifton's biography and circumstances.

Rather than adopting commercial products like Martin and Clifton, the essayist and journalist John McPhee deliberately shaped his writing software in self-conscious imitation of his long-standing habits of organizing projects. Since 1984 he has relied on a suite of custom-designed applications centered upon an obscure IBM-compatible editing program called Kedit. He recounts the story of a friend at Princeton University who listened to his description of how he filled notebooks and journals, transcribed and typed and retyped the entries, cut them apart with scissors, and then pasted them back together again. "Howard [J. Strauss] wrote programs to run with Kedit in imitation of how I had gone about things for two and a half decades," McPhee recalls.[40] Chief among these is one called Structur [*sic*], which "explodes" individual notes into discrete files that then become the basis for future writing in Kedit, and another program, called Alpha, which performs the reverse function, consolidating and collapsing an array of different files into a single file with internally ordered segments that McPhee can edit and revise as he sees fit. McPhee, of course, is particularly known for his interest in the convergence of the natural world and built infrastructure, in books such as *The Control of Nature* (1989) and *Uncommon Carriers* (2006). In the essay from which I have quoted here, "Structure," he writes about a group of intangible software programs that serve to emulate the structures of his own thought, in turn allowing him to shape and give structure to his extended writing projects. Structur and Alpha are one-of-a-kind writing instruments created in a single author's image: to know the software is to know something of the mind of the writer, however obliquely.

When tens of thousands of people shared, tweeted, or embedded the clip from George R. R. Martin's *Conan* interview, it was doubtless with the sense of casual novelty—"get-a-load-of-this"—that characterizes so much of the social Web. But there was also something more at work—the sense, however intuitive or instinctive, that this was something worth knowing, specifically worth knowing about this author and his writing. Not that WordStar explained anything, exactly. But the contrast between the inaccessible recesses of Martin's mind—knowable only to him, at best only in partial and fragmentary form—and the banality of the instrument through which he channels its expression is, I believe, a large measure of what accounts for the fascination with the technical details of his computer system. We can read about what it was like to use WordStar, even going back to original documentation; we can look at screenshots or video capture of the software in operation; we can experiment with an emulated version to get a more hands-on feel; we can even find an old Kaypro or Osborne or DOS PC and fire it up for ourselves.[41] The black box is cracked open, if only a little: not a muse descending, just George R. R. Martin pecking away with his vintage software to create the story-world that immerses millions of people.[42]

WordStar fit. It complemented Martin in the same way, perhaps, as Jack Kerouac's famous typing of *On the Road* on a 120-foot roll of paper or Auster's cigarettes and Olympia typewriter. "The best writers," concludes Tom McCarthy, "have always understood that to write is to both grapple with, and to some extent, allegorize the very regime of technological mediation without which writing wouldn't exist in the first place."[43] The technological regime McCarthy is speaking of here is writing's *interface,* by which I mean not only what is literally depicted on a screen (menus, icons, and windows) but also an interface in the fuller sense of a complete, embodied relationship between a writer and his or her writing materials—the stance and poise and "feel" invoked by William Dickey. In other words, McCarthy is speaking of what we earlier termed *materiality,* the materiality of both word processing and of writing more generally. This materiality often has implications for interpretation, and it always has implications for preservation and documentation, for history and for memory. This is the scholar's art. But materiality also grounds us. It demystifies. Materiality is where—is how—our knowing begins.

ONE

WORD PROCESSING AS A LITERARY SUBJECT

Not long after he began work in earnest on the mechanical tabulating machine that would be called the Difference Engine, the nineteenth-century British inventor Charles Babbage found himself a regular guest at the table of Samuel Rogers, a poet who lived in a lavishly rebuilt house in London's St. James overlooking Green Park. The two had struck up a friendship through Rogers's brother, a banker, with whom Babbage had financial dealings. Rogers's breakfasts had long been an institution among London's literati, a place where Coleridge might talk with J. M. W. Turner as Walter Scott passed a pot of jam to Byron. Babbage, of course, was no poet, though from time to time the idea of writing a novel had crossed his mind as a means of financing his long-deferred work on another project, the Analytical Engine—the device whose design foresaw many of the principles of a universal machine that would be articulated by Alan Turing a century later.

At one such breakfast, however, Babbage the polymath found himself discoursing not on his engines, or mathematics, astronomy, or indeed any of the numerous other topics with which he was acquainted, but on composition. How *did* a poet work, he inquired: Did one start with a fast, rough draft of the whole and then go back to revise, or is the process sentence by sentence, line by line, laboring over each until it was just so before moving on to the next? His host employed the latter method: Rogers ventured that he had never once written more than four (or maybe it was six) lines of verse in a single day, so fastidious was he about the perfection and polish of each. It was said that Robert Southey, by contrast, flew through his drafts, turning out poetry and prose at prodigious rates and going back to revise only after he had a thick sheaf of pages. And so it goes, a bit of literary table talk of the sort writers have been engaging in for centuries. Still, here we have the individual generally credited with anticipating the concepts of modern computation chatting with some of the foremost literary figures of his day about the nuts and bolts of writing and revision.[1]

Word processing would seem an obvious literary subject, for table talk or otherwise. Many of us must imagine that its present-day ubiquity was somehow preordained, the trajectory of its uptake as smooth as the convex curve of a classic CRT screen. And indeed, word processing's standard narrative possesses an overwhelming sense of inevitability. Typically one of the very first moves journalists or technical writers would make in introducing readers to the topic was to align it with the long history of prior writing technologies. References to Sumerian cuneiform or monks in scriptoria or Gutenberg's printing press suddenly abounded in the computer industry. "Thousands of years ago, people put their thoughts down on clay tablets," began *Byte* magazine's December 1984 review of WordPerfect. "Modern authors have the word processor."[2] These contemporary descriptions sound a lot like today's TED Talks and other Silicon Valley disruption scenarios: "A word processor is, quite simply, the most amazing thing that has happened to writing in years," begins the author of a column in the August 1983 issue of *Writer's Digest.*[3] Ray Hammond, just one year later in the preface to a handbook about word processing addressed specifically to literary authors and journalists, agrees: "The computer is the most powerful tool ever developed for writers."[4]

The freedom and flexibility that word processing apparently afforded—what Michael Heim experienced as bliss—seemed so absolute that it was hard to conceive of the technology as even *having* a history apart from the long series of clearly inferior writing utensils leading up to the present-day marvel. Yet, however inevitable the development of word processing might have seemed in the history of writing writ large, serious literature—*belles lettres*—was another story. Hammond acknowledged that among creative writers, "computers have a terrible image, one which at best bores writers and at worst terrifies them."[5] The exception should not be surprising. John Durham Peters reminds us that according to the best evidence we have, writing was invented as a mnemonic device for bookkeeping and calculation, not as a surrogate for speech. And he notes that books—let alone anything we might care to characterize as *belles lettres*—have never been the mainstay of printing, which gravitated overwhelmingly toward instrumental texts like bills, records, indulgences, and reference works.[6] Likewise, scholars such as Peter Stallybrass and Lisa Gitelman have demonstrated that the mainstay of printing in the West since Gutenberg has been so-called "job printing" (a nineteenth-century term), as opposed to codices and books, with perhaps the exception of the Bible itself.[7] And according to John Guillory, it is the intra-office memo (not the novel or the sonnet) that stands as the "humblest yet perhaps most ubiquitous genre of writing in the modern world."[8]

Approaching word processing as a specifically *literary* subject therefore means acknowledging that we seek to concern ourselves with a statistically exceptional form of writing that has accounted for only a narrow segment of the historical printing and publishing industry. Moreover, for the literati, word *processing* was invariably burdened (not buoyed) by the associations carried by the gerund in that compound term, which could function as a foil for nothing less than humanity itself. "The writer, unless he is a mere word processor, retains three attributes that power-mad regimes cannot tolerate: a human imagination, in the many forms it may take; the power to communicate; and hope," opined Margaret Atwood in a 1981 address to Amnesty International.[9] When conceptualist Kenneth Goldsmith—who has built a reputation on his verbatim transcriptions of traffic reports and newspaper dailies—declares, "I used to be an artist; then I became a poet; then a writer. Now when asked, I simply refer to myself as a word processor," he implies that he is merely part of the zeitgeist; but in fact the comment provokes us precisely because of word processing's associations with mechanization, automation, and repetition; bureaucracy, productivity, and office work.[10]

Novelists, dramatists, poets, and essayists nowadays set preferences and manage files more often than they change ribbons and zing carriage returns. Yet neither the designers of word processors nor the inventors of the typewriter envisioned literary writing as the inevitable application for their machines. The typewriter was initially conceived and marketed as an aid to the blind, deaf, and motor-impaired, and for taking down dictation. Hammond calls it a "desperately limited tool . . . totally linear in operation, frustrating a writer's attempts to mold a piece of writing as a whole."[11] (For Hammond the computer is much closer to working with paper and pencil than a typewriter is, a conclusion we have also seen in Robert Sawyer and Adam Bradley's observations about WordStar.) Nonetheless, typewriters would enjoy early and notable associations with literary culture and creativity, as catalyzed by Twain's celebrity and most fully embodied by T. S. Eliot's typing of *The Waste Land* a generation later, a poem in which a typist (that is, a *type writer*) figures prominently.[12] Hannah Sullivan has shown how typewriting during this period was quickly assimilated into individual authors' workflows, though still functioning primarily as a site of revision and correction as opposed to free-form composition (most writers continued to draft longhand).[13] Some writers learned to type up manuscripts for themselves, numerous others employed secretaries and assistants—an often gendered dynamic that, as we will see in Chapter 7, was reproduced decades later with the advent of word processing. Typewriting also became an important part of the representational fabric of modernism through figures as diverse as H. N. Werkman, Gertrude Stein, and Bob Brown.[14]

Darren Wershler, who has given us perhaps our best book on typewriting, enumerates the iconic imagery that quickly took hold: close-ups of the hammers and keys, inky letterforms steadily stamped behind a ribbon, or else (aurally now) the staccato chatter of the mechanism to signify productivity.[15] The totality of this image complex was as absolute as it was in some sense unexpected—as Adam Bradley astutely notes with regard to fiction writers, "it is a testament to their success in harnessing the typewriter's capacity as a tool for creative expression that we most often think of typewriters in the nostalgic light of the hard-boiled writer cranking out pages by dim lamplight."[16] (Ernest Hemingway took to staging photographs of himself at his typewriter in which he appears to be looking into the face of his muse like a toreador staring down a bull.) Yet Bradley also reminds us that the association between typewriting and literary productivity "was a mastery achieved only over time, and with many casualties among those authors unable to adapt."[17]

Cinema has furnished some especially vivid portrayals of those casualties. Consider the typewriters (equally monstrous and mundane) that tortured William Lee in *Naked Lunch* (1991) or John Turturro's character in the Coen brothers' *Barton Fink* (also 1991); or the way in which Stanley Kubrick used a typewriter to record the deteriorating mind of Jack Nicholson's character in the film adaptation of *The Shining* (1980). (The scene where Shelley Duvall confronts her husband's prodigious output of pages, all containing repetitions of the same phrase, is a restaging of Truman Capote's famous put-down of that other Jack—Kerouac—"It is not writing. It is only typing.")[18] It would seem that word processing can't help but make for a poor cinematic prop by comparison, facilitated as it is by mass-produced machines made of plastic, not metal, the bold rattle of the keys replaced by muted clicks and clacks, the backlit screen but a pale counterpart to the existential whiteness of the page.[19] "The idea of taking everything and cramming it into this little electronic box designed by some nineteen-year-old in Silicon Valley . . . I can't imagine it," David Mamet once declared.[20] Computers thus seem far removed from the most iconic renditions of literary authorship: more fax machine than fountain pen, we might say. "I am surprised by how much I like my computer," admits Anne Fadiman. "But I will never love it. I have used several; they seem indistinguishable. When you've seen one pixel you've seen them all."[21]

By contrast, evidence that the typewriter is now a consummate literary subject (and object) is seemingly everywhere today: typewriter appreciation pieces, some thoughtful, many merely wistful, are a mainstay of journalists and essayists;[22] the Internet routinely spawns lists of famous authors and their favorite typewriters;[23] hipsters carry typewriters into coffee shops and set them up next to their espresso (and then post pictures to social media);[24] Steven Soboroff's collection of typewriters belonging to everyone from the Unabomber to John Updike regularly tours and is exhibited for charity;[25] Tom Hanks, a well-known aficionado, bankrolled an iPad app called the Hanx Writer that imitates the look and especially the *sound* of a typewriter on a touchscreen;[26] and when Hanks wrote an op-ed for the *New York Times* to launch the Hanx Writer, he lamented that audiences at Nora Ephron's *Lucky Guy* (in which he had recently starred in his Broadway debut as tabloid journalist Mike McAlary), got only the subdued clickety-clacks of word processors in the on-stage newsroom, rather than the hardboiled racket of vintage typewriters.[27] As tellingly as any of these, when Manson Whitlock's New Haven typewriter repair shop closed after his death at the age of 96, the story was covered in the *New York Times.* "I don't even know what a computer is," he was quoted as saying in 2010.[28]

The ubiquity of computers and word processors has clearly allowed us to retroactively buff and varnish the typewriter's aura of authenticity. Thus typewriter collector and scholar Richard Polt has collected numerous such instances as I have been describing into an attractively produced book entitled *The Typewriter Revolution,* whose opening manifesto enjoins readers to hold public "type-ins" and otherwise promote typing's tactility as an alternative to the seductive distractions of the "data stream."[29] Just as importantly, however, we should acknowledge that vectors of influence can be bidirectional and recursive, and otherwise follow strange loops—a phenomenon evident in the work of artist Tim Youd, who stages public performances in which he retypes, word for word, page by page, classic American novels on vintage typewriters in a place associated with the book in question.[30] For each such project, Youd sources the same original make and model of typewriter that the author used to write the book. The novels are then retyped in their entirety on a single page of typing paper—that is, when Youd gets to the end of the page, he simply replaces it in the rollers and starts typing on it again from the top. The results are striking: dark, abstract Rorschach blots that bear some resemblance to Concrete Poetry, though the physical page itself is left in tatters, a latticework held together only by what cumulative negative space has remained intact in and around the inky palimpsest of typewritten letterforms. While this work might seem like the ultimate expression of typewriter fetishism, I would suggest it is just as much marked by word processing for the way in which Youd's retyping reimagines books as data, reflecting an aesthetic (and anxiety) that is all about the "processing" of words, the commodification of canon and tradition, and the sheer vastness, scope, and density of written text in the present.

God looked down at the writers and said, 'I haven't done anything for these people for a long time, hundreds of years, so I'm going to make up for it,'" Frank Conroy is quoted as saying during his tenure as director of the Iowa Writers' Workshop.[31] Of course attestations of divine providence tend to lack historical nuance, to say nothing of technologies such as the typewriter, correction fluid, acid-free paper, and the ballpoint pen (biro).[32] How, then, should we fix and formulate a literary history of word processing in a larger history of writing technologies? The first narrative to resort to is that of technological progression, or perhaps what Jay David Bolter and Richard Grusin would come to call a "remediation": a new medium defining itself by assimilating and appropriating the associations and affordances of a previous medium.[33] Many observers of technological change have been tempted by this pattern. When Margaret Atwood, no naïf she, suggests that the Internet is still only "at the early typewriter stage," we are clearly meant to assume that it will one day attain the word processing stage.[34] In this progression, then, the word processor follows the typewriter, which followed the ballpoint pen, which followed the fountain pen and goose quill, and so on. Some of the relevant technological history would also seem to encourage such a view. A 1960 program named

Expensive Typewriter ran on the TX-0 or "Tixo" computer then installed in MIT's Lincoln Laboratory.[35] The joke was that the state-of-the-art TX-0 (whose commercial incarnation, the PDP-1, would cost upward of $100,000, stand six feet high, and weigh over half a ton) was really nothing more than a (very!) expensive remediation of what typewriter historian Bruce Bliven Jr. had just a few years earlier called that wonderful writing machine.[36]

Undeniably, literary authors often understood their word processors as exactly this kind of remediation, at least initially. Doubtless they would have been encouraged by the fact that there is a basic visual continuity between a personal computer and a typewriter—a keyboard and a screen behind and laterally perpendicular to the keyboard, much as a piece of paper comes curling out of the rollers. The conversion narrative wherein a writer testifies to the power and potential of word processing—never to touch a typewriter again—thus became one of the most commonplace by-products of how the technology was assimilated. "I've got a magic typewriter," declared science fiction novelist Larry Niven in a 1980 issue of *Writer's Digest.* "The word processor program moves blocks of type around (any size); writes over a line that isn't perfect, leaving no trace; erases words or lines or paragraphs in a second or two; and when I'm ready it prints as pretty a manuscript as you ever saw, with justified margins and no visible corrections."[37] The prolific fantasist Piers Anthony switched to a computer in 1985 only *after* his wife, a computer engineer, convinced him that the keyboard could be reprogrammed to re-create the custom Dvorak layout he insisted on using with his manual typewriter.[38] Then there is Jorie Graham: "I still use it like a fancy typewriter."[39] José Saramago: "What I do on the computer is exactly what I would do on the typewriter if I still had it, the only difference being that it is cleaner, more comfortable, and faster. Everything is better."[40] Amos Oz: "The word processor is, for me, nothing but a typewriter, only you don't have to use Typex to erase or correct a mistake."[41] Joan Didion, commenting on the IBM Thinkpad she was using at the time of a 1996 interview, likewise stated, "I just use it like a typewriter." But she immediately complicates her own response, adding, "Before I started working on a computer, writing a piece would be like making something up every day, taking the material and never quite knowing where you were going to go next with the material. With a computer it was less like painting and more like sculpture, where you start with a block of something and then start shaping it."[42]

Not all writers choose to use a computer, even today. The list of those who do not is long, and includes some of our most celebrated: Paul Auster and Don DeLillo, Cormac McCarthy and Joyce Carol Oates, to name but a few. "I need the sound of the keys, the keys of a manual typewriter," DeLillo once declared in an interview, choosing to speak (intentionally?) in short, staccato sentences: "The hammers striking the page. I like to see the words, the sentences, as they take shape. It's an aesthetic issue: when I work I have a sculptor's sense of the shape of the words I'm making. I use a machine with larger than average letters the bigger the better."[43] So *pace* Didion it is the typewriter and *not* the computer that allows him to "sculpt" his prose. And like Henry James, he relishes the aural dimension of the experience.

Wariness and critique of media and technology is of course one of the hallmarks of DeLillo's fiction. But Cormac McCarthy, who has enjoyed the benefits of a long-term residency at the Santa Fe Institute (one of the world's premier centers for advanced scientific research), has likewise done all of his writing on a light-blue Lettera 32 Olivetti that he purchased in a pawnshop in Knoxville for $50 in 1963. (It fetched a quarter of a million dollars at auction when McCarthy replaced it with another Olivetti.)[44] Meanwhile Paul Auster's Olympia typewriter is perhaps one of the most iconic writing instruments in the world. Since acquiring it in 1974, he tells us in the book he eventually coauthored with Sam Messer, "every word I have written has been typed out on that machine."[45] That may be true, but it is also a fact that Auster writes first in longhand and types his prose only afterward. And he regards this division of effort as fundamental. The reasons are resolutely somatic: "Keyboards have always intimidated me. I've never been able to think clearly with my fingers in that position. A pen is a much more primitive instrument. You feel that the words are coming out of your body and then you dig the words into the page."[46] The typing that then follows is a mechanical act, but no less essential to the process: "Typing allows me to experience the book in a new way, to plunge into the flow of the narrative and feel how it functions as a whole. I call it 'reading with my fingers,' and it's amazing how many errors your fingers will find that your eyes never noticed."[47] As for upgrading, Auster is adamant: "There is no point in talking about computers and word processors," he states, and goes on to relate his horror of pushing a wrong button and losing an entire day's work—or worse, an entire manuscript ("I knew that if there was a wrong button to be pushed, I would eventually push it").[48] Similarly, Joyce Carol Oates also hand-writes and then types—but not with a word processor.[49] Although she typed her earliest books directly, the typewriter has since become for her "a rather alien thing—a thing of formality and impersonality."[50] Her practice dovetails with Auster's description of his typing, wherein he acknowledges the bodily toll it takes—strained back, aching neck—because the only way to get clean typescript after revisions is to start over again from scratch.[51] But though they parallel one another in their workflows—first longhand and then typewriter—Auster and Oates regard their relationship to these instruments differently, and have adopted them for different reasons. Ironically, Oates—known not least for her prodigious output—is routinely accused of just the kind of extreme productivity that word processing's detractors once feared would overtake the literary establishment.[52]

But while some writers have foregone computers altogether, many more use them alongside other writing technologies, including longhand, verbal dictation (itself once recognized as a form of word processing), and the manual annotation and correction of hard-copy drafts. Kazuo Ishiguro maintains two writing desks, one whose surface is slanted for longhand and the other with a decade-old computer unconnected to the Internet.[53] Annie Proulx writes first drafts by hand and then revises on a computer;[54] Gloria Evangelina Anzaldúa typically composed directly on her Macintosh but edited her final drafts by hand.[55] Toni Morrison told a group of creative writing students at Swarthmore that she writes exclusively with pencils on legal pads, but in a more in-depth exchange in the *Paris Review* it emerges that computers and even voice dictation figure in her process as well.[56] Michael Ondaatje is notorious for the dense handwritten drafts he produces, but he also dictates the text to a digital recorder from which it is transcribed by an assistant into a digital document, which Ondaatje then proceeds to edit further.[57] Similarly, J. K. Rowling is often touted for working longhand because she wrote the first *Harry Potter* novel that way in an Edinburgh coffeehouse—and she still does write longhand—but since her second novel she has also worked with a computer.[58]

The reality, of course, is that *every* writer's individual habits and practices are deeply personal and idiosyncratic, and it is difficult, if not impossible, to extract patterns in support of generalizable conclusions—beyond the intense intimacy and commitment that the act of writing invariably demands. Some writers dictate aloud. Some write longhand and then type their work on a typewriter or computer. Some compose at the keyboard but then print out their work for handwritten revision. Others don't need the hard copy. Some writers print everything out, mark it up, and then retype it themselves. Others hand it off to an assistant. A few still revise by (literally) copying and pasting strips of text.[59] Some writers find the computer alienating, intimidating. Others see it as an intimate, profoundly unmediated experience. Some writers value the slowness of the pen. Some value the speed of the keyboard. Some chafe at the labor of retyping, others embrace it. Some writers are enamored by the small rituals of the process, the changing of the worn-out ribbon, the bright bell of the carriage return. Others are unsentimental. Some writers require an absolutely specific instrument—or setting, or time of day, or slant of light, just so. Others write on anything and everything, anytime, anywhere. Work your way through "The Art of Fiction" and its kindred features in the *Paris Review*—what George Steiner, in his own "Art of Criticism" interview, describes as "the largest collection of insight into this of any publication"—and you can find accounts describing all of these and more.[60] The one inescapable conclusion is that our instruments of composition, be they a Remington or a Macintosh, all serve to focalize and amplify our imagination of what writing is.

Word processing hovers uneasily between the comfortably familiar and the encroachingly alien. Rather than a reimplementation or remediation of typewriting, I prefer to think of it as an ongoing negotiation of what the act of writing means. (To the point is one recent marketing slogan for Microsoft Word: *More Than Words.*)[61] The very term "word processing" is increasingly difficult to utter unselfconsciously. Even "computer" is beginning to seem cumbersome alongside "laptop" (a specific form factor for a computer) or the trademarked brand names of individual products. "I wrote that on my iPad" sounds natural; but gesture toward the same device and say you've just written something on a computer (or a word processor) and it sounds wrong, the sort of locution one might expect from someone who still has a landline and answering machine. Nowadays we walk around with supercomputers in our pockets—but we still prefer to call them by a vestigial name from a prior telecommunications epoch. What does it mean, then, to recapture our sense of what word processing once was at a moment when we *text* each other with our . . . phones?

The big histories of the Information Age now emerging are of little help with this question. There is scarce mention of word processing in books such as Walter Isaacson's *The Innovators* or James Gleick's *The Information.*[62]

Perhaps this is because the technology lacks a single obvious "genius" figure to personify it. Yet I suspect it is also because word processing is widely perceived as belonging to the realm of *application* instead of innovation. To this way of thinking a word processor is merely an arbitrary instantiation or just one particular configuration of the universal machine that is a modern digital computer. The technorati have often shown surprising disdain for word processing on exactly these grounds. Ted Nelson, for example, visionary author of the book *Literary Machines* (1980) and founder of the Xanadu project, has frequently inveighed against programs like WordStar and Word that are based on the WYSIWYG model. For him, these represent the triumph of a fundamentally conservative vision. "A document," he laments, "can only consist of what can be printed."[63] Jay David Bolter, a classicist who was an early advocate for computers as writing tools, rendered much the same verdict, concluding that word processing was "nostalgic" in its respect for the aesthetics of print.[64] In this view, the promise and potential of newer, more experimental modes of electronic writing—including nonlinear hypertext, a term Nelson himself coined and has popularized throughout his career—is at odds with a technological paradigm whose highest achievement lies in mimicking the appearance of something that might have come from Gutenberg's own press.[65]

Scholarly interest in the history of electronic literature has similarly gravitated overwhelmingly toward those authors who sought to reimagine our definitions of the literary through branching, multimodal, and interactive narratives or poetic compositions. "A story that changes every time you read it," in the words of one of the form's most accomplished practitioners, Michael Joyce.[66] The most important platforms for this kind of experimentation were HyperCard (first released by Apple in 1987), Storyspace (initially co-designed by Joyce with Bolter and John B. Smith, and developed since 1990 by Eastgate Systems), and then the Web itself.[67] Interactive fiction—so-called "text adventures," which had a brief commercial vogue in the early 1980s through the success of a company named Infocom—have also increasingly been explored by scholars.[68] Lori Emerson, among the best of recent academic critics, has carefully detailed the ways in which poets like bpNichol, Geof Huth, and Paul Zelevansky each leveraged the programmable capabilities of the early Apple II line of computers to craft innovative on-screen textual compositions.[69] The texts thus produced are indeed striking, harbingers of a new aesthetic intimately tied to the procedural capabilities of digital media but also knowingly reaching back to Concrete

Poetry, Surrealism, and other well-documented movements. Yet the Osborne 1, which debuted in 1981 (the year the Apple II became the best-selling computer on the consumer market), was likewise an important platform for writing, as we have already seen. One motivation in my labeling of word processing as a literary subject is to balance the preponderance of critical and historical attention already devoted to those relatively few writers who, as Emerson relates (after John Cage), viewed the computer as a labor-*making* device—allowing for bold but sometimes rarefied experiments—rather than as the far more commonplace labor-saving device it was for most users. And yet, as we will also see, the boundaries between such supposedly polarized uses and users are porous; academic focus on the avant-garde has risked obscuring our recollections of a much wider, albeit more quotidian, history of writing on screens.

Researchers in literacy, composition, and rhetoric have been exploring and assessing the impact of computers on writing for at least the last four decades. I have already mentioned the contributions of Daniel Chandler and Christina Haas, who together with such pioneering scholars as Gail Hawisher, Eric Johnson, Andrea Lunsford, Joel Nydahl, Cynthia Selfe, John Slatin, and others helped launch the field that has produced the most sustained, thoroughgoing investigation of exactly how word processing and other digital writing modalities—right up to present-day Wikis, blogs, and social media—shape the art and act of composition. The resources these scholars put in place are crucial to understanding the emergence and evolution of the technology. In 1983 Bradford Morgan, an English professor at a remote technical college in North Dakota, founded the *Research in Word Processing Newsletter*.[70] The first issues were produced on a CPT 8000 dedicated word processing machine that was available to him on the campus, and then mimeographed for distribution. Another colleague, James Schwartz, soon joined Morgan to produce the *Newsletter*—known especially for its bibliographies of popular and scholarly work about word processing—each month for the remainder of the decade. The existence of such a publication seemed "inevitable," according to Morgan.[71] In the introduction to the first issue he wrote, "The word processor not only saves times [*sic*], conserves labor, and solves problems, but it also reinforces the traditional mission of writing programs." The subscriber list grew to hundreds, including academics, journalists, and creative writers (Piers Anthony among them). Likewise in 1983, the University of Chicago Press published the first edition of the *Chicago Guide to Preparing Electronic Manuscripts*, noting that word processing meant that "with cooperation between author and publisher . . . the elusive keystrokes can be captured and reused."[72] In 1985 the Modern Language Association took the unusual step of offering an official recommendation to its membership on the choice of a word processor: Nota Bene, a program created by a Yale doctoral student named Steven Siebert who was frustrated at the inability of existing software to adequately handle the notes and citations so critical to scholarly writing.[73] The gradual uptake of computers within English departments also gave rise to another subtle circumstance: unsuspecting creative writing instructors would one day find a desktop machine (with all the latest software) provided as standard furnishing for their campus offices—a very literal realization, if you will, of what Mark McGurl, in reference to the institutionalization of creative writing programs in academia, terms the "program era."[74]

Occasionally textual scholarship—the discipline that studies the material history of the transmission and transformations of literary or other texts—has yielded some attention to the minute particulars of word processing, as has been the case with Adam Bradley's work on Ralph Ellison or Doug Reside and the libretto for Jonathan Larson's Broadway musical *RENT.* (Reside has applied forensic computing techniques to the original diskettes Larson used in his computer, thereby recovering alternate versions of some of the lyrics and MIDI compositions.)[75] Experience teaches us that we are likely to see more of this kind of work, which depends on precise knowledge of individual writers and their personal computing history.[76] A textual scholar would know, for example, that *The Waste Land* not only featured a typist as one of its figures but that Eliot's actual typing of the poem—on three different typewriters—proved the key by which Lawrence Rainey unlocked the history of the text and accurately reconstructed the different episodes' order of composition. (Doing so required document sleuthing of the sort usually practiced by the FBI.) Would such a coup have been possible if Eliot wrote the poem on a succession of Mac PowerBooks? Scholars interested in questions such as these for literary manuscripts that now exist only as document folders on hard drives or data in the "cloud" will one day have to come to terms with the particulars of different operating systems, software versions, and hardware protocols, as well as the characteristics of a variety of different hard-copy output technologies, from dot matrix and daisywheel to inkjet and laser printer.[77] All of these problems and possibilities depend not just on our knowledge of "computers" or "word processors," but also on our knowledge of very specific products and technologies.

Twenty years ago, just at the moment of the Web's apotheosis, Sven Birkerts first gazed at the spinning globe in the corner of the old Netscape Navigator browser and counseled his readers to "refuse it."[78] In a much more recent essay he meditates on Joseph O'Neill's use of Google Earth (whose default view is similarly the planet as seen from space) as a literary device at the end of his 2008 novel *Netherland*. Birkerts finds the literary image conjured by the technology compelling, so much so that he briefly considers installing the software on his own computer; but once again he opts to refuse it, reaching out to click but then withdrawing his hand: "The fact that such a power is available to the average user leaches from the overall power of the novel-as-genre," he concludes.[79] Say what one will, but this seems to me a dim view of that genre's contemporary relevance if maintaining its vitality somehow depends on keeping readers an arm's length from the very subjects it seeks to encompass.

Part of the problem with commentaries such as Birkerts's is a kind of flattening—a sweeping of all nuance and distinction, all attention to the minute material particulars of individual circumstance—under the banner of epochal terms like the Information Age. But what do such terms really mean to us here in the day-to-day world of the present? Joyce Carol Oates may not use a word processor, but she is on Twitter, with over 100,000 followers.[80] Toni Morrison writes longhand but also uses a computer (and, as we know from the 2007 advertising campaign, reads books on an Amazon Kindle). Paul Auster writes longhand but employs an assistant to do his computer work. The same is true of Michael Ondaatje. Regardless of each of their individual attitudes toward word processing and its place in their workflows, there is no easy sorting of any of these writers into analog or digital binaries, into those who have either embraced "it" or refused it.

In 1985 John Barth sat for his *Paris Review* interview with George Plimpton. "Do you think word processors will change the style of writers to come?" Plimpton asks. "They may very well," Barth replies, and continues: "But I remember a colleague of mine at Johns Hopkins, Professor Hugh Kenner, remarking that literature changed when writers began to compose on the typewriter. I raised my hand and said, 'Professor Kenner, I still write with a fountain pen.' And he said, 'Never mind. You are breathing the air of literature that's been written on the typewriter.' So I suppose that my fiction will be word-processed by association, though I myself will not become a green-screener."[81] One way to take this is as mystification and nothing more, a throwaway statement wherein the air exuded from the typewriter (or word processor) swirls about the more earnest endeavors of the fountain pen, like vapors from an atomizer. The comment is perhaps that, but it is also something more I think. For a severe media determinist like Friedrich Kittler, it would in fact be taken as a statement of remarkable self-awareness, one in which the writer frankly acknowledges the paucity of human agency amid a mechanized regime of keys, chips, ribbons, rollers, and signals. "Writing in the age of media has always been a short circuit between brain physiology and communications technologies—bypassing humans or even love," he flatly declares.[82] (Kittler is contrasting the mechanized imprint of the typewritten page to the tradition of writing love letters by hand—readers familiar with his analysis will understand that we are brushing the surface of his whole cosmology.) Here John Barth is marking his own dominant impression of the technology, colored by alien associations and the threat of sublimation: one *becomes* a green-screener, bathed in that monochrome glow. And indeed, Barth himself *did* become one, the evidence of his manuscripts revealing him as assimilated by the time he was writing *The Last Voyage of Somebody the Sailor* some five years later.[83] Moreover, he integrated the poetics of electronic writing into his fiction, in novels like *Coming Soon!!!: A Narrative* (2001), which imagines Johns "Hop" Johnson as a fictional protégé version of himself captivated by hypertext, as well as in short stories like "Click" (1997) and "The Rest of Your Life" (2000). None of this should be very surprising given Barth's interest in literary experimentalism of all sorts, not to mention the foundational conceit in his Cold War fable *Giles Goat-Boy* (1966) that the manuscript of the novel had been compiled from reels of mainframe computer storage tape—certainly one of the earliest instances of a computer having *any* kind of voice or role in serious fiction outside of science fiction.

No less notable is the story of Hugh Kenner himself, Barth's interlocutor in the dialogue above. Kenner was a critic of the first rank, a distinguished scholar who studied under Marshall McLuhan and then devoted a career to Ezra Pound and other major figures of British and American literary modernism. He also, however, maintained a longtime interest in computer-generated poetry, wrote columns and reviews for *Byte* magazine, and in 1984 published a user's guide to the Heath/Zenith Z-100 computer, a competitor to the IBM PC noted at the time for its color graphics. In the introduction to the user's guide, he mentions building his first one from a kit: "I . . . turned it on, pressed the keys, and saw letters and numbers jump to life on the screen. It worked!" He also states he'd gotten it to help with his writing, and indeed it does ("automating the drudge-work"), though not until after the subsequent addition of a printer.[84] This publication is little remembered today, but just a few years after it appeared Kenner would publish a still well-regarded set of essays entitled *The Mechanic Muse* (1987), demonstrating how Eliot, Pound, Joyce, and Beckett each responded to technological change in their own era.

Given that Kenner was a student of McLuhan, it is perhaps best to read the exchange about the "air of the typewriter [and word processor]" as an impromptu extension of the thesis of *The Gutenberg Galaxy:* that language, mobile and recombinant, is now literally at our fingertips, pervading our structures and systems of thought, and reconfiguring ours senses. Whatever new twist the green-screeners portend, Barth resigns himself to his fate. But air—the ether—can also be understood as a medium of sorts, as it was for much of the nineteenth century.[85] In this respect Kenner and Barth become anticipatory voices of some of the more influential branches of media studies today, those embracing networks and layers of coded or scripted artifice as the material base of a new, decidedly nonvirtual reality—a reality that also (nowadays) circulates throughout the air itself via wireless radio and cellular signals, and that even then intimately tied to networks of power, capital, and control in a global system where identity and commerce intermingle in data flows that are, algorithmically, indistinguishable from writing.

Word processing may thus be a literary subject, but word processing also shapes and informs literary subjects—the persons who inhabit the system (and economy) of literature, green-screeners or otherwise. As both Haas and Chandler remind us, writing is a medial process, characterized by the author's relationship to an ever-expanding array of tools and surfaces. "Technologies cannot be experienced in isolation from each other, or from their social functions," is how Chandler puts it. "Our use even of a pen necessitates the complementary use of related technologies (such as ink and paper) no less than does our use of a word processor."[86] We have seen that George R. R. Martin disconnects himself from the Internet while writing, but that does not mean he is isolated from other kinds of networks and dependencies: How, for example, does he maintain that DOS machine? Where do spare parts come from? How does he get files from the computer's hard drive to his publisher? Who converts them from WordStar's native .ws format, and how? Or does he print them out? If he does, on what kind of printer? And whose job is it then to rekey the text so it can pass to layout and production?[87] (This is to say nothing of Martin's reliance on that most elementary of modern networks, the power grid.) The reality of these contingencies and arrangements, their consequences and dependencies, are all part of what it means to "know" that a particular writer uses a particular writing tool, word processing or otherwise.[88] Writing, in other words, is *never* off the grid. It is *always* about power—a "power technology," as Durham Peters calls it.[89] He means both the legalistic and societal power that writing embodies and encodes, as well as its even more fundamental capacity to project language through and across basic physical categories like space and time.

A literary history of word processing must therefore acknowledge not only the hybrid, heterogeneous nature of both individual persons and their personalities, but also the highly complex scene of writing (and rewriting) that we observe today, one where text morphs and twists through multiple media at nearly every stage of the composition and publication process. The history I offer here thus largely and willfully resists generalizations and sweeping conclusions; it highlights instead the stories of individuals, it pays heed to the *difference* different tools and technologies actually make, and it reveals how attitudes and assumptions can sometimes change over the span of even just a few years. It also, I would hope—constructively, even joyfully—extends our imagination of what writing is by illustrating the variety of ways in which *all* manifestations of that activity coexist and cohabitate with technology. Take, for example, Neal Stephenson and his epic *Baroque Cycle.* His colophon to the three books (just shy of 3,000 published pages of scientific and historical fiction) captures what I mean. He tells us there that the manuscript was drafted longhand, using a succession of boutique fountain pens. Stephenson then transcribed the text to his personal computer system using the venerable Emacs program and typeset it himself using TeX, the computer typesetting system to which Donald Knuth (perhaps our most famous living computer scientist) devoted nearly a decade of his career to perfecting. His publisher, however, wanted the manuscript set using what was then the industry standard, QuarkXPress, so Stephenson next took it upon himself to put together a conversion program written in LISP.[90] Point for point, the brand names of the fountain pens—Waterman, Rotring, Jorg Hysek—balance out the technical particulars of the software programs and computing languages that follow. Stephenson's fountain pens are, of course, beautiful, precision instruments; but his hand-tooled LISP program is beautiful in its own way, and every bit as integral to the writing act.

Or else consider Jonathan Franzen, who in recent years has repeatedly spoken out against the pernicious influence of social media.[91] It would be easy to cast him as a neo-Luddite and killjoy, a refusenik after Birkerts. But witness too this extraordinary passage, worth quoting in full for the sheer grit of it: the self-knowledge, the recall of technical details, and especially the deep, multisensory intimacy of the relationship to technology it exposes:

> I bought my first computer in 1989. It was a noisy metal box made by Amdek, with a paper-white VGA monitor. In good codependent form I came to appreciate the noise of the Amdek fan's hum. I told myself I like the way it cut out the noise from the street and other apartments. But after two years of heavy use, the Amdek developed a new, frictive squeal whose appearance and disappearance seemed (though I was never quite sure of this) to follow the rise and fall of the air's relative humidity. My first solution was to wear earplugs on muggy days. After six months of earplugs, however, with the squeal becoming more persistent, I removed the computer's sheet-metal casing. Holding my ear close, I fiddled and poked. Then the squeal stopped for no reason, and for several days I wrote fiction on a topless machine, its motherboard and tutti frutti wires exposed. And when the squeal returned, I discovered I could make it stop by applying pressure to the printed circuit-board that controlled the hard disk. There was a space that I could wedge a pencil into, and if I torqued the pencil with a rubber band, the corrective pressure held. The cover of the computer didn't fit right when I put it back on; I accidentally stripped the threads off a screw and had to leave one corner of the cover sort of flapping.[92]

No virtual realities here, only the minute torques and tolerances of the everyday: relative humidity, rubber bands, and stripped screws, their shavings of low-grade steel no doubt collecting somewhere inside the burlesque cavity of the machine's exposed innards. It's a prose passage whose only real reason for existing is to advance the theme of his essay ("scavenging"), and yet it also reminds us that all of *this* is what writing really is, these details no different in their way from those that inform the re-creation of the scriptorium so marvelously rendered by Umberto Eco, or from the care and concentration with which a connoisseur restores the delicate moving parts of a vintage typewriter.

Of course sometimes the details really do make all the difference. Here is Nicholson Baker (who once made a living as a word processor; that is, a professional operator of them), describing the writing of *A Box of Matches* (2004), which he did like the novel's narrator with his journal, early in the morning and in the dark: "I had a couple different laptops because they were not all that dependable, and one of them had a slider bar. I could slide the screen brightness down to almost nothing, so I was sitting in complete darkness. The screen would have just the tiniest hint of phosphorescence and a faint crackle of static electricity. I thought, This is an option Dickens did not have."[93]

TWO

PERFECT

On March 20, 1981, the *New York Times* op-ed page carried a brief item musing that historians and biographers might soon be in for "slim pickings." What precipitated this pronouncement was a report that Jimmy Carter, just months out of office and hard at work on his memoirs for Bantam, had lost several pages of text after hitting the wrong keys on his brand new $12,000 Lanier word processor. Word processing, the *Times* speculated, threatened to put historians—"those bloodhounds of the paper trail"—out of business. "Archivists," meanwhile, "will be deprived of words scratched out, penciled in and transposed with wandering arrows. They will have to make do with electronically perfect texts."[1]

Perfect. No other word so encapsulates the aspirations as well as the anxieties that accrued around word processing. We remember WordPerfect, of course, the software that rose to dominate the market after WordStar (from the mid-1980s through the early 1990s), but there were also now largely forgotten programs like LetterPerfect and Perfect Writer.[2] Etymologically, "perfect" comes to us from the Latin *perfectus,* by way of Old French; the original Latin meaning encompassed the idea of completeness, the state of fullness or of being finished. The familiar hand gesture, index finger bent to touch tip of thumb, other fingers stiffened in accent, is a way of visualizing that condensed state of ripeness, the attainment of closure and completion. In word processing's parlance, "perfect" was meant to connote a finished document, flawlessly formatted and printed, cleanly and clearly expressed. The sense of completion, meanwhile, also spoke to the desire for efficiency and productivity, a zeal for getting the job done.

Could a single invention both improve the quality of the work being performed and accelerate the pace at which it was completed? It seemed too good to be true, yet this dual ideal—the quest for flawless efficiency and effectively flawless results—was present from the very inception of word processing as a concept and technology.[3] In 1974 a consultant named Walter A. Kleinschrod produced a report on word processing for the American Management Association. "The touch of a button," he wrote, "triggers a perfect final copy."[4] A second AMA text echoed and embellished this language: "The typist can push a button and the machine, tirelessly and flawlessly, chugs out a perfect draft—once, twice, a hundred times if required, without fatigue or the errors fatigue can bring."[5] Whatever else it was then, word processing was also a form of futurism that, as Thomas Haigh deftly notes, was of a piece with a push-button lifestyle and, indeed, food processing, which was introduced to consumers by Cuisinart at just about the same time.[6] As word processing continued to mature and also transition from dedicated stand-alone systems to early personal computers, this original ideal lost none of its appeal. "With a word processing system," wrote computer scientist Ivan Flores a decade later in 1983, "you can actually produce a perfect document."[7]

Error was the archenemy of perfection: typos, misprints, mistakes, caked-up layers of correction fluid, and patches of paper rubbed raw by erasures—all of these disrupted the smooth, homogeneous surface of an impeccably presented text. So obsessed were early office managers with the pernicious influence of error on efficiency that entire studies were commissioned; the habits of typists were scrutinized at the most minute levels. It was known, for example, that a secretary making a mistake in the first few lines of the page was likely to remove it and feed in a fresh sheet and start anew; a mistake introduced farther down the page, however, would be attacked by the eraser. It was known how long the rubbing and retyping would take, along with the erasure and correction of any carbons associated with the original page. And, of course, it was known how much this all cost—in paper, in carbons, in erasers, and above all, in time.[8]

Word processing promised to change all that. "Errors," proclaimed the Kleinschrod report, "are no longer sins of incompetence, destroyers of confidence."[9] Indeed, they were so "easily rectified" that "first-time perfection" was no longer demanded, marking an astonishing shift in attitude. "Recording on magnetic tape is just as easy as typing at your fastest speed," assured a roughly contemporary manual prepared by IBM's Office

Products Division in support of its most advanced word processing product at the time, the Magnetic Tape Selectric Typewriter (see Chapter 8). "Typographical errors are corrected merely by backspacing and typing over the incorrect character. A perfect tape is created which gives you a perfect copy . . . and no erasing!"[10]

The ideal of perfection, so seductive and seemingly emancipatory in an office setting, signaled something different in literary contexts, an area the AMA consultants mentioned briefly in 1974 as an example of word processing's as yet unexplored applications—private and "creative," as opposed to functional and organizational.[11] Some literary authors were not bothered by these connotations and indeed embraced them, adopting the language of "perfect" for themselves. The flurry of fingers across increasingly sensuous keyboards, the instant gratification of the glowing letterforms on the screen, the effortless ease and efficiency of corrections and revisions, and the polished look of the final printed document—all of these conspired toward a common aesthetic and experience readily summarized in the optimism of Russell Baker: "The wonderful thing about writing with a computer instead of a typewriter or a lead pencil is that it's so easy to rewrite that you can make each sentence almost perfect before moving on to the next sentence."[12] Stanley Elkin, who as we will see in Chapter 7 was using a Lexitron word processor by the end of the 1970s, was even more enthusiastic in his evaluation: "The word processor enables one to concentrate exponentially; you have absolute command of the entire novel all at once," he opined, and then continued, employing his pet name for the machine: "You can go back and reference and change and fix and . . . so in a way, all novels written on the bubble machine ought to be perfect novels."[13]

Elkin's statement is especially notable for its appreciation of word processing's complex refashioning of a writer's sense of the text, its search functions affording the user total access and total recall. For other authors, however, word processing's most immediate intervention was to reinforce the importance of the manuscript itself, cultivating an appreciation for their writing as a thing in the world, the proverbial well-wrought urn. John Updike, keynoting a 1988 computer science conference at MIT, opined that word processing made the production of "perfectly typed" text "almost too easy."[14] Recall also Larry Niven and his "magic" typewriter, the one that "writes over a line that isn't perfect," leaving no trace and printing "as pretty a manuscript as you ever saw."[15] The conceit that writing could somehow be perfected was thus an important part of what word processing promised. Yet for both Updike and Niven that perfection lay just as much in the finished appearance of the text as in the perennial search for *le mot juste.* Sometimes these two different understandings of what literary perfection meant became conflated. Poet and self-help author Peter A. McWilliams, in his *Word Processing Book* (1982), narrated the tale of how John Keats, window-gazing in the company of Henry Stephens, came to compose the proverbial line "A thing of beauty is a joy forever," by trying out spoken versions on his friend before hitting on exactly the right formulation. "What think you of that, Stephens?" "That it will live forever."[16] For McWilliams, a writer as gifted as Keats is essentially a word processor incarnate; his mind alone is enough to harness the vast array of imponderables poetic composition demands. The rest of us, however, must rely on lesser devices, which is to say the assistance of mechanical contrivances. "Word processing machines allow for maximum flexibility in alteration, changes, correction, revision, and expansion," McWilliams explains. "After all this processing of words has taken place, the word processing computer will print out as many copies as you like, letter perfect."[17]

We see here the collapse of the superficial appearance of the document with its more ineffable qualities as McWilliams moves us, in the space of a page, from the deepest recesses of Keats's fertile mind to the veneer of a hard-copy printout. The ideal of perfection thus becomes closely tied to the supposed dematerialization of the written act: a perfect document is one that bears no visible trace of its prior history; indeed, it is as though the document did not have a history, but rather emerged, fully formed in its first and final iteration, from the mind of the author. Transferring this concept to the AMA's milieu, every business executive signing his name to letter-perfect copy after however many prior drafts produced by the unseen hands of a secretary at her keyboard became a Keats incarnate, the most mundane communications unsullied by any trace of error, happenstance, or hesitancy of thought.[18]

In literary circles the pretense of perfection was just as often grounds for suspicion and anxiety. John Barth recalls that 1981 was the year the graduate program in creative writing at Johns Hopkins University received the first writing samples self-evidently prepared by the applicant on a word processor. "I was impressed by its virtually published look," he admits.[19] But Barth showed the sleek, seemingly professionally produced pages to a colleague, a science fiction writer. He "took one suspicious look at the justified right-hand margins, the crisp print and handsome typefaces, and said, 'This is terrible! They're going to think the stuff's finished, and it only looks that way.'"[20] This was to become a pervasive sentiment, repeated over and over again: journeyman authors would be tricked into thinking their work was better or more polished than it really was, and—what was worse—hardworking writers who were busy paying their dues at the typewriter with Snopake, scissors, and glue would be squeezed out of the market. When *Writer's Digest* ran a 1982 cover story under the title "Writing Made Easier with Personal Computers," there was a storm of protest in the next month's letters to the editor column. "*Typing* made easy with personal computers, not writing," read one.[21]

Editors and agents were likewise often skeptical. It is true that early dot matrix printouts were genuinely difficult to read, and so submission guidelines would routinely include prohibitions against them. But the belief that word processing meant cutting corners also widely obtained. At the outset of his 1984 book, Ray Hammond found it necessary to remind his readers that "the computer has no power to write words."[22] Some writers went so far as to actively conceal the fact that they were using a word processor, lest they raise suspicions that the work was somehow automated or otherwise inauthentic. In *Writer's Digest* a novelist named Karen Ray narrated her editor's anxiety that the computer would inhibit her creativity, and her subsequent decision to disguise the fact that her manuscript had in fact been prepared on a word processor by selecting fonts approximating the appearance of typescript and feeding single sheets of paper into her printer rather than using a tractor feed. "The best way to write with word processors may be . . . secretly," she concluded.[23] Gish Jen, describing her own first experiences with a computer, echoed these anxieties: "Maybe there was something dronelike about working on a mainframe; maybe sitting at stations and terminals really was at odds with the play at the heart of creative work." She describes fellow students at the Iowa Writer's Workshop, working at just such mainframe terminals (taking advantage of the ease and convenience the system afforded) deliberately doctoring their printouts—adding autograph annotations, even rumpling the corners—before sending them out for consideration. ("Real writers," their instructors insisted, used pencils or typewriters.) Yet for Jen, the personal computer, specifically the Apple, overcame these prejudices: "Computers coaxed out of me an expansiveness the typewriter never did," she says. "What came out . . . was not further from the human heart; it was closer. It was looser, freer, more spontaneous—more democratic too."[24]

Anne Fadiman acquiesced to the word processor, as we have seen, but she also claimed to be able to detect the "spoor" of it in books: "The writers—no longer slowed by having to change their typewriter ribbons, fill their fountain pens, or sharpen their quills—tend to be prolix."[25] This fixation on seemingly unseemly length—closely related to the other common concern that word processing made writing too easy, something also expressed by Updike as well as Barth's Johns Hopkins colleague—likewise receives a refreshing counter from Jen, who underscores the presumption inherent in consigning one's words to a medium supported by a non-renewable source: "I was not a person who would have looked at a ream of paper and thought, 'Sure, that is mine to fill up.' But I turned out to be a person who could keep moving a cursor until I'd filled up one ream, then another."[26]

The anxieties manifested in all quarters of the literary establishment, as the tide of word processing quickly took hold, had surfaced in science fiction since at least the late 1950s as part of the genre's general fixation on artificial intelligence and the automation of society. Fritz Leiber's *The Silver Eggheads,* originally a story in *Fantasy and Science Fiction* magazine (1958), later expanded into a novel (1961), imagines a future in which writers are celebrities, focusing full-time on the maintenance of their extravagant public images while digital machines—"wordmills"—do the work of creating fiction. Eventually the writers revolt.[27] Michael Frayn, meanwhile, gives us in *The Tin Men* (1965) the William Morris Institute of Automation Research, one of whose many dozens of departments is devoted to using computers to generate newspaper stories from recombinant storylines ("Paralyzed Girl Determined to Dance Again," "Child Told Dress Unsuitable by Teacher," and so on). Along the way we meet a struggling novelist, Rowe, who plots his books by first penning the glowing reviews he expects them to receive. His reliance on the recycled language of literary criticism is obviously meant to mirror the activity of the computer scientists and their machines. At the end of the novel we learn that *The Tin Men* is in fact the fruit of the Institute's first attempt to automate the production of fiction.[28] Stanislaw Lem's "U-Write-It" is a 1971 speculative fiction that posits a "literary erector set," which furnishes a would-be author with "building elements," namely strips of paper containing snippets from past masters—all very much like Tristan Tzara's dada composition, not to mention then-contemporary experiments in the French OULIPO group.[29] Word processing similarly evinced the specter of something like Italo Calvino's Organization for the Electronic Production of Homogenized Literary Works, a sinister (and cynical) outfit possessed of powerful computerized engines for churning out best-sellers biometrically calibrated to capture a reader's attention.[30]

Arthur C. Clarke may have had any or all of these predecessors in mind when he sat down at his own word processor (to be detailed in Chapter 3) to write a 1986 story for *Analog* magazine entitled "The Steam Powered Word Processor."[31] In it the Reverend Charles Cabbage, frustrated with composing sermons week after week—all of them just variations on one another, after all—undertakes to build the titular engine, which eventually draws the unwanted attention of the archbishop of Canterbury. Clarke relishes much detail on the operations of his pseudomachine, which he dubs the Word Loom, creating what amounts to an early instance of the genre called steampunk (the term did not yet exist). But the Word Loom, laboring at capacity, eventually wrests itself to pieces in the western transept—a victim of its own prodigious labor as the follies of the hapless Cabbage are displaced and magnified by its enormous coal-fired bulk. Most telling, though, is the afterword, which relates the fate of the only surviving copy of Cabbage's opus *Sermons in Steam*, shelved in the British Museum. The book, upon examination, is clearly printed with ordinary letterpress, we are told, with the exception of a single leaf, pages 223 and 224: "An obvious insert. The impression is very uneven and the text is replete with spelling mistakes and typographical errors."[32] These are, of course, the very *imperfections* that word processing (and spell-checking) is supposed to eradicate; yet here they are, ironically presented as proof positive of this one leaf's mechanical provenance. This irony is the fulcrum of the story: just as the creation of the steam-powered word processor ultimately demanded far more labor and attention from the reverend than the biweekly crafting of his sermons—thereby defeating its purpose, as the final cataclysm attests—here these minute mechanical imperfections function as a microcosm of the folly of seeking to automate the word.

John Varley was one of the few science fiction writers to continue using a typewriter, an IBM Correcting Selectric.[33] In 1985 he indulged in a mock-serious epistolary dialogue with his editor, Susan Allison at Berkely Books, in which he demanded that all forthcoming editions of his works include a statement attesting that the text was created "using only natural ingredients: The purest paper, carbon typewriter ribbons, pencils, ballpoint pens, thought, and creativity. . . . Not a word ever phoned in via modem."[34] One trusts the irony is intentional (carbon typewriter ribbons, let alone ballpoint pens, as "natural" ingredients?). This faux testimonial, when printed, was to be accompanied by a self-designed sigil depicting antique typewriter keys amid a flourish of ribbon. The piece was called "The Unprocessed Word," and elsewhere in it Varley offers the tale of a fellow writer identified only as "DT" who, not satisfied with the recently released MacWrite, supposedly acquires programs entitled MacConflict and MacClimax. The stories quickly begin to (literally) write themselves: "Now, in today's mail, comes MacFirstline, but I don't think I'll run it. I think I'll kill myself instead."[35] Here, as in Clarke's story, the implications are clear: the ostensible perfection of the printed text becomes a stand-in for a whole set of much deeper anxieties related to authenticity in the writer's craft—originality and creativity, truth and beauty, all those Romantic shibboleths of old. If imperfection can become the measure of authenticity (true as much for the output of the Word Loom as for the typos and eraser rubbings of more prosaic workplace texts), then the more efficient the act of writing becomes, the more suspect it is rendered, until we reach the kinds of limit cases limned by Clarke, Varley, and their predecessors—fully automated algorithms and apparatuses for generating text.

Many readers will recall an exemplar that was earlier than any of these, from Gulliver's visit to the Academy of Lagado in Jonathan Swift's *Gulliver's Travels* (1726). There he encountered what was also a kind of word loom, a device that consisted of combinations of words stamped on dies atomized in a system of wires and pulleys, its rods and levers worked by a professor's diligent pupils: "By his contrivance, the most ignorant person, at a reasonable charge, and with *a little bodily labour,* might write books in philosophy, poetry, politics, laws, mathematics, and theology, without the least assistance from genius or study."[36] Of course this mechanism, too, must be revealed as flawed: like the Reverend Cabbage's steam-powered text riddled with typos, the Lagado machine produces only occasional "broken sentences," which the head academician has not yet found time to improve upon.

Swift's invention has been recognized as a "word processor" by previous critics, and so it is—but not just in the overt sense of its being a mechanical contrivance for the creation of prose. Elsewhere in the passage, Swift relates the method by which the output is read aloud by an apprentice and dictated to a scribe for a permanent record, therefore anticipating a far more comprehensive scene of word processing—encompassing, as we will see in Chapter 7, precisely such relays of dictation and inscription, composition and rendition, found in the hyperrationalized office environs that were the purview of the American Management Association. Above all, however, Swift's writing about the writing machine anticipates the late twentieth-century anxiety that the "perfection" of writing—which is to say, optimizing the means of its production—would lead to nonsense. Like the Reverend Cabbage's *Sermons in Steam* then, or like the output of Lem's U-Write-It or what we can presume to call the "MacBooks" of Varley's literary confidant, the proceedings of the Academy of Lagado are volumes best consigned to the Library of Babel. The Library of Babel, of course, is the centerpiece of Jorge Luis Borges's great fable of the universe as a library, containing everything on its endless shelves that has been or could ever be written, in all languages, sense and nonsense. Borges was himself a librarian; but it is to computers and the Internet that the story has been applied as parable over and over again.[37]

Find a best-seller from a name-brand author and (especially if the book has a foil-embossed cover) check to see how many times the word *perfect* appears in the blurbs. What exactly is that word doing adorning something so subjective (and resolutely middlebrow) as a specimen of genre fiction? Is the book a "perfect" confection like a soufflé, or is it perfect like a piece of mass-produced merchandise, executed impeccably and unimpeachably to spec? In one of the most influential literary endorsements ever tendered, Ronald Reagan helped catapult Tom Clancy to celebrity by pronouncing his first published novel, *The Hunt for Red October* (1984), "the perfect yarn."[38] Clancy wrote his first draft on an IBM Selectric; always the forecaster of technological trends, however, he gave an Apple computer running WordStar a cameo appearance—his hero Jack Ryan uses it as he pecks away on just such a monograph as might be published by the Naval Institute Press, the small Annapolis-based house specializing in naval history that would in fact acquire Clancy's own manuscript.[39] The book found its way thence to readers inside the nearby DC beltway, and eventually all the way to the Oval Office. Clancy was soon able to afford his own Apple IIe computer and had already traded up to a Macintosh by his third book, his fame and career well launched.[40] About writing, and indeed about computers, he would remain always unsentimental: "If your objective is to write a book, get a computer and write the damn book," he told an audience much later in life. "It's a lot easier than you realize it is."[41]

Few best-selling authors have been more polarizing than Clancy. His fans are many—David Foster Wallace was reportedly among them—but his detractors are legion, seeing in his books (which continue to appear posthumously) something very close to the realization of Calvino's OEPHLW. But it is James Patterson, whose occasional nickname in the press is "The Word Processor," who perhaps more than any other writer today seems to embody the very ethos of word processing.[42] First and foremost, Patterson is prolific. He writes series books. With established characters. Set formulas. Short chapters. And short sentences. The books are purely consumptive—page-turners, read once and then discarded, a new one (for there is always a new one) rotated into place on the nightstand or the tablet. Most audaciously of all, he openly works with coauthors, typically crafting extended outlines (dozens of pages) that he then hands off to one of a half-dozen or so regular collaborators for fleshing out in a first draft—which Patterson then edits and revises, the process iterating until the book is done.[43] Video footage shot in 2012 shows rows of shelves in his home office laden with manuscript pages for writing projects in various states of completion, dozens of them, each neatly stacked with a large-print cover sheet denoting title, coauthor, and current state of the draft.[44] There are reportedly no less than three employees at Little, Brown and Company devoted exclusively to handling this output.[45] Patterson's working methods may resemble a master artist with a clutch of pupils in a Renaissance atelier. But *Time* magazine called them "an affront to every Romantic myth of the artist we have."[46]

It is equally impossible to discuss Patterson's career without invoking statistics and financials.[47] A typical year might see a dozen new Patterson titles published, some fourteen to fifteen million individual units shipped in all. He's reportedly sold upward of 300 million copies of his books. Total entries on the best-seller list: fifty, give or take. Total books published: north of 100. His reported annual earnings in one recent year: $94 million. Total earnings: $700 million. How does this compare to authors like Stephen King, John Grisham, and Dan Brown? Patterson outsells them all—not individually, but collectively. Perhaps the most astonishing statistic of all: one in seventeen hardcover novels sold in the United States in one recent year was a Patterson book.

Although he has his share of critics, fellow best-sellers like King among them, in interviews the Word Processor is unapologetic about his success. But he also retains perspective about his own work: "I'm less interested in sentences now and more interested in stories," as he put it on one occasion.[48]

Patterson's career dovetails with the culmination of decades of changes in the bookselling industry.[49] By the mid-1980s the consolidation among former rival publishing houses was well under way in New York, as the chain stores increased their dominance in the retail market—often at the expense of independent booksellers.[50] The result was unprecedented stress on the publishing lists. Between 1986 and 1996 (just prior to Patterson's ascendency), 63 of the 100 best-selling books in the United States were written by just six authors: Tom Clancy, Michael Crichton, John Grisham, Stephen King, Dean Koontz, and Danielle Steele.[51] They were all authors capable of delivering a perfectly consistent product to stock the tables at the front of the store.

Word processing in and of itself was the catalyst for none of this. There is no simple causal relationship between these structural changes in the bookselling business and the rise of word processing in the personal computing industry. That said, the two phenomena did largely coincide. When Gore Vidal declared in 1984 that "the idea of literature is being erased by the word processor," he had something specific in mind by way of institutions, canons, and traditions.[52] Tellingly, Vidal offers his remark in the context of an encomium for the life of Logan Pearsall Smith, the essayist and grammarian who labored in obscurity for much of his career, dedicating himself (*pace* Patterson) to the craft of sentences. "As a writer," notes Vidal, "Logan himself was very much [of the] school of America's own (now seldom read) Emerson who was at his best in 'the detached—and the detachable—sentence.'"[53] Eventually Vidal gets round to laying the blame for Smith's unrequited legacy at the feet of his pursuit of . . . yes, "perfection" (Vidal uses the word), which Smith, nobly in Vidal's view, prized far beyond commercial success; this being "the inevitable fate," Vidal tells us, "of one who has been denied not only the word processor but the Apple home computer in which to encode Thoughts."[54]

The commodity status of the literary is neither new nor news, of course—books, even great books, have always also been saleable products, and many of the famous authors depicted in the mural in the Barnes and Noble superstore café sweated over their sales figures. But it is easy to understand why the kind of push-button automation word processing seemed to promise—giving any hack the ability to place a letter-perfect manuscript into the hands of a publisher—came to epitomize the fears of many working writers. Such writers, the backbone of the venerable midlist whose sales had heretofore sustained the industry, found their position in jeopardy as the independent bookstores gave way to the chains, the backlist was all but entirely eclipsed by the frontlist, the great houses were consolidated, and the books themselves became properties in transmedia conglomerates with brands like "Tom Clancy."

As for James Patterson? What kind of word processor does he have? Surely it must be a mighty one! But the novelist who more than any other functions as the embodiment of everything satirists from Jonathan Swift to Italo Calvino and Arthur C. Clarke have sought to skewer doesn't use one. He works his stacks of manuscripts longhand.[55] How perfect is that?

Jimmy Carter was able to retrieve his deleted prose with assistance from the manufacturer (ironically the Lanier was branded the "No Problem" word processor).[56] But what can it mean that some thirty-five years ago this incident rose to the attention of the *New York Times*? The op-ed, written at a moment when people were still very much in the midst of transitioning to the new technology, put its finger on one of the defining aspects of this new kind of text: its smooth, seemingly seamless perfection, its crisp, luminescent characters—the beauty and regularity of the digital word.

The peril and allure of an electronically perfect text derived from several different qualities of word processing. One, as we have already seen, is the physical testimony of the hard copy, unblemished by any trace or remnant of the messy work of composition. (This would seem to be what the *Times* had in mind.) There is also the widespread association of computers with automation and homogenization.[57] As early as 1986 one observer was already able to note, "The history of word processing is the story of the gradual automation of the physical aspects of writing and editing."[58] And as we have also already seen, automation had become a target for literary satire from a variety of quarters. I have further attempted to suggest that such associations compounded many authors' anxieties at a moment when increasing numbers of them felt vulnerable owing to structural changes in the publishing industry. But to fully understand the prevalence of "perfect" as a trope around word processing, we must also look to the material qualities of computation itself, to what was happening on the other side (as it were) of what Iris Murdoch once prosaically referred to as the "glass square."[59]

First and foremost, the glass square was luminous. It glowed. We take the luminosity of our screens for granted now, thinking in terms of battery life when we bother to think about brightness and lumens at all, or designing purposefully nonbacklit screens to reduce eyestrain. At the time, however, the glare of all those brilliant ray-traced phosphors cast some long, lyrical shadows. "Word processing is a dry distance of a label for what is, more accurately, writing with light," is how one early commentator put it.[60] Of course, "photography" also means literally light-writing, but the technical particulars of word processing are very different from those of either chemical or even digital photography.[61] Pressing a key on a standard computer keyboard means completing a circuit. The character input is then routed to the computer's central processing unit, which in turn typically does several different things with the signal thus received, such as storing a representation of that character on some version of magnetic, optical, or solid state media and directing the display to make the character data visible as the typographic representation we see on the screen—originally by firing a pattern of electrons from a glass tube to strike a phosphorescent surface (the so-called cathode ray). Thus Andrei Codrescu, on the occasion of Andrew Kay's death, remarked, reminiscing about his first computer, "The Kaypro . . . let you write with light on glass, not ink on paper, which was mind-blowing. It felt both godlike and ephemeral."[62] And John Updike, describing an evening walk in his boyhood hometown of Shillington, Pennsylvania, for the *New Yorker* writes, "The raindrops made a pattern on the street like television snow, or like the scrambled letters with which a word processor fills the screen before a completed electric spark clears it all into perfect sense."[63]

In fact the technical sequence I have just described is a vast simplification, making no mention of device drivers, font libraries, file systems, virtual memory, physical hardware interfaces, or numerous other variables. Moreover, as Sean Cubitt has shown, light—the actual physics of waveforms and radiation—is not an alternative to or the transcendence of materiality, but rather it is the very epitome of materiality. Printing and engraving technologies have long depended on the use of light as a medium through their manipulation of texture—thus techniques including mezzotint, aquatint, and lithography helped create the conditions for the rasterized Cartesian grid of the CRT (and subsequently the LCD and plasma) display, a function of what Cubitt describes as the arithmetical "triumph of enumeration over continuum."[64]

An aesthetics of luminescence in turn had direct implications for how writers thought about and articulated their experience of the medium, particularly in terms of speed and freedom or flexibility. "Working with light on a screen rather than marks on a page, I find that I can noodle and doodle and be much more spontaneous," said Russell Banks. "The faster I can write, the more likely I'll get something worth saving down on paper. From the very beginning, I've grabbed onto any technology that would allow me to write faster—a soft pencil instead of a hard pencil, ballpoint instead of a fountain pen, electric typewriter instead of manual."[65] Implicit also in Banks's remarks is a drive toward the dematerialization of inscription, a notion we have already traced back to Michael Heim's very early observations about word processing. Over and over again, writers reflected upon the speed of word processing: "I worked from a very early age on a typewriter which I was given—a little portable—so I was in one sense prepared for the computer because I never wrote by hand. It proved to be a godsend actually because I was able to write much more rapidly on it and was able to complete *The Tunnel* in a year, writing about half of it in that last year on the computer," commented William Gass.[66] Similarly his colleague Stanley Elkin told *Time* magazine in 1981, "If I'd had [the Lexitron] in 1964, I'd have written three more books by now."[67] Sometimes speed was assessed even more functionally, as by romance author Robyn Carr: "I haven't timed my typing but I think I'm up to 100 words per minute," she claimed after switching to a CP/M-based system called the Burroughs Redactor III.[68]

In 1983 Merv Griffin interrupted studio guest Michael Crichton, who was there to promote a nonfiction book on the computer revolution. "What is a word processor?" he demanded. "Do you talk at a machine and it writes your book?" "No, no," replies Crichton. "A word processor looks like a typewriter but it has a screen instead of a sheet of paper. And when you type, the words appear on the screen. They are just electrical impulses on the screen. Which means that you can move around on the screen, change what you've written, pull blocks of text, put them elsewhere. You have complete freedom." As Crichton talks, he gestures in an animated fashion, using his hands to suggest the bounded space of the screen, the horizontal lines of text, and the dynamic interplay of the features.[69] While Michael Heim might have explained the box in the air that Crichton's hands were so busily describing by reference to electronic text's supposedly purely "symbolic" character, I prefer Daniel Chandler's characterization: "suspended inscription."[70] Suspended inscription means that the stored record of a text is separate from whatever the medium or surface on which it is ultimately printed or inscribed in more palpable form. When one writes with a pen, creating and composing a text is coterminous with the work of inscribing it; and it is the same with typewriting, the press of a key initiating a simple act of mechanical leverage that sends the type bar hurtling toward the page, its kinetic energy thus impressing the inky fabric of the ribbon it encounters in its path onto the paper behind it in the embossed shape of a letterform. (As Ivan Flores notes, with a manual typewriter, the harder you hit the key, the darker the impression.)[71] But word processing is different. As Flores puts it, "The quest was for some way to break this relation so that something could be done between the time a key is hit and the information is printed."[72] This is a powerful observation, but we can go even further. With Flores's help, we can see that word processing's suspension of inscription is in fact a suspension both temporal *and* locative in nature. In other words, there is a gap or delay between the act of writing the text and rendering it in its documentary form; moreover, the record of the text and its documentary instantiation occupy physically distinct media and surfaces.

Every writer who made the transition to the new technology would feel the consequences of this shift. Text was stored on a disk or on (or in) some other media until such time as it was sent to a printer, but in the interim it would hover just out of reach behind the glowing glass screen in the writer's gaze. Text blinked and flickered, brightened and darkened; sometimes it even left phosphor cells burned into the glass. These new writing surfaces—some square and some rectilinear, some flat and some convex, the letters green or amber on black, then (improbably) grey on blue, and (eventually) black on white—is what afforded the user the illusion of real-time control over a document.[73] Time and again writers were struck by the instantaneousness—there is no other word—of their new word processor. A press of a key and the document changed before the user's eyes, not just simple backspaces to fix a typo but deleting or moving whole fields of text or changing a word throughout an entire manuscript. Robyn Carr again: "If I have to change a character name—and I've had to on my first two books—I can just code the change into the machine and it will automatically be changed on the disk," enabling the revised text to appear (as Carr goes on to note) on the printed hard copy.[74] Word processing thus emerges as a combination of the indefinite suspension of inscription *and* the allure of real-time editorial intervention—in stark contrast to the typewriter, where writing and editing were of necessity mechanically separate operations.[75] In effect, the writing surface becomes a Möbius strip, with the writer both writing and not-writing at the same time—which is to say, writing in multiple locations simultaneously, one text made of light and another stored indefinitely prior to printing onto yet another (even more durable) surface. Word processing was thus the simulation *and* the suspension of writing—"writing" and "not-writing"—instantaneously manifest and yet potentially endlessly postponed. This is ultimately what Heim and so many other converts were celebrating.

Unsurprisingly, Continental philosophy and theory tried its best to account for these newfound qualities of writing. Friedrich Kittler used the example of word processing and specifically WordPerfect to launch the argument in his most famous refutation of the myth of digital transcendence, drawing attention to the difference between human- and machine-readable writing by noting that the actual word "WordPerfect" is too long to be faithfully rendered under the software's own native DOS regimen of eight-character file names.[76] For Kittler, word processing marked a definitive break with prior writing technologies because words stopped being mere signifiers and become executables instead: "Surely tapping the letter sequence of W, P, and Enter on [a] keyboard does not make the Word perfect, but this simple writing act starts the execution of WordPerfect."[77] And it does so without ambiguity, perfectly predictable each and every time. Jacques Derrida, who in 1986 traded in his Olivetti electric typewriter for what he referred to thereafter as *le petit Mac* (it was photographed for *Circumfessions*), also pronounced upon word processing, most notably in a 1996 interview devoted to the subject. Derrida mentions his surprise at a word processing program prompting him when his paragraphs got too long; after some initial hesitation he opted to respect the constraint, "so submitting . . . to an arbitrary rule made by a program I hadn't chosen."[78]

What is perhaps most striking about many of these early accounts of word processing, however, is simply how much they had in common with one another, even when they disagreed. Consider these two testimonials:

> Poets do have to make changes, but they cannot think so; they must think that the next word and phrase will be perfect. At times, and these are the happiest, they have the feeling that words are being given to them with absolute finality. The word processor works directly against this feeling; it tells you your writing is not final. And it enables you to think you are writing when you are not, when you are only making notes or the outline of a poem you may write at a later time. But then you will feel no need to write it.

> It's a different kind of timing, a different rhythm. First of all, you correct faster and in a more or less indefinite way. Previously, after a certain number

> of versions, everything came to a halt—that was enough. Not that you thought the text was perfect, but after a certain period of metamorphosis the process was interrupted. With the computer, everything is rapid and so easy; you get to thinking you can go on revising forever.

The first is from the poet Louis Simpson writing in the *New York Times Book Review* to warn his fellow travelers away from the seductions of word processing; the second is from Derrida's interview.[79] Simpson sees the word processor as beguiling, whereas Derrida is more interested in the difference it makes. Yet both organize their remarks around the open-ended nature of revision and both question the conceit of perfection. Nor was it always easy to predict a given writer's loyalties and predispositions. The conservative columnist and essayist William F. Buckley Jr., for example, was a vocal early adopter and wrote a sharp response to Simpson that amounts to taking him to task for mystifying writing in ways Derrida would have no doubt appreciated.[80]

For others, however, the future of writing was already on the wall. The Czech media philosopher Vilém Flusser, working at the same time as Heim, yoked word processing and electronic document technologies to automation, believing it was only a matter of time before humans would become superfluous to the very act of writing, which machines would be able to execute far more efficiently.[81] In particular, Flusser distinguished between what he termed "inscription" and "notation"; the former is characterized by instruments for digging and incising, for gouging writing *into* a surface; the latter by brushes and pigments or guiding ink through channels for writing *on* a surface.[82] In this formulation (as Russell Banks also seemed to intuit) the word processor is the culmination of centuries of technological refinement that began with brushing ink onto parchment using featherweight quills. But there is also a strong relationship between the effort that writing requires owing to the resistance of the surface, and the writing's long-term durability. Letters carved into granite last longer than ink stains on paper, which last longer than pixels illuminated on the screen. For Flusser, a truly momentous occurrence in the history of writing was the advent of the practice of firing soft clay tablets in the oven to harden them, thereby transmuting a soft and pliable writing surface into an unyielding one for posterity.[83] Word processing could be said to do much the same, allowing the malleable structures of computer memory (we call it software, after all) to harden into hard copy when the writer is ready. The kiln of the digital age is thus the laser printer. Its beams of light give us letter-perfect copy.

Anne Rice got her first computer, an Osborne 1, from Macy's on credit.[84] She used WordStar and quickly realized it was doing something profound to her. "Well, I think once you really get used to a computer and you get used to entering the information from that keyboard, things happen in your mind, I mean, you change as a writer," she claimed in a 1985 CBS radio interview after writing *The Vampire Lestat*.

> You're able to do things that maybe you never would have thought of doing before. I think your mind has to meet that challenge. I think it takes about two years to realize what this thing can really do so that you begin to use its whole potential. Because what it does is it forces you to come up to it. It says, I can do anything you want now—so think, what do you want me to do?[85]

This unadulterated freedom—or at least the supremely powerful perception of such—has everything to do with the material characteristics of word processing just described.

In another interview a few years later Rice would go even further, calling her subsequent *Queen of the Damned* (1988) "the first book in which I really used the computer as the pure poetic tool it is capable of being."[86] Like Stanley Elkin, she describes being able to move effortlessly across the full expanse of her text, backward and forward with near simultaneity, editing and revising at will. She contrasts the experience to the typewriter, and the accumulation of a large and "ponderous" (her word) draft, the necessity of having to retype whole chapters after introducing even minor changes. "You're dealing with a mechanism, with labor, and all of that's swept away by the computer," she exults. "You're no longer making the mechanical compromises that move [the writing] away from poetry."[87] And thus, the inevitable conclusion, as ominous in its own way as a dynasty of immortal vampires: "There's really no excuse for not writing the perfect book."[88]

THREE

AROUND 1981

In 1981 (the year the *Times* reported on President Carter's mishap with the Lanier) neither "mouse" nor "window" were terms most people would have associated with computers. The Lanier itself was one of the early standalone word processing systems then on the market, along with the IBM Displaywriter, the Lexitron, the CPT 8000, and the Wang, to name some of the more popular models and brands. The first-generation of Z-80 and Intel 8080 eight-bit microcomputers, typified by Cromemco and IMSAI, had meanwhile given way to so-called integrated personal computer systems from Apple, Tandy, Commodore, and Osborne. IBM was only just getting into the PC business that year, and Microsoft had just recently moved from Albuquerque to Washington State with a contract to produce something called a Disk Operating System (their primary competition was the rival operating system CP/M). Disks themselves were literally floppy—5¼- or 8-inch black squares that actually waggled up and down; the smaller 3½-inch floppies (which were not floppy but rigid) did not yet exist. Hard drives existed, but were seen only rarely on the sorts of machines most consumers would buy. *Time* magazine was still more than a year away from declaring the personal computer its "Machine of the Year." *Popular Electronics* had featured the Altair 8800—usually considered the world's first microcomputer—on its cover in January 1975, but in 1981 was still a year away from renaming itself *Computers and Electronics*. *Tron* and *War Games* had not yet made it to movie theaters, and William Gibson had not yet coined the term "cyberspace" in his fiction. *Wired* magazine and the World Wide Web, meanwhile, were both a decade away, give or take. In the arcades, *Ms. Pac-Man* was queen of the quarters. The "Internet" was a messy amalgamation of government and academic research networks, useful mainly for exchanging files or something called "electronic mail." These networks crisscrossed with commercial services like The Source and CompuServe (America Online did not yet exist). Bulletin boards—not part of the Internet as such—were an alternative ecosystem, attractive to the hobbyists and enthusiasts who also sometimes still assembled their own computers from kits. Users were accustomed to acquiring new programs by manually typing them from source code printed in books, newsletters, and magazines. In 1981, Cold War tensions between the United States and the Soviet Union were some of the most heightened on record and would get worse before they would get better. In March of that year President Reagan would come close to dying from a bullet fired by John Hinckley Jr.; in April the Space Shuttle *Columbia* lifted off for the first time; in July Sandra Day O'Connor was nominated to the U.S. Supreme Court, and Charles, Prince of Wales, married Lady Diana Spencer; in August MTV went on the air for the first time, launching the network with "Video Killed the Radio Star" by The Buggles; in October Anwar Sadat was assassinated and the Dodgers beat the Yankees in the sixth game of the World Series.

The beginning of the 1980s was not the beginning of word processing, either in the workplace or in the home. As we will see in later chapters, there are a number of earlier exemplars, and a writer just beginning to use a word processor in 1981 had arguably already missed the chance to be called a genuine early adopter. Nevertheless, 1981 was about the time word processing entered public awareness at large and became a topic of conversation and debate in the literary world as elsewhere. An article published that year in *The Economist* claimed that 500,000 people were "already" using one.[1] Len Deighton recalls a lunch from that era at which the London-based mystery writer H. R. F. Keating quipped, "It used to be that when writers got together they talked about money: now they talk only about word processors."[2] *Time* magazine, meanwhile, ran a piece called "Plugged-In Prose," with the following lead: "Every month more writers are discarding their pencils and typewriters for 'word processors,' technical jargon for small computers with typewriter-like keyboards, electronic screens for scanning and manipulating text, units to store information, and high-speed printers."[3] The key development was the rise of the integrated systems, which made personal computers attractive to people without technical backgrounds. These systems provided novices with all of the hardware they needed in a single package, including CPU, monitor, tape or disk storage, and often a printer and even software. Retail computer stores also began appearing—clean, well-lit spaces that encouraged window shopping and kibitzing with the sales staff. Computers, in other words, had come of age as consumer electronics.

New software was released almost daily. Though WordStar was quickly recognized as the leading program of its kind, there were literally scores of alternatives on the market, with choice dependent on not only features and capabilities but also compatibility with what were generally mutually incompatible host systems. In our own era, when one word processor has held onto the dominant market share for so long, it is difficult to appreciate the challenges this entailed. A detailed buyer's guide published in a 1983 issue of *Writer's Digest* compares some three dozen different programs across a matrix of over two dozen variables and features, including (besides price and compatibility) the availability of block commands like copy, move, or delete, search and replace, and file backup as well as options for form printing, pagination, superscript and subscript, proportional spacing, underling and emphasis, word counting, and "screen display same as printed copy."[4] Word processing also spawned ancillary software genres, notably spell-checkers and thesauri (not built into many programs) but also typing tutorials as well as programs for creating indices, tables of contents, footnotes, and outlines.

While the abundance of choice may seem empowering in retrospect, it was also a significant obstacle to getting started. Stewart Brand's *Whole Earth Software Catalog*, an offshoot of the legendary counterculture publication *Whole Earth Catalog*, put its finger on the problem: "For new computer users these days the most daunting task is not learning how to use the machine but shopping."[5] Charles Bukowski (who wouldn't begin using a computer in earnest until he got a Macintosh for Christmas in 1990) nonetheless captured the moment in a poem written circa 1985 called "16-bit Intel 8088 chip."[6] Laced with references to brand names such as Apple, Commodore, and IBM, it includes this observation:

> both Kaypro and Osborne computers use
> the CP/M operating system
> but can't read each other's
> handwriting
> for they format (write
> on) discs in different
> ways.[7]

The poem concludes by contrasting the fundamental irreconcilability of all of these artificial systems with the natural world that unchangingly, unknowingly coexists with them.

As with other constituencies, authors coped with the vagaries of the computer marketplace as best they could. Flip through an issue of *Writer's Digest* from around 1981 and you will see page after page of advertising for computers, word processors, and word processing software alongside images of manual and electric typewriters and fountain pens—not least because the ads for the new technology often used the latter to convey a sense of trust and continuity with the past. A wrong choice on the part of a first-time buyer—who might arrive at the local computer store primed with good or bad advice from friends and the latest industry gossip gleaned from computer magazines, only to first be subjected to a hard sell from the sales staff—could mean thousands of dollars wasted or worse. Steven Levy, a technology journalist, summed up the nature of the decision: "I compare using a word processor to living with somebody. You go into it with all kinds of enthusiasms, and things are wonderful. Then, you see other word processors promising more. More features, friendlier style. The question is, is it worth tossing over a relationship in which you've invested months for a word-transpose toggle, an indexing function you'll use maybe twice, and a split-screen capability? A choice of a word processor is a major life-decision, and no-one can afford (in terms of time, money, or emotional capital) to play the field."[8] This account, as casually sexualized as it is earnest, makes clear that a word processor was not only the most vexed and complicated but also in many ways the most *intimate* piece of software many consumers would acquire. Word processors externalized individual conceptions and presumptions about the act of writing, forcing users to confront—in a literal feature matrix—what was essentially a model of their own personal writing practices.

The stakes could sometimes be even higher. 1981 was also the year the U.S. Centers for Disease Control first recognized the existence of acquired immune deficiency syndrome (AIDS). Judy Grahn, a poet living in California and identifying as both lesbian and feminist, was writing using a short-lived word processing system manufactured by the Exxon Corporation. The book she was working on was not poetry; it was *Another Mother Tongue: Gay Words, Gay Worlds,* which when published in 1984 was widely recognized as a pioneering work of gay history and historiography (also fused with autobiography). Grahn had begun using the Exxon 510 as part of a concerted effort to finish and publish the book as quickly as possible:

"The Exxon system would automatically repaginate and renumber footnotes, so it was the most time-saving thing you can imagine. . . . Time meant everything to me as I wanted to get the book out to a generation of Gay men who were coming down with AIDS and HIV positive diagnoses; the Right wing fanatics were telling them ugly stuff about punishment and retribution, and my book refuted all that."[9] The unlikely existence of an Exxon word processor was a by-product of the energy giant's 1980 take-over of Zilog, maker of the successful Z-80 microprocessor, which rivaled the popularity of the Intel 8080 as the foundation of many first-generation computers.[10] Grahn concludes, "The word processor cut at least a year from my writing time."[11] But in as much as word processing was regarded as transformative by so many upon their first encounter with luminescent letters on a glass screen, it is also clear that the reality of contending with those early systems was often more mundane. Users were expected to come to grips with stacks of recondite manuals in ways that seem unthinkable now; they banded together in users' groups, started up homegrown newsletters to rival the newsstand glossies devoted to personal computing, and generally learned to hack around and figure things out for themselves. Such was the nature of computing at a moment before the popular advent of the Internet, before the graphical user interface, before even the general availability of hard drives.

These kinds of experiences—anything but perfect—are the focus of this chapter. Computers were indeed "moving in," as *Time* magazine would put it the following year, and they were not shy about it, demanding not only power outlets but also dedicated furnishings and other accessories, even sometimes their own room in the house. As one of the contributing writers to *Time* was to put it: "Computers were once regarded as distant ominous abstractions, like Big Brother. In 1982 they truly became personalized, brought down to scale, so that people could hold, prod, and play with them."[12] Certainly 1981, the year that saw the release of such systems as the IBM PC and the Osborne 1, was at least as much of a watershed in that regard. While some writers recoiled at the prospect of a computer under their roof, others took the leap, discovering an affinity for the technology (warts and all) in ways they could not have anticipated.

Over the course of four articles written between November 1981 and June 1982 for the newsstand magazine *Popular Computing*, none other than Dr. Isaac Asimov (then sixty-one years old) recounted his coming to grips with a word processor; together these articles constitute an unusually detailed self-portrait of a writer's encounter with the technology.[13] He begins by narrating the arrival of the machine on May 6, 1981, at his thirty-third-floor New York City apartment. The system was one of the most popular ones on the market, a TRS-80 Model II manufactured by Tandy and sold through its Radio Shack retail stores. It was accompanied by a daisy wheel printer and Radio Shack's own word processing software, Scripsit.[14]

The beginning is not auspicious. Ray Bradbury, a famous contemporary, is said to have once declared: "With a book tucked in one hand, and a computer shoved under my elbow, I will march, not sidle, shudder or quake, into the twenty-first century."[15] Asimov, by contrast, describes his stance thus: "head high, eyes flashing, fists clenched, and brain paralyzed with fear."[16] He also expresses skepticism that the technology can do anything to improve his already prodigious output, roughly a book a month. Nonetheless, he endeavors to honor the experiment, contemplating the sealed boxes and imagining them filled with "arcane incunabula."[17] He asks us to accept that he has stacked them in a corner of his writing office and then practices navigating the space until he can be assured of avoiding any accidental contact with them, even in the dark. But that is merely prolonging the inevitable. A few days later a technician, dispatched from Radio Shack headquarters in Fort Worth, arrives to set everything up. Asimov's wife, Janet, makes an appearance at this point, suggesting that the computer take pride of place in the living room instead of the back office. The machine is powered on and the technician types some lines with Asimov watching over his shoulder. "Words and sentences appeared, and parts were then erased, substituted, transferred, inserted, started, stopped."[18]

Dr. Asimov is given homework: manuals to read, workbooks to complete, tape-recorded tutorials to listen to. He confesses in his diary that night: "'I'll never learn how to use it.'"[19] Time passes. The computer looms as an ominous presence in the otherwise tranquil domestic space of the apartment, dominating its surroundings: "I had, by now, developed the habit of flinching when I passed the computer corner, throwing up my arm as though to ward off an attack."[20] Gradually, however, Asimov lets his determination get the better of him and begins to experiment with the machine. On June 14 he has the typescript of an article to revise. He transcribes and edits it with

Scripsit. And somehow, without discernible reason or explanation, everything (so to speak) clicks:

> I will never know what happened. The day before I had been as innocent of the ability to run the machine as I had been while it had still been in its original box. A night had passed—an ordinary night—but during it something in my brain must finally have rearranged itself. Now, there I was, running the machine like an old hand. In making my corrections, I could even use my right hand on both the "repeat" and the cursor arrows, without looking, and that little blinking devil jumped through every hoop in sight.[21]

Thus empowered, he immediately turns to revising a waiting novel in the same fashion. He whistles while he works. Perhaps channeling Twain, he declares out loud to his wife: "'All it takes is grit, determination, a sense of buoyant optimism, and good old Yankee know-how.'"[22]

This story of a "word-processor" (as the title of the original *Popular Computing* article has it rendered) is also, let us not forget, being told by a master storyteller. It delights with the image of the patriarch of the scientific literary imagination, author of the *Foundation* trilogy and giver of the Three Laws of Robotics, humbled by the mute pile of consumer electronics in the corner of his apartment. It narrates a Eureka moment then scripts a pitch-perfect montage of happy productivity, very near cinematic in its elements; all of it wrapped within the self-effacing charm that no doubt endeared Asimov to his readers.

Initially he admits only modest gains. He is already a blazingly fast typist, claiming 90 words per minute, so he doubts the machine can make him any faster. But he does note that the keys are easier to depress and make less noise, which allows him to hear the television while he works, a first. The most salient observations, however, concern revision. As has been noted, Asimov's output was prodigious. He described himself as a "one-man book-of-the-month club," and there is not much hyperbole in that statement.[23] Asimov had grown used to simply "barreling along," as he puts it: "When an article requires, let us say, ten pages, I end up with ten uncrumpled pages," he says.[24] Errors—typos—were the copy editor's problem. The writer's job was to produce the copy:

> The result is that any page I type may smoke slightly from the heat generated by the speed, but it is also garbled. The word "the" is spelled in various

> fashions—“eht,” “eth,” “teh,” “th e” and so on. These are distributed, with fine impartiality, randomly over the page. I make few distinctions between “seep,” “seen,” and “seem” (or “esem” for that matter), and I am quite apt to mention “the button of the flask” when I am referring to that part of the flask at the opposite end from the top.[25]

The result, he finds, is that he receives frequent last-minute phone calls from his copy editor pleading for him to reveal the meaning of phrases such as “snall paint,” which is really “small print.”[26] But then something curious happens. The more he works with the word processor, the more he finds himself compelled to proofread and correct his own typos and errors:

> Bang goes the “F1” and the “u” and the “F2” and “cold” suddenly becomes “could” and no sign exists that it was ever anything else. I send the cursor flying, up and down, left and right, and all the “cart”s become “cat”s, and all the “hate”s become “heat”s (or vice versa). What’s more, commas go zooming in by the thousands and interrogative sentences which, in the old days, had question marks attached at the rate of one in three, now have one inserted with loving care every time.[27]

What’s happening here is the redistribution of labor. Asimov, as author, is performing work that he would have previously and cheerfully left to his copy editor. In his memoir, *I, Asimov,* published some years later, he insists that he used the Tandy for “one job and no more—the preparation of manuscripts.”[28] Short pieces of under 2,000 words—and it must be said there were many of these—he composes directly on the computer; but he continues to produce first drafts of longer works on his trusty typewriter, taking comfort in the sight of the rapidly accumulating sheaf of pages. The computer is the venue for producing clean copy.[29] “Manual corrections make the manuscript look messy. . . . My editors will stand a little messiness from me, but with everyone handing in clean copy that has been corrected invisibly on the screen, I’m afraid my messiness would stand out and give editors the subliminal notion that my writing was poor simply because it was messy.”[30] But Asimov, for all of his attentiveness to the impact of the technology on his writing process, does not question whether he is in fact now spending *more* time with each manuscript by doing this new kind of textual work. In the pages of *Popular Computing,* he was even less equivocal: “I end up with letter-perfect copy and no one can tell it wasn’t letter-perfect all the time," Asimov tells us. "Then I have it printed—br-r-rp, br-r-p, br-r-p—and as each perfect page is formed, my heart swells with pride. . . . So it's not a question of speed after all, but of perfection," he concludes. "And I hope the copy editors appreciate the new me."[31]

The Talisman is a novel about a boy named Sawyer who has to undertake a long and arduous journey, much of it on his own, to retrieve the magical McGuffin of the title to cure his movie star mother of cancer or a cancer-like disease. There's more to it than that of course, and along the way there are dangers and adventures aplenty; but the most important thing to know is that the story is set in not one world but two. There is the United States, across which the twelve-year-old Jack Sawyer must travel from Maine to the Pacific; and then there are the Territories, a kind of mirror world, whose geography more or less conforms to that of the American interior (albeit at an altered scale) and whose inhabitants are "twinners" of people Jack meets in this world. The Territories has its own currency and its own customs, its own politics and its own problems. It even has its own language, in which Jack finds himself natively fluent. He travels there by "flipping," a maneuver at first accomplished with the aid of a mysterious juice (actually malt liquor given to him by his spirit guide, a stereotypical "magical Negro" character named Speedy). But Jack soon realizes that the hooch is just a placebo—the ability to flip has always been innate within him, as it is for many of his allies and the chief villains of the story.

By the time it appeared in 1984, this novel was already one of the most anticipated (and most hyped) of the decade. The first printing was a massive 600,000 copies.[32] Its authors were Stephen King and Peter Straub, who were something of literary twinners themselves: at the time of their collaboration, both were broadly typed as horror novelists. King had just finished *Cujo* (1981), which was preceded by *Carrie* (1974), *Salem's Lot* (1975), *The Shining* (1977), *The Stand* (1978), *The Dead Zone* (1979), and *Firestarter* (1980), among others. In other words, the bedrock of the King canon had been well laid, and he was indisputably a publishing phenomenon. Straub was more of an acquired taste, more grown-up, the writer one turned to when wanting something besides rabid dogs, vampires, and the apocalypse. Books like *Julia* (1975, which King had blurbed), *Ghost Story* (1979), and *Shadowland* (1980) had all been well-received by the critical establishment as well as by fans. The two writers had become friends after meeting in

London and had reportedly decided to write a book together as early as 1977 or 1978 but didn't get going on the project in earnest until several years later. (They worked for different publishers and had preexisting commitments they had to fulfill first, and, in any case, the ideas took time to gestate.) When the joint contracts for *The Talisman* were signed with Viking and Putnam in September 1981, the advance was reportedly one of the largest the industry had ever seen. All that remained was the writing.[33]

By then the book had been plotted in some detail. There was a thirty-plus-page outline, even some first passes at dialogue and scenes.[34] There was also broad agreement between the two regarding the narrative arc, characterization, and themes. "This is a cheerful and fairly unpretentious story of a good kid with a lot of guts trying to save his mother's life—and experiencing every wonder in the world (and out of it) along the way," King wrote in a synopsis.[35] But King lived in Maine and Straub in Westport, Connecticut, a five- or six-hour drive away. How then to exchange work at the pace their creative energies demanded? King and Straub had both already taken note of word processing's increasing popularity among their fellow writers, and had separately been considering buying systems of their own. It seemed like a good investment, and both were successful enough to pay the price tag. In what would surely have been a leap of faith, however, they also decided that the technology would somehow facilitate the long-distance collaboration that was now at hand. King wound up with his celebrated Wang (see Chapter 4), which he acquired after a visit to the company's headquarters in nearby Lowell, Massachusetts. Straub shopped around and talked with other writers, but eventually settled on an IBM Displaywriter 6580.[36] He was familiar with IBM products and had been using a Selectric typewriter to finish his then-current novel, *Floating Dragon.* The price tag wasn't cheap: $14,000. The salesman gave Straub assurances that he and King, despite their different word processing systems, would be able to send electronic files back and forth using something called a modem.[37] In other words, one could almost say they would be able to flip to each other's Territories.

Straub immediately began using the Displaywriter to finish off *Floating Dragon,* which was about two-thirds complete.[38] "Almost as soon as I began I realized how much work that thing was going to save me," he recalls.[39] When it came time to start in on *The Talisman,* King spent several days in Connecticut. The two took turns in front of Straub's new machine. "He'd sit down for a little bit and write something, and then I'd sit down.

He'd look over my shoulder. Now that was a great way to start," Straub recalls.[40]

The Displaywriter was a relatively new product in 1981; IBM had introduced them in June of the previous year. Despite the company's pioneering involvement in office word processing, this was—incredibly—the first IBM word processing product to feature a full-size video display screen. And despite the price tag and the increasing availability of personal computers (including, as of August 1981, the IBM PC), the Displaywriter found a market, at least for a time. (It was the first commercial word processor with a spell-checker.)[41] The journalist William Zinsser, who had previously written the best-selling book *On Writing Well,* opted for a Displaywriter and then used it to write (and typeset) a popular book entitled *Writing with a Word Processor* (1983).[42] The Canadian novelist Marina Endicott also used one, as did the American humorist Bruce Feirstein (he wrote *Real Men Don't Eat Quiche* with it).[43] Endicott recalls learning the machine while working for the Ministry of Education in Toronto, noodling around with fiction when the job permitted. "I'm pretty sure I wouldn't be a writer at all except for the IBM Displaywriter," she still believes.[44]

The system had no hard drive. Using it required first "booting" it with a program stored externally on an 8-inch disk. (Despite the elephantine dimensions, each of the disks stored only about 280 kilobytes, nowhere near enough for a full novel.) The actual word processing program was known as Textpack. After a minute or so of clicking and clacking from the boot disk, the IBM logo would appear in phosphorescent green in the lower right-hand corner. The screen itself was 80 by 20 characters. A user would begin a document by selecting a "typing task" from the main menu. The Displaywriter was compatible with several good-quality IBM printers (one of which, the 5215, used the same golfball technology as the Selectric) and—crucially—it allowed for modem connections. Depending on the particulars of the actual hardware Straub and King were each using, transmission rates would have been between 300 and 1,200 baud, or between 30 and 120 characters per second—so no more than around 1,000 words a minute even under the best of circumstances. Still, it was faster than driving back and forth between Bangor and Westport, and faster by orders of magnitude than the U.S. mail. "Steve could actually see the lines going across his screen," Straub recalls. "I couldn't on mine. I had to call it up later, after it was all in. But Steve could see things as they came in, which I thought was something."[45] Indeed it must have been: watching text scrolling down a video screen in one's home office as it was being transmitted over a telephone line.

But this solution wasn't without glitches. As was the case with the mirror-world of the Territories, not everything always transferred in quite the same way. "The IBM is a very adaptable machine, and by fooling around with the codes on mine, I could make it possible for 99% of our stuff to go through without a hitch," Straub remembers.[46] He elaborates:

> There were differences in protocol, in underlying language between the IBM and the Wang systems. So quote marks didn't come across the same way, italics didn't come across the same way, all these little embedded codes were different from machine to machine. So we had to work out a secondary language for all these things that caused glitches in the manuscript. So let's say for example instead of quote marks we used a pound sign. Then when we got the material then you did a global search and replace. For a pound sign you put quotes and for whatever the other thing was you put italics. There were a lot of those codes but once we got kind of into the swing of things it all went pretty easily.[47]

Straub, who also began subscribing to computer magazines around that time, found these rituals appealing: "I liked the idea that I was learning my way through a rather deep forest and was succeeding in putting things together. Later on when it became a little easier I kind of missed all the old mumbo jumbo that you had to do."[48]

Twinning and flipping are easy metaphors to borrow from the novel to foreground the practical difficulties of collaboration and technological compatibility. But so tidy an analogy can only go so far. An August 1983 letter from King to Straub describes King's meeting with their editor, Alan Williams at Viking, who had various changes to suggest; King agreed with most of them and in the letter tells Straub he will have an assistant "produce a second draft from my Wang discs," and then he and Straub can meet to go over it.[49] Such a seemingly insignificant detail attests to a scene of writing that is invariably messier and more complicated than any simple medium-specific account would suggest, in this case involving not only King and Straub's creative imaginations and their modems, but also the advice of an editor, the labor of a paid assistant, disk storage, hard-copy printouts, postal correspondence, and in-person meetings. All of these modes of communication and interaction were thus part of what it meant to write *The Talisman.* As was also the case with Asimov, word processing took its place in and among existing work habits and networks—and it reconfigured them to varying degrees—but it never simply replaced them.

Around 1981 Amy Tan was living in San Francisco and working as a medical journalist.[50] Her first novel, *The Joy Luck Club* (1989), was still years away. On the job she used a TRS-80 to write and file her stories. Tan ruefully dubbed the machine "Bad Sector" for the number of times its disks would return that dreaded error message when she was saving or retrieving a file.[51] (A bad sector message could mean different things, but none of them good: maybe a physical flaw on the fragile surface of the disk itself, maybe a hopelessly corrupted file.) "Happened a lot back then," Tan remembers.[52]

A self-described "geekette," she followed with interest developments in the burgeoning home computer industry—much of which was centered just south of the city in Silicon Valley.[53] That was the year Adam Osborne debuted the Osborne 1, the first genuinely portable personal computer. What made it portable was not so much its weight (it came in at a hefty 23½ pounds), but the fact that its core components—keyboard, screen, processor, and disk drive—were all integrated within a single, self-contained unit that was transportable like a carrying case. Because the globetrotting executive was seen as a key demographic, the dimensions of the case were carefully calibrated to fit under a business class airline seat. The Osborne 1, as we have already seen, also came bundled with WordStar, and its relatively low cost, ostensible portability, and above all self-contained all-in-one design made it a popular choice for a number of writers. Tan would have had no way of knowing, of course, but this is the machine Ralph Ellison would shortly purchase, as would a young writer named Michael Chabon, as yet unpublished.[54]

The Kaypro II debuted the following year, at the West Coast Computer Faire held in San Francisco. The brainchild of Silicon Valley entrepreneur Andrew Kay, it had many similarities with the Osborne 1, chief among them the integrated case design. It was a few pounds heavier, but its screen was almost twice the size (Ellison, like many Osborne users, would eventually get a separate external monitor to avoid having to peer into the notoriously tiny screen). And like the Osborne, the Kaypro was based on the Z-80 chip and its attendant CP/M operating system. The computer came bundled with an application suite dubbed Perfect Software, which included the word processor Perfect Writer. (This was eventually replaced by WordStar in the start-up bundle, and it had briefly been preceded by yet a third word processor, Select.) The Kaypro was also popular with writers. Arthur C. Clarke got one. So did Andrei Codrescu. "It was incredibly hip," he remembers in a 2014 piece on the death of Andrew Kay. "And it made writing something very different."[55]

Amy Tan would buy her Kaypro II in 1983, presumably purchasing it at the standard retail price of $1,795 (the same as the Osborne).[56] She had not yet taken an active interest in writing fiction, but she had transitioned from medical writing to freelance corporate communications work for firms like Pacific Bell, AT&T, and IBM. Her Kaypro II, and Perfect Writer in particular, thus saw heavy use. Indeed, Tan, a self-described workaholic, found herself burning out, a victim of her own success. Nonetheless, at the very height of this period she found time to help start a Kaypro users' group called Bad Sector, the same name she had unceremoniously given to her first computer.

Users' groups were one of the fixtures of early computer culture. The first and most famous of them all was the Homebrew Computer Club, which met in the auditorium of Stanford University's linear accelerator. Adam Osborne was a member, and at one of those now quasi-legendary gatherings Steve Wozniak had demonstrated his prototype for what eventually became the Apple II computer. But users' groups were all very "homebrew"; they tended to coalesce organically, their members finding each other through notices tacked up in computer shops (or on virtual bulletin boards), ads in newsletters, and word of mouth. Typically they were tied together by an interest in a common system or product. Members would swap tips and help one another troubleshoot. Software was demoed and doubtless copied and exchanged, sometimes on the down-low. Yet manufacturers often worked closely with the users' groups, giving members opportunities to preview products that were still under testing and give feedback. Industry reps would frequently turn up at the meetings. Computer bulletin boards and early online networks existed, but these face-to-face gatherings furnished the most vital form of community.

By the time of its founding in 1984, Bad Sector was a relative latecomer to this scene: there were several other Kaypro users' groups already active in the Bay area, including BAKUP in Oakland and another called KUG, which was simply the Kaypro Users' Group. Those meetings would sometimes attract upward of a hundred people. The founders of Bad Sector, however, were motivated by the desire for something on a smaller, more intimate scale. They also wanted to have some fun, to geek out in style. Alongside Tan, the founding members of the group were the technical writer Ray Barnes and the photojournalist Robert Foothorap, who was then stringing for *Time* magazine and had photographed Wozniak and Jobs with their Apple I prototype. "From the Bay area there were two national stories of merit," Foothorap recalls. "One was computers and the other was HIV."[57]

The fullest account of Bad Sector's activities comes to us from the journalist Ben Fong-Torres, also a member, who often wrote for *Rolling Stone.* In 1985 he contributed an article about the group to *Profiles,* a handsome, well-appointed organ published by Kaypro itself.[58] In some detail Fong-Torres recounts the goings-on at a typical monthly meeting, held in this instance in the living room of Ray Barnes's Haight-Ashbury apartment and attended by some eighteen people; three of them, he notes, were women, including Tan. "People had questions; invariably, others had answers," Fong-Torres wrote. "Or at least clues."[59] The origin of the group's colorful name also soon became manifest: "One person got a 'bad sector' message when he formatted a disk. When he tried to find the bad sector, he was greeted with: 'No bad sector.' 'I'd throw the disk out,' said Barnes. But he also offered a suggestion: 'If you have the new CP/M you can format on either drive. It's a tick and goes away. No big deal.'"[60] The meeting proceeded in that vein: "Foothorap offered two disks of friendly tips on Perfect's Writer, Calc and Filer software; Barnes mentioned a new Kaypro Users' Handbook that fleshes out Kaypro's own thin manuals. A guest told the group about a section of an independent guide to Perfect software 'that I found invaluable, three pages of undocumented commands.' Barnes suggested he photocopy those pages so that 'Bad Sector' can make them available to anyone interested. Agreed."[61]

Tan, meanwhile, offered to host a tutorial for a subset of the group specifically interested in Perfect Writer and word processing. Though little remembered today, Perfect Writer was widely used in the Kaypro community. It was notable for its ability to split the screen in half and display two different documents simultaneously, something not even WordStar could do at the time (this functionality was a result of Perfect Writer's roots in the powerful Emacs software used for text editing on mainframe systems). "Perfect Writer is at its best with long or complicated documents," the *Whole Earth Software Catalog* noted.[62] One can thus understand the attraction for fiction writers. Fong-Torres then quotes Tan at length on how the group got started: "Someone told me, 'I know someone who bought the same computer as you; you might want to make a connection.' My interest was to have someone to go to whose computer I could borrow if mine broke down. And that was Robert Foothorap. I told him I was interested in starting a group for support purposes, and to exchange information. I didn't want something like BAKUP which is so big that I'd wind up spending 20 hours a week taking phone calls. I wanted something simple and fun."[63]

Fun was unquestionably a key part of it. "Amy set the atmosphere," Barnes recalled, "through her love for chocolate, liqueur and coffee."[64] Indeed, Fong-Torres treats the appearance of a sumptuous Black Forest cake (as opposed to a demo from the attending Kaypro rep) as the centerpiece of the evening. Tan herself described it all thus, in the shorthand of a tweet: "We had reps from Kaypro, the WP software at meetings, free Kaypro mugs. B&B. chocolate decadence. Grew from 5 members to 100."[65] But the eventual size and popularity of the group proved its undoing. It folded not long after Fong-Torres finished his article, after a run of only about a year and a half.

Ultimately Bad Sector is just one more piece of an individual writer's history. Certainly Amy Tan's involvement in the group must be placed in proper proportion to other events marking her often tragic or turbulent biography, from childhood family trauma to her later struggles with Lyme disease. Still, most accounts of Tan's early life move straight from her unfulfilling career as a journalist and technical writer to the submission of her first short story, "Endgame," to the Squaw Valley writers' workshop in 1985 and its subsequent publication in *Seventeen* magazine. At the very least, however, the Kaypro users' group she cofounded furnished an important social outlet during this interregnum. The meetings, often in her and her husband's home—their street address was the one listed in the directory of users' groups in *Profiles* magazine—afforded her access to a new kind of creative community. The story of Bad Sector complicates a narrative in which technology, through the purgatory of a technical writing career, serves only as a foil for Tan's subsequent accomplishments as a novelist.[66] Moreover, it was an unusually prominent leadership role for a woman amid the culture of early home computing: "Everybody else was pretty much male tinkerers," recalls Foothorap.[67]

Was the Kaypro—with its self-contained setup, the keyboard unfolding from the case to reveal the waiting screen—a beacon for someone contemplating a career as a fiction writer? It's hard to imagine that it wouldn't have been. (By her own account, Tan has since written all her books on computers, starting with that same Kaypro.)[68] Regardless, the story underscores just how deeply Tan was invested in the early culture of word processing. She was an expert user, completely up to date on current developments and insider news. "The computer is my life," was what she told Fong-Torres at the time. "It's great finding people as fixated as you are."[69]

Readers who scrutinized the acknowledgments in Arthur C. Clarke's novel *2010: Odyssey Two* (1982), the long-awaited sequel to *2001: A Space Odyssey* (1968) that had supposedly brought him out of retirement, would have encountered the information that it was written on something called an "Archives III" microcomputer running WordStar, and that its manuscript had been conveyed from Colombo, Sri Lanka (where Clarke had lived since the 1950s), to New York on a 5¼-inch disk. The text's final words are these: "Last-minute corrections were transmitted through the Padukka Earth Station and the Indian Ocean Intelsat V."[70]

Clarke had acquired the suggestively named Archives computer sometime in the second half of 1981, when he was about a quarter of the way through the initial draft of *2010* (he had bought an Apple II at about the same time, but used it mainly for games rather than for writing).[71] The Archives III would have retailed for some $8,500; it ran the popular Z-80 CP/M combination, but befitting its name it also featured a five-megabyte hard disk drive, uncommon for its day. Clarke had an assistant transcribe the text of an existing manuscript into the machine (which he christened "Archie," once his own nickname) and proceeded from there, an instant convert: "I was in exactly the same position as an Egyptian scribe who had spent his life carving inscriptions on granite—and suddenly discovered ink and papyrus."[72] Clarke was thus using a word processor a full half decade before writing his affectionate send-up of the technology, "The Steam-Powered Word Processor." He testifies with the now-familiar litany: that it removes the "drudgery" from writing, and thereby improves the quality of his writing; and that every manuscript can be a "perfect" production, effortlessly output by the printer. "I can honestly say I have never touched a typewriter since that day," he concludes.[73]

Sri Lanka, of course, was not the usual locale for a prominent British-born science fiction author—but Clarke had made it his home in no small part due to its suitability as a base for his lifelong interest in scuba diving and underwater exploration. He disliked travel and rarely left the island, becoming an influential public figure and patron there and in 1983 establishing a science and engineering institute, the Arthur Clarke Centre for Modern Technologies. The location also undoubtedly contributed to the fact that Clarke quickly learned to use a modem in conjunction with his computer, which allowed him to access early online news and data services. After entrusting the fragile floppy containing the complete manuscript of *2010* to international mail, he employed the modem technology to transmit a text file containing additional revisions to New York, thus yielding that final, dramatic detail resonant with the imagery of the giant radio-telescopes at the beginning of the book and film. This electronically encoded stream of bytes, which undertook its own adventurous voyage through Earth's atmosphere and telecommunications infrastructure, was named ODYCOR, or Odyssey Corrections.[74]

Archives, Inc. was but a short-lived entrant in the volatile personal computer industry, and a couple of years later Clarke was in the market for a new system. IBM had just sent him one of their PCs, perhaps hoping for an endorsement. Instead he selected a Kaypro II, in part because a science writer researching a book on personal computing wanted to interview him by modem, and Kaypro offered software to make that possible.[75] Clarke, meanwhile, was also in conversations with MGM and director and screenwriter Peter Hyams about the movie adaptation of *2010.* Hymans considered Clarke's involvement in the project crucial, but Los Angeles and Sri Lanka were twelve hours apart in international time zones, thus making regular phone calls not only expensive but inconvenient. Clarke quickly realized that the modem could become a "time shifter" (his phrase) and that each of them could write long-form communiques while the other slept.[76] Hyams duly acquired a Kaypro of his own, and he and Clarke began an extended electronic correspondence, using the Kaypro II machines and a piece of communications software called MITE to transmit back and forth text files that they would then store on diskettes with their home systems—essentially the same method that (doubtless unknown to them) Straub and King were employing to write *The Talisman.* The resulting correspondence dates from September 16, 1983, to February 7, 1984, and is collected in a book that the two planned almost from the start of their collaboration, entitled *The Odyssey File* (1984).[77] It recounts discussions about casting, plot alterations, set design, and many other aspects of the production, as well as conversations about mutual acquaintances, current affairs, and baseball scores. This, of course, sounds very much like a typical back-and-forth on electronic mail, a term that in fact Clarke used to describe the process. But email in the sense we know it was beyond their reach; although the "Internet" existed in its nascent form as the ARPANET, it was not generally accessible to the public, making a science fiction author (and British national living in Sri Lanka) and a Hollywood film director unlikely candidates for exceptions.

Clarke describes the transmission process that quickly became routine for both of them: "When I get up in the morning, I switch on my machine and tell it to call Peter's office number," he writes. "I then take charge, just as if I were sitting at the keyboard in Culver City [Los Angeles], and ask it to list all the files that are stored on its No. 2 floppy disk, designated B. . . . Peter's machine then swiftly lists the names of all its current files, and displays them to me in numerical order: 'PH49 PH50 PH51. . . .' I look at the last entry and see if it's a new number—in other words, a file I've not already received. If it is, I type 'SEND B:PH51,' tell my own machine 'RECV B:PH51,' and sit back while the file comes through."[78] Transmission took about a minute per page. He then used WordStar to open and read the files, a practice that facilitated the production of the book Clarke and Hyams wrote together to document the correspondence. Very early on Clarke writes: "I'm one step ahead of you. I'm instantly Wordstarring and printing out our immortal prose. . . . Since you will also be accumulating printouts it can be continuously edited from your end and would be ready about the same time as the movie."[79]

The material accumulates rapidly as the correspondence progresses, the medium becoming second nature for both of them. Hyams: "I started out with a beautiful disk. It was so clean you could eat off it. Now . . . every day it is getting more and more cluttered with PH's and ACC's," he wrote, referring to the naming convention they adopted for their respective files.[80] Clarke, meanwhile, notes that he cannot see the text as it is transmitting but wonders if it might be possible to print it in real-time: "There's an Echo option I've not investigated."[81] There were, of course, glitches. Eventually, for example, one of Clarke's floppy disks fills up: "Here's something that may make your flesh creep," he writes to Hyams after the fact. "When I tried to save this file, for the first time I got the message FATAL ERROR—DIRECTORY FULL."[82] And one time Clarke's Kaypro seemingly dies: "The 1 amp fuse at the back had blown, presumably through a voltage surge."[83] He replaces it and urges Hyams to install a surge protector of his own.

It is impossible to read these mundane mentions of blown fuses and finicky diskettes without mentally juxtaposing them to the striking renditions of computers in both *2001* and *2010;* there is HAL's iconic red lamp, of course, but there are also the scenes of Dr. Chandra floating weightless, deep within HAL's crippled processor core, sliding crystalline shards of memory in and out of their sheaths in order to debug and reboot the creation that is his life's work. Clarke himself was sufficiently absorbed by the minutiae of the modem process that he wrote up detailed instructions for the use of the MITE program, which he published as an appendix to *The Odyssey File.* Readers of the book would thus not only enjoy the behind-the-scenes look at the film's creation, they would receive the benefit of a detailed tutorial in the use of a piece of software they might own themselves. Alas, Clarke's earnest and meticulous instructions were out of date by the time the book went to press: "Steven's [his assistant] latest dispatch informs me there is a new and much improved version of MITE waiting for me in Los Angeles. I hope it does not differ too greatly from the program in which I have invested so much blood, sweat, and tears."[84]

Clarke would subsequently claim that computers, particularly WordStar, had brought him back out of retirement by restoring pleasure to the act of writing. It "doubled" his production with a "quarter" of the effort.[85] While there is undoubtedly something genuine to the sentiment, it shouldn't be accepted at face value. As his biographer notes, Clarke had "retired" on other occasions before.[86] Nonetheless, there is no mistaking the joy with which Clarke approached personal computing. Ray Bradbury narrates his friend's excitement browsing in a computer shop on one of his trips to Los Angeles when the film version of *2010* opened.[87] Clarke returned home to Sri Lanka with a newly minted Kaypro 2000, an early laptop design (weighing in at a comparatively svelte twelve pounds and looking more like something from the set of *The Empire Strikes Back* than Hyams's movie, it was also the first Kaypro to run MS-DOS). Clarke in fact collected a number of different machines, and was thereafter frequently photographed with a computer in the background. (Besides the Kaypro, photos show him seated with a Commodore Amiga and various Apple systems.) Later he became permanently confined to a wheelchair as a result of post-polio syndrome. At the time, however, he had this to say: "If I'm eventually incapacitated I'll get wired into a word processor—and then, Isaac, look out!"[88]

When a writer of Andrei Codrescu's caliber invokes an image like "writing with light on glass" and comments of his own Kaypro II that it "felt both godlike and ephemeral," it is hard to perceive anything but a stark break between word processing and earlier technologies of inscription.[89] How can ink and the messy clatter of ribbon or typing ball, to say nothing of carbons and correction tape, possibly compete with the allure of a word processor? The glare would be too strong. The reality, however, was often different, as the stories of the authors in this chapter show: Asimov making his grudging Yankee accommodation with the machine, or Clarke undertaking his meticulous documentation of his modem's idiosyncratic protocols, or Tan or Straub acknowledging the appeal of the technology even as they wrestle with its imperfections. Word processors may have saved writers work, but they also created new kinds of labor, whether it was sweating over manuals and tutorials, or taking on responsibility for aspects of the writing process previously left to others, or just troubleshooting and experimenting. "With microcomputers," wrote a columnist for *Writer's Digest* in 1981, "you have to be a tinkerer as well as a writer."[90]

The experience of the enigmatic Manhattan-based author Harold Brodkey, who purchased a DECmate computer in late 1981 or 1982, offers a vivid contrast to Codrescu. The Digital Equipment Corporation, best known for its high-end mainframe and business computers, had introduced the "Mate" as its entry into the burgeoning personal computer market. Initially, as Brodkey told *PC Magazine,* he was unimpressed, finding that it encouraged him to write hastily and sloppily.[91] "Those little flickers of fire are not prose," he said of the characters on the screen.[92] He returned to longhand, but the investment was too great for the computer to sit unused. And his most important project was no farther along than it had been. Brodkey, who published his fiction in the *New Yorker,* was widely regarded as one of the most promising writers in America; the contract for his novel had first been signed in 1964, and many doubted that the book, then titled *A Party of Animals,* would ever appear. It was eventually published (at least in part) as *The Runaway Soul* in 1991. A visitor to his Upper West Side apartment during this period described the scene thus:

> Harold has raised expectations so high . . . that of course he had to introduce roadblocks in his path. He bought a computer. But this was still the era when a computer filled a whole room, when only industry and spies owned them, when one had to master a whole new method of writing, of programming. Harold invited me to see the machines humming and buzzing in one room, which someone from IBM was teaching him, day after day, week after week, how to operate. The entire long, sprawling manuscript

> would have to be transferred to the computer. Only then could it be properly analyzed for content, repetitions, inner consistency, and flow.[93]

Computers had moved in, as *Time* said; but making space for them—making space in the creative process, to say nothing of cramped Manhattan apartments—was not always easy. By the time the *Paris Review* came calling in 1991, this same room was described as housing "a collection of computer equipment worthy of a bond-trading room."[94]

Brodkey died of complications from AIDS in 1996. Back in 1982, however, at about the time Brodkey was first coming to grips with the DECmate, the journalist James Fallows wrote a column for the *Atlantic* entitled "Living with a Computer." In his case the computer was an Optek Processor Technology SOL-20 that he had acquired in early 1979. Fallows takes nothing for granted as he details its operations: "When I sit down to write a letter or start the first draft of an article, I simply type on the keyboard and the words appear on the screen," he relates to his readers.[95] Fallows subsequently remarked on the extent to which programming and "tinkering" were part of the day-to-day experience of word processing at the time: "It was like in the era of the first cars," he recalls. "If I hadn't paid attention to how these things fit together I wouldn't have been able to make it work."[96]

Such accounts serve to complicate simple distinctions between the kind of avant-garde techno-experimental writing often celebrated by academic critics, on the one hand, and conventional writing—mere word processing, dull and uninspired from a technical standpoint—on the other. Even for writers such as Fallows or Straub, who weren't seeking to self-consciously generate avant-garde texts, the degree of hacking, improvisation, and problem solving the early systems required meant that the borderlines between innovation and convention, novelty and standard operating procedure, were frequently blurred in practice. When Arthur C. Clarke and Peter Hyams zipped messages back and forth to one another halfway across the planet and through multiple time zones using temperamental modems that had to be nursed through power surges and other glitches, they were conscious of doing something altogether very new, regardless of the fact that the final form their texts took was a mass-market paperback (and a movie tie-in at that). Writing, as Brodkey's visitor observed, now sometimes meant something like programming, the algorithmic manipulation and analysis of text. Certainly that wasn't the experience of every author; but for some, text indeed became something more akin to code, to be broken down and built up again in heretofore unthinkable ways.

Michael Chabon, meanwhile, had a much simpler problem. It was 1985, Chabon was twenty-two at the time, and he was at work on what would become his first novel, *The Mysteries of Pittsburgh.* His Osborne 1, which he had purchased three years earlier, had remained in flawless working order. "It never crashed, it never failed, and I loved it immoderately," he says.[97] The problem was, Chabon couldn't reach the keys. (More about that in a moment.) The piece in which he provides these details was published in the *New York Review of Books* some twenty years later; there, the staid genre of the literary bildungsroman becomes laced with particulars of personal computer technology that furnish the framework for self-reflection. Near the climax of the essay, for example, Chabon narrates sliding a diskette into his "B" drive and then pausing: "Was this really the kind of writer I wanted to become?"[98] It's a neat set piece: initializing a disk marks the initiation of a writer's career.

But back to that problem: Chabon couldn't reach the Osborne's keys because the computer—all twenty-five pounds of it—was balanced atop a chest-high workbench that was the sole furnishing in the basement room in Oakland, California which he had claimed for his writing. So, as Chabon tells it, he dragged a steamer trunk over to the workbench, placed a folding chair on top of the trunk, and then sat down. "I found that if I held very still, typed very chastely, and never, ever rocked back and forth, I would be fine."[99]

Writing with light—from a folding chair balanced atop a steamer trunk in a dingy basement room. One could do worse for an image of the precarious perch many writers were to find themselves on as they sat down in front of their first word processor in or around 1981.

FOUR

NORTH OF BOSTON

In a black-and-white photo of Stephen King in his Bangor, Maine, office taken in 1995, we see him in profile in a swivel chair, feet propped on his desk, Corgi dog camped underneath. He has a legal pad on his lap and a pencil in his hand. One of the best-selling novelists in American history is in his element. In the background, just part of the clutter, is a computer. More precisely, it is a Wang System 5 word processor, complete with keyboard and a built-in monochrome screen.[1] The System 5 was marketed by Wang as a "standalone"—it was intended for the customer who was interested in purchasing the company's famous word processor rather than one of its complete office systems consisting of multiple workstations on what we would today call an intranet. It relied on a brace of 8-inch floppy disk drives and was typically mated to a Diablo impact printer. Its molded enclosure was a study in space-age curves, straight out of *The Jetsons*.

The picture is by the literary photojournalist Jill Krementz, and appears in a book of her author portraits called *The Writer's Desk* (1996). The concept behind the book is a simple one: to capture writers at work, or more precisely in their workplace (all fifty-five of the images that make up the book are interior shots, though one, the photo of William F. Buckley Jr., is in the backseat of a car). The book would have appealed to a certain type of buyer, the one who enjoys a glimpse into the secret, supposedly solitary habits of authors. (King himself has toyed with this kind of tradecraft—for example, in his novel *Misery* when the stricken novelist Paul Sheldon confounds the terrifying Annie Wilkes with his deliberately obfuscating talk of "Webster pots.") Much of what you see in Krementz's photos is what you would expect: if not Webster pots (which don't really exist), then stacks of papers, books on shelves. Typewriters of course, old and new, manual and electric. There are pipes and cigarettes, and there are cats and dogs; in one or two cases there are children, though (interestingly) never spouses. There are rugs and lamps and wastebaskets. And then there are the desks. Some are improvised and impromptu, like Ross MacDonald's lap desk; some are suitably grand, great reassuring slabs that no doubt serve to ground the writer, landing strips for high-flying thoughts; some are utilitarian and some are unmistakably unique, like Rita Dove's *Stehpult,* a handmade German-style standing desk.

Though the book is called *The Writer's Desk* and indeed showcases the desks prominently, of course it isn't really a collection of images about furniture, or even the spaces the furniture fills: it's about the authors themselves, or more precisely it's about the juxtaposition of the authors—their faces and bodies—with the physical settings they inhabit. Kevin Kopelson meditates on this same subject in *Neatness Counts,* noting that the desk is the "stabilizing center" of whatever room it occupies, both metonymy and synecdoche for its owner's identity.[2] On one level the impulse here is unabashedly voyeuristic, as John Updike acknowledges in the introduction, losing no time in comparing what's captured in these images to the "beds of notorious courtesans."[3] We cluck our tongue at the messiness, or else sigh with satisfaction as we contemplate the cozy study, all just as we might have imagined. Each image evokes its own counterpoint, a single frozen moment of unusually intimate access set against the backdrop of the writer's published work and public reputation.

Only ten of the writers in Krementz's book are, like King, photographed with a computer visible anywhere in the frame. Amy Tan has moved on from her Kaypro II to a Mac PowerBook, possibly the same model Mona Simpson is photographed with. Veronica Chambers and Cathleen Schline also both have laptops. Roy Blount Jr. has an IBM-compatible with a great big monitor. Edwidge Danticat has a Macintosh desktop. John Updike leans over a terminal of seemingly indeterminate manufacture, though it too is a Wang product. John Ashbery works at a typewriter, but has a PC alongside. The computers are perhaps the most jarring of the workplace paraphernalia that pop up here and there in the photos, but like the phones and fax machines that are also occasionally visible, they are a reminder that writing is a business and industry as well as an art and a calling. The images containing computers are perhaps especially compelling because they expose the writer's relationship to such a quotidian commodity. So-and-so's prodigious, hand-tooled desk is all her own, but look! *Amy has the exact same computer I do.* (So it was that in 2012, J. K. Rowling—often touted as refusing to use a computer—created a product boost by declaring that her MacBook Air had "changed her life.")[4] Paradoxically, perhaps, the juxtaposition of these well-known authors with mass-produced consumer electronics does as much to humanize them as the lines on their faces, the angle of their eyes.

We see the author's software less often than we see their computers. Half of the time the computer is shot obliquely or from behind. In several of the other photos, glare and lighting prohibit a look at what's on the display. There may or may not be a trace of text visible on John Ashbery's screen (this seems appropriate). Only two authors uninhibitedly reveal their displays: Russell Banks, who has a file menu visible, and Roy Blount Jr., who appears, puckishly, to be watching a thumbnail-size video of *Butch Cassidy and the Sundance Kid* (1969)—the text accompanying his photo wants to know, "Why write, when you can watch a movie on your typewriter?"[5] (Today it is not uncommon for writers to deliberately disconnect their computer from the Internet to avoid the seductions of Netflix and everything else.) As for the authors whose screens we can't see, perhaps they're just faking it, posing for the camera, the computer not even powered on—more so than the typewriters or legal pads it seems, the computers invite us to distrust the candor or spontaneity of the image. If the authors don't show us their screens, what are they hiding? The desk*top* is clearly not the same as the desk: while the latter is the nominal subject of each image and the primary surrogate for the individual writer's identity, the digital desktop is, with just a couple of exceptions, oblique or occluded—literally screened.

In 1995 when Krementz's photo of Stephen King was taken, his Wang word processor had been around for nearly fifteen years—ancient and obsolete by any technological standard. How then to explain its ongoing presence in what looks to be a relatively small and modest workspace? As we saw in Chapter 3, King bought it to write *The Talisman* with Peter Straub. But it quickly became integral to the workflow in his busy home office. Marsha DeFilippo, King's longtime personal assistant who came to work for him in 1986, recalls that some of her first assignments consisted of using the Wang to key in the then-typewritten manuscripts of *The Eyes of the Dragon* and the *Tommyknockers* so they could be mailed to Viking on disk.[6] But when the Krementz photo was taken, the venerable Wang would have been thoroughly outclassed by the PowerBooks and x86 PCs King's colleagues were then acquiring. (Wang Laboratories had in fact filed for bankruptcy three years earlier, a severe blow to the small industrial city of Lowell, Massachusetts, where the company was headquartered, some thirty miles north of Boston.) Supplies for the printer, not to mention the 8-inch disks, would have been increasingly hard to come by. Was King still using it, and if so why? Perhaps the Wang had become something like a talisman itself.

King was quick to grasp the potential for adolescent humor in the particular brand of word processor he had chosen; while working with Peter Straub, he recalls retiring to his study to "pound" on his "big Wang," and then, "I would call up Peter and say, 'It's ready.' And then I would send him what I had pounded."[7] But there is also no question King was intrigued by what it would mean for his writing. In his introduction to his 1985 short story collection *Skeleton Crew* he writes about the effortless way in which words, sentences, even whole paragraphs could be blipped in and out of existence:

> In particular I was fascinated with the INSERT and DELETE buttons, which make cross-outs and carets almost obsolete. . . . I thought, "Wouldn't it be funny if this guy wrote a sentence, and then, when he pushed DELETE, the subject of the sentence was deleted from the world?" Anyway, I started . . . not exactly making up a story so much as seeing pictures in my head. I was watching this guy . . . delete pictures hanging on the wall, and chairs in the living room, and New York City, and the concept of war. Then I thought of having him insert things and having those things just pop into the world. Then I thought, "So give him a wife that's bad to the bone—he can delete her, maybe—and someone else who's good to maybe insert."[8]

Writers are used to playing god, but now the metaphor was literalized. Characters lived or died at the touch of a button. Whole worlds could be born (or obliterated) with a few volts of electrical energy. These ideas became the basis for a short story he wrote soon after acquiring the Wang. Titled simply "The Word Processor" when it was first published in *Playboy* in 1983, it was reprinted as "Word Processor of the Gods" in the *Skeleton Crew* collection as well as adapted for the television series *Tales from the Darkside* in 1984.[9] Likely the first extended fictional treatment of word processing by a prominent English-language author written in a realist manner, it is worth examining more closely.[10]

The basic premise follows from King's musings above. Richard Hagstrom is a schoolteacher and writer, none too successful (his first novel was a flop), but he keeps at it, gamely plugging away in a converted shed behind the house. He is estranged from his indolent and unattractive wife, Lina, who, when not playing bingo or binging on sweets, mocks his literary pretensions; their one child, Seth, is a dull and apathetic teenager and similarly estranged from his father, able to tolerate only the company of the fellow miscreants in his punk rock band. But we soon learn that Richard's life could have turned out very differently. His high school sweetheart (Belinda) ended up marrying his bully of an older brother (Roger) because Richard retreated from a confrontation with him; but that couple's son—Jon—is (improbably) a sensitive and precocious teenage whiz kid, in short everything Richard longs for in an offspring and that Seth is not. When the story opens, Belinda and Jon and the loutish Roger have all just died in a van wreck with a drunken Roger behind the wheel.

Jon, however, has left behind an unlikely birthday gift for his Uncle Richard: "At first glance it looked like a Wang word processor—it had a Wang keyboard and a Wang casing."[11] Jon has built a Frankenstein machine hacked together with (upon second glance) parts from not only Wang but also IBM and Radio Shack, as well as a Western Electric telephone, an erector set, and a Lionel electric train transformer. Richard has coveted a word processor for years: "I could write faster, rewrite faster, and submit more" (312). Of course simply buying one would have been out of the question: "The Radio Shack model starts at around three grand. From there you can work yourself up into the eighteen thousand-dollar range," Richard observes (312). Jon's creation is heavy and ungainly and ugly, but Richard sets it up in his study anyway, symbolically displacing his electric typewriter, which gets relegated to the top of a file cabinet. When he turns the word processor on, it displays a message: HAPPY BIRTHDAY, UNCLE RICHARD! JON, the green-tinged words "swimming up" out of the darkness of the heretofore empty screen (311). Richard, understandably, is shaken: "Christ," he whispers, and almost shuts it off. But he doesn't. (Of course he doesn't.)

While the story that follows reads as little more than stock *Twilight Zone* fare (and is in fact reminiscent of an episode from the show, as we will see in Chapter 7), it nonetheless manages to capture much that is worthy of our attention about the newness of the electronic word, its exaggerated detail offering us a time capsule of how the technology was received and perceived very close to when it first entered public consciousness. We have seen that the word processor introduced a new intermediary element—a literal screen—between the writer's fingertips and the printed page. This screen—cool, opaque—signified ultimate possibility, a kind of heterotopia, the setting and stage on which the computer could flaunt its otherworldly powers, and an irrefutable reminder—always right there, squarely before our eyes—of the computer's alien otherness. A screen placed language in suspended animation. On a screen, words could be deleted and inserted, searched and replaced, eventually formatted, even illuminated, like the medieval rubrications of old. ("They took one look at the screen and saw the magic we could do with words," a Wang salesman is said to have commented.)[12] Language became pliable, malleable—in a word, writing became *processed.*

Writing became processed, but it also became executable. This is the command that first seizes the attention of Richard Hagstrom as he studies the machine: EXECUTE. On the Wang, the key was double-size. "It wasn't a word he associated with writing; it was a word he associated with gas chambers and electric chairs . . . and, perhaps, with dusty old vans plunging off the sides of roads" (313). In fact, however, the specter of "execution" has been associated with writing since the Bible: "The letter killeth, the spirit giveth life."[13] What explains such a formulation? Part of it, of course, is the *Logos,* the presence of the divine word in Christian theology. But unlike a person in conversation, a word on a page is silent and unyielding, immutable and impassive. As Walter J. Ong has noted, "The spoken word is always an event, a movement in time, completely lacking in the thing-like repose of the written or printed word."[14] Speech is associated with presence and life, writing (and books) with absence and death. ("In yon dark tomb by jealous clasps confined," as Oliver Wendell Holmes once put it.)[15] The appreciation of the spoken word as an event—Ong points out that in Hebrew the same word is used for both—is crucial for grasping the import of EXECUTE, for it means that writing is taking on the characteristics of oral performance. Indeed, Ong describes digital communication as a "secondary orality," aiming to capture with this term something of the contradictions of a script that is both written but also not-written, seemingly endlessly mutable and changeable. Computers thus make the written word *actionable.* EXECUTE was the juice, the lightning, the scroll in the forehead of the golem (to invoke the old Jewish legend). The digital word was not only a commodity to be milled and processed in some vague industrial sense, but also a conveyance laden and latent with potential energy; functional, programmable, a kinetic happening that could be unleashed with a keystroke or command. Little wonder that the parable of the sorcerer's apprentice is often conjured in discussions of computing: When the machine executes, it does our bidding (or so we hope), sometimes irreversibly (or so we fear). This is Richard (and surely King himself) coming to terms with the *eventuality* of electronic writing.

MY BROTHER WAS A WORTHLESS DRUNK, Richard types, and presses EXECUTE; the words appear on the screen. Richard is impressed: "Whether it would store information in the CPU still remained to be seen, but Jon's mating of a Wang board to an IBM screen had actually worked" (314). (In fact, no such keystroke was necessary to transpose input text to screen when using an actual Wang, further suggesting the extent to which King was struck by the concept of "executing" in his imagination.) Gazing at the text hovering on the screen in front of him, Richard associates the word processor with another oracular relic, this one the source of a painful childhood memory—his Magic Eight-Ball toy, smashed to pieces by his miscreant brother, Roger: "It wasn't nothing but a cheap, shitty toy anyway, Richie. Lookit there, nothing in it but a bunch of little signs and a lot of water" (314). On one level, this association calls to mind the proverbial black box, the technology that "just works"—we don't know how, we don't know what's inside, we don't need to know. But unlike the Eight-Ball, hermetically sealed until that traumatic moment, Jon's word processor messily exposes its innards, displaying the cannibalized bits and pieces of its components from the get-go. Indeed, the casing is literally cracked open: "not gently, either; it looked to [Richard] like the job had been done with a hacksaw blade" (307). Yet Richard, of course, knows nothing about the thing's actual operation despite his eyeballing of the components. For most users, a word processor would have seemed very much like a Magic Eight-Ball, a black box (or orb), whose proclamations were delivered up—swimming up out of darkness on an empty screen—with a rough shake and anxious peering.

Idly, Richard types a sentence describing the photograph of his wife that hangs (watchfully) on his studio wall, then rejects it and hits the DELETE key. Suddenly her picture disappears from the wall along with the words on the screen. All material traces of its presence have been effaced, undone, smoothed over, as if they never were. The hook the picture hung on is gone, and the wall where the hook would have been screwed in is clean and unblemished. Here is what, as we have seen, struck so many early adopters as one of the essential differences between word processing and their previous writing implements. The eraser on the pencil leaves its shavings, ink leaves stains, the white-out brush deposits its rough film; a manuscript page always bore the scars that had been inflicted on it in the course of the rough surgery of revision. But words on the screen vanished instantly, utterly, if indeed they had ever really been there at all. Like some modern-day self-healing superplastic, the surface simply sealing itself over an incision, or else like water (perhaps inside a Magic Eight-Ball), momentarily disturbed, now smooth and placid once again.

Thus begins the process by which Richard begins to deliberately and self-consciously copy edit and revise his own life. There is a catch, however: just as the genie grants only three wishes, the word processor has a built-in constraint. Its jury-rigged parts and transformers begin to burn themselves out, cooking up faster and faster each time Richard switches the thing on. Soon the ominous word OVERLOAD begins to flash on the screen. So after a few tentative forays (deleting and inserting the picture, giving himself a bag of gold) Richard gets down to business. MY SON IS SETH ROBERT HAGSTROM, he types. Then presses DELETE. Awful! Has Richard just murdered his own son? In one sense he has—pressing the fateful key seeming little different from pulling a trigger. But there is a difference, King wants to suggest; the strange new technology of the word processor offers a moral escape hatch. Murder, you see, leaves a body, evidence—a smoking gun. But like his wife's picture disappearing without any material trace that it had ever existed, Seth is simply removed from the world. His shoes are not piled up in the hall in the house (Richard checks), and the basement rec room is not a slovenly rehearsal space for his punk band. He can't be *murdered* because he never was. In short order, Richard proceeds to INSERT Jon, DELETE Lina, and then INSERT Belinda.

"How *could* a machine do such a thing?" (317), Richard wonders. How indeed? When the word processor finally burns itself out from OVERLOAD, Richard turns around to find Jon peering over his shoulder at the smoldering remains. The two converse with an easy rapport, as though they have been father and son always. And in a real sense they have, since Seth and Lina are not no-more but instead *never really were*. How could a machine do such a thing? It could because King has grasped something essential about the ontology of writing on the screen, something expressed more fully just a few years later when Michael Heim wrote about word processing. As we have seen, Heim notes that the transition from writing on paper to writing on the screen is not merely one of material substrate, but involves the redefinition of writing from inscription to the abstract realm of algorithmic symbol manipulation.[16] Unlike the messy physical word of diacritics and erasures, the "cross-outs and carats" King mentioned in his *Skeleton Crew* introduction, symbols are abstractions. They are always either there or not there, and they are never ambiguous. For most users, there are no messy remainders with symbols, no smoldering wires or fragments of the Eight-Ball, no eraser shavings, no palimpsest of text still faintly visible underneath the revision. Seth and Lina and the other elements of Richard's life can be blipped in and out of existence because Jon's creation somehow confers the magic, godlike power to treat them as symbols, temporarily elevating Richard's life to an empyrean away from the messy material world of history and lived experience.

One final plot detail serves to dramatize this effect: Richard, having INSERTED Belinda and Jon back into his life with the machine just seconds away from its terminal OVERLOAD, has time to "execute" perhaps one last instruction. What will that instruction be? That the word processor has been completely debugged and will work flawlessly forever? That he becomes a brilliant and renowned novelist? Or maybe only long life and happiness with his new family? But he types nothing. "His fingers hovered over the keys as he felt—literally felt—all the circuits in his brain jam up like cars grid-locked into the worst Manhattan traffic jam in the history of internal combustion" (324). Richard, in short, is suffering from information OVERLOAD just like the machine with which he is now so closely identified. Electronic, industrial, and neurological descriptors collide with one another in this short passage. Having manipulated the most basic living symbols available to him, the nuclear family unit, he is paralyzed with the possibilities of anything else. He stops. He freezes. He cannot *process*. And the circuits burn out for good.

The writer's revenge is an old story, despite the newfangled technology at the center of this one. And it's not so much fiction that we have here as a Romance, in the original sense in which Nathaniel Hawthorne meant it ("somewhere between the real world and fairy-land, where the Actual and Imaginary may meet, and each imbue itself with the nature of the other").[17] Everything is perfect at the end of King's tale, right down to the aroma of cocoa that comes wafting in from the house on the crisp night air. As D. H. Lawrence put it, there are never any muddy boots in a Romance.[18] No muddy boots here either (Seth's "ratty tennis shoes" in King's text), and no troubling reminders and remainders—except one. As Richard turns to leave the shed with Jon, he takes one last look at the scorched components of the word processor: "Delete it," Richard commands in the final line of the story (325), thereby consigning the word processor itself, the last mute material remainder of his revisions, to the scrap heap.

Today Krementz's photograph of King is most widely seen on the cover of his nonfiction treatise, *On Writing* (2000), which he completed while recovering from near-fatal injuries after being struck by (as it happens) a van on a roadside in rural Maine. In that book he recalls the physical agony of his first session sitting upright in front of his word processor trying to write during a long convalescence. (King reverted to longhand for his next novel, *Dreamcatcher* [2001], as a result: "This book was written with the world's finest word processor, a Waterman cartridge fountain pen," he tells us in the afterword.)[19] In *On Writing* he offers a response of sorts to Krementz's focus on writers' desks, describing his own relatively modest one, which he keeps pushed into a corner as a reminder about perspective. "Life isn't a support system for art," he says. "It's the other way around."[20]

Meanwhile, on Stephen King's official website, visitors can explore an animated rendition of King's "office," a virtual quasi-three-dimensional walk-through replete with memorabilia from his career.[21] One navigates by—awkwardly—using a kind of faux control panel displayed at the bottom of the browser. Eventually we find our way into a re-creation of King's study, and there on the desk (which is not in a corner) sits a computer clearly branded as an Apple with Microsoft Word open on the screen.

We can move in closer: the document on the screen is labeled "Letter from Stephen regarding *Under the Dome* and *The Cannibals*." We can read the text. It appears at first that King is merely rehearsing the composition history behind his novel *Under the Dome* (2009), which began life in the late 1970s as a manuscript of the same title (now lost), was then resuscitated as a manuscript (typescript, actually) called *The Cannibals* (aborted and also lost), before King finally returned to and completed the thousand-page novel many years later. But the typescript for *The Cannibals* was not lost after all, as we learn from reading the "letter": "So, for your amusement, and as an appetizer to *Under the Dome,* here are the first sixty pages or so of *The Cannibals,* reproduced, warts and all, from the original manuscript which was dredged up by Ms. Mod [Marsha DeFilippo] from a locked cabinet in a back room of my office. I'm amused by the antique quality of the typescript; this may have been the last thing I did on my old IBM Selectric before moving on to a computer system."[22] King was working on *The Cannibals* while filming *Creepshow* in Pittsburgh with George Romero in 1982, so the timing here accords with his acquiring the Wang in preparation for writing *The Talisman* with Straub. But the manuscript isn't being presented solely for our amusement or as a teaser for the forthcoming book; reading the text of the letter to the end, we learn that there had been suggestions on the Internet that King had lifted the idea of a whole town covered over by a mysterious dome from the 2007 *Simpsons Movie,* which features just such a plotline. King posted the first sixty pages of *The Cannibals* in September 2009, and another sixty pages a month or so later, thereby clearly establishing the originality of his main idea. He thus uses a virtual surrogate of his office computer as the Potemkin platform for a digital facsimile of a typewritten manuscript, its "warts" (cross-outs and carats, the very stuff vanquished by the word processer) serving as seemingly irrefutable evidence of its pre-digital provenance, some two-and-a-half decades before the *Simpsons Movie.*

Stephen King has always been interested in exploring new writing tools and new platforms, manifesting a consistent interest in e-books and Internet distribution, for example. He has authored several works available originally only or still exclusively in digital form, including "Riding the Bullet" (2000), an as-yet unfinished epistolary novel called *The Plant* (2000), and "UR" (2009), which does for the Amazon Kindle what "Word Processor of the Gods" did for the Wang. Today King favors Apple computers, laptops mostly, and Microsoft Word, and sometimes a program called Final Draft for screenplays. This is the stock response he gives when a question about technology comes up, as it occasionally does with interviewers or fans. The answer disappoints some: "I envisioned him using something a little more high-end, maybe some specialty software that's out there," says one fan online.[23]

Yet King has also always consistently demystified the act of writing; this, more than anything, is the message he seeks to bring home in *On Writing,* which he subtitles *A Memoir of the Craft.* Like many professional authors, he works according to a strict schedule, seating himself at his desk (in the corner) every morning and not leaving until he has met his daily quota. "One word at a time," is his preferred explanation for how he does it. Word processing fit comfortably with this routine, smoothing the rough edges of revision, keeping drafts organized, allowing him to lay down word after word, page after page, chapter after chapter, the 8-inch disks of the Wang dutifully consuming and storing it all, the occasional soft churning sounds of the drive—presumably inaudible over the loud rock music King favors while working—serving to mark the progress.

For a writer as consistently thoughtful about his tools as King, it is not surprising that a great deal of what is in "Word Processor of the Gods" anticipates the more deliberate and sustained considerations of the digital word by Heim and a host of later critics struggling to come to grips with the transition from page to pixels. Above all, we should recognize the story's awareness of the strange new ontology of word processing, the way it lifted written language into a symbolic, procedurally actionable realm, coupled with the inscrutable opacity of the physical apparatus working the magic. Like a Klein bottle, King's word processor, visible in an angle never captured by Krementz, is reflected in the convex (curved like the Eight-Ball) mirror of the Wang's monochrome screen by the story he first wrote there.

It was in a spirit of novelty similar to King's when he wrote "The Word Processor" that on or about March 13, 1983, John Updike sat down in front of the Wangwriter II just installed in one of the upstairs writing rooms of his home in Beverly Farms, Massachusetts, and did something rather unusual.[24] Instead of reaching for a pencil, as was his habit with poetry, he composed the first version of the lines that would eventually be published as follows directly on the softly glowing (yet somehow insistent) screen newly emplaced in front of him:

INVALID.KEYSTROKE
Wee.word.processor,.is.it.not
De.trop.of.you.to.put.a.dot
Between.the.words.your.nimble.screen
Displays.in.phosphorescent.green?

Your.cursor—tiny.blinking.sun—
Stands.ready.to.erase.or.run
At.my.COMMAND.to.EXECUTE
Or.CANCEL:.which? The.choice.is.moot,

So.flummoxed.are.my.circuits,.met
This.way.by.your.adroiter.set.
I.cannot.think.Your.wizardry
Has.by.some.ERROR.cancelled.me.[25]

Judged against Updike's oeuvre, "INVALID.KEYSTROKE" is a slight effort, but the year he composed it (1983) is the year Adam Begley, Updike's biographer, identifies as the pinnacle of the author's literary career.[26] By then Updike had published the third of his Rabbit books, *Rabbit Is Rich* (1981), which had garnered in quick succession the National Book Critics Circle Award, the National Book Award, and the Pulitzer Prize for Fiction; his short stories, essays, reviews, and poems were a mainstay of the *New Yorker* and other magazines, and he was fast at work on *The Witches of Eastwick,* which was to become perhaps his best-known novel with its 1987 film adaptation. There could be no other candidate to play the role of a paternal dean of American letters: Tom Wolfe, with whom Updike suffered an ongoing feud, had not yet produced any fiction; Pynchon and DeLillo lacked Updike's broad public platform and appeal; David Foster Wallace was just then writing the senior thesis at Amherst that would become his first novel. One could, in fact, do worse than to name Stephen King as Updike's greatest rival for the attention and affection of the American reading public.

With regard to adopting word processing, however, Updike was neither a trendsetter nor a holdout. He was instead what he so rarely was otherwise: merely typical. Just two years previously he had been quoted in *Time* magazine saying, "I am not persuaded that the expense and time it takes to learn the machine would be worth it. I'll stick to my manual, as I have for 20 years."[27] Nonetheless, 1983 was a year of transitions (he and his second wife, Martha, having just moved to Beverly Farms), so perhaps the time seemed right.[28] He told Roger Angell, his editor at the *New Yorker,* that it would surely change his writing but he didn't know how.[29]

As the lines of the poem suggest, Updike, like King, is initially struck by the sense of mastery he possesses, fingers hovering over the keys, the machine willing and waiting to unleash its awesome, unfathomable powers in response to his every whim. Yet, like Richard Hagstrom, he finds himself paralyzed by its seemingly limitless possibility—or else humbled by his own human foibles in the face of such electronic wizardry. Indeed, Updike would never fully get over this feeling; a decade later, in an essay entitled "Updike and I," he wrote of "John," who, sitting down in front of the "blank-faced word processor," wonders if his alter-ego will desert him: "Suppose, some day, he ["Updike"] fails to show up?"[30] Yet he would also speak of the Wang as "dazzling" and attest to having "mystical experiences" with it.[31] A "delicate opacity," like "a very finespun veil" was the very essence of the machine.[32] Even so, Updike was attentive to the material particulars of the hardware in the composition of the poem: EXECUTE and CANCEL were both actual keys on the Wang keyboard; and the monitor indeed displayed its text in green (this is technically known as P1 phosphor, and was extremely common; amber and white were also used for monochrome displays).

Displaying a dot between words was likewise a feature of the Wang and some other early word processing systems, but it is of course a much older convention. It is properly known as an *interpunct,* and Paul Saenger has demonstrated that it was commonplace in the ancient world prior to the introduction of vowels into the Phoenician alphabet.[33] Nicholson Baker (who himself used a Wang word processor as an office temp) also makes the connection to *scriptura continua,* telling us specifically that the convention was thus revived by the new technology.[34] But there is also a typographic subtlety we must take note of: In the poem as printed, and indeed in the typescripts at the Houghton as well, those oh so *de trop* dots do not float midway between the top and the bottom of the line as they originally did when displayed in phosphorescent green upon Mr. Updike's nimble screen. Rather, they are printed at the bottom of the line, as ordinary periods. The center dot would have been present on Updike's display as a formatting code, much like tab symbols, hard returns, and other such marks; but these were never actually output to the printer, and there was no easy or obvious way to do it, certainly not for a novice user. What we see, then, is Updike substituting an approximation of a special formatting code with an ordinary punctuation mark, a gesture that speaks at once to his sensitivity toward the unique characteristics of the medium he is working in as well as the limits of his personal know-how.

Updike played with the text of his tetrameter apostrophe, doubtless editing on the screen but also printing at least five hard copies and annotating one of them (it appears to be the first) heavily by hand. At one point, for example, he also had this:

> The.mind.is.just.a.set.of.sparks
> Composed.inscrutably.of.quarks
> And.so.are.you,.you.dazzling.thing.
> I.touch.S.I.N.G;.you.sing![35]

Besides the comparison between human mind and electronic brain (a theme to which he would return in subsequent fiction) we see evidence of further indulgence in the typographic effects that mark the poem, the act of typing "S I N G" and the machine's soaring execution of his desires thus alphabetically made to coincide. But ultimately Updike exercised the DELETE key, and discarded that additional stanza. Off the finished poem then went to the *New Yorker,* where it was rejected in short order; the letter, signed by Howard Moss and dated March 24, reads in part, "Though God knows this is timely, something kept us from taking it—I'm not sure what."[36] Referring to the digital provenance evoked by the dots between each word, Moss notes "the idea of the poem and the way its done being one, maybe it's that after the first few lines, you've got it."[37] The piece was instead placed in an annual of light verse edited by Robert Wallace.

Updike's relationship to the screen would remain ambivalent. "Upright on a green screen the words look quite different from the way they do flat on a piece of paper," he told *PC Magazine* in 1984. "The Gutenbergian ethos will be sorely missed, at least by me."[38] Indeed, he acquired a reputation for being something of a Luddite, as cemented by his 2006 exhortation to American booksellers to "defend your lonely forts," those few remaining outposts of gravitas and contemplation amid the "anthills" of electronic media.[39] Nevertheless, over the course of his career Updike would produce his share of writing about computers and digital technology. There is *Roger's Version* (1986), a novel that explores the nature of computation through the device of a theologian's search for an algorithmic equation for god. On the strength of that book he received an invitation to keynote an MIT computer science conference two years later, which resulted in an essay, "A Writer's View of the Computer Laboratory." His 2004 novel *Villages* features a protagonist who develops "DigitEyes," a digital drawing tool. And besides "INVALID.KEYSTROKE" he has also written the poems "Death of a Computer" (2004, published 2009; unlike "INVALID," it was initially composed longhand), and "Birthday Shopping" (2007), about browsing a big-box retailer for a new computer to buy. Meanwhile, in 1997, when such exercises were still very much a novelty, he participated in a writing experiment for Amazon.com in which he supplied the beginning and ending of a short story completed—via the Internet—by others in the middle. The stunt was widely covered in the media, and Updike wrote about it himself for the *New Yorker.*[40]

As for "INVALID.KEYSTROKE," it was not reprinted in any of Updike's poetry volumes, nor to the best of my knowledge anywhere else. I myself first encountered it in typescript in the reading room of the Houghton Library. Yet the story of its composition turned out to be memorable enough for Updike to use it to introduce a 1991 gathering of essays he titled *Odd Jobs,* where he also notes that the word processor had made the prospect of taking on such occasional pieces—the odd jobs of the collection's title—all the more seductive. It is there that Updike also tells us that he composed this poem "to" his word processor "on" his word processor.[41]

In many respects Updike and Stephen King are both typical of authors writing about their experiences using their first computer or word processor. Such pieces constitute almost their own genre. "STILL VERY MUCH LEARNING TO THINK ON THIS MACHINE," Russell Banks writes in all caps at the beginning of a document that is a kind of stream-of-consciousness exploration of his word processor's capabilities. "STRANGE EXPERIENCE, UNFAMILIAR MIXTURE OF SPEED AND SLOW-DOWN."[42] Terrence McNally, from a file dated June 10, 1988, named NEWLIGHT: "This is the 22nd line. After I finish it and two more, the screen should begin to move upwards and I will only be seeing the last 25 lines."[43] In *Foucault's Pendulum* (1989), Umberto Eco begins his plot with an electronic file found on the word processor of one of his protagonists. The machine is kabbalistically named Abulafia ("Abu"), and Eco treats us to several exuberant pages of what is clearly his own authentic celebration of the technology: "If you've written a novel with a Confederate hero named Rhett Butler and a fickle girl named Scarlett and then change your mind, all you have to do is punch a key and Abu will global replace the Rhett Butlers to Prince Andreis, the Scarletts to Natashas, Atlanta to Moscow and lo! you've written war and peace."[44] And later:

> There, indiscreet reader: you will never know it, but that half-line hanging in space was actually the beginning of a long sentence that I wrote but then wished I hadn't, wished I hadn't even thought let alone written it, wished that it had never happened. So I pressed a key, and a milky film spread over the fatal and inopportune lines, and I pressed DELETE and, whoosh, all gone.
>
> But that's not all. The problem with suicide is that sometimes you jump out the window and then change your mind between the eighth floor and the seventh. "Oh, if only I could go back!" Sorry, you can't, too bad. Splat. Abu, on the other hand, is merciful, and grants you the right to change your mind: you can recover your deleted text by pressing RETRIEVE. What a relief! Once I know that I can remember whenever I like, I forget.[45]

The similarity to King's remarks in the introduction to *Skeleton Crew* is striking, and testifies to a commonality of the experience by writers who otherwise are very different.

In one passage from an essay written later in his life, Updike characterizes his word processor ("a term," he quips, "that describes me as well") as the last in a succession of writing instruments that originated with crayons and colored pencils.[46] But it would be a mistake to interpret this genealogy of machines as merely progressive; instead, they coexisted, literally side by side in his office, with different tasks associated with each. He described the scene for Jill Krementz in her book:

> An oak desk bought at Furniture in Parts in Boston twenty years ago is, along with a metal typing table and an old manual Olivetti, where I answer letters and talk on the phone. An olive-drab steel desk, a piece of retired Army equipment bought over thirty years ago in Ipswich, is where I write by hand, when the fragility of the project—a poem, the start of a novel—demands that I sneak up on it with that humblest and quietest of weapons, a pencil. . . . The third desk, veneered in white Formica, holds the word processor where everything gets typed up and where many items . . . are composed.[47]

These details orient us toward what is in fact a very complex writing environment, with texts originating in various media and migrating back and forth between them in the course of their revision—longhand and typescript, hard copy and disk. (We catch a glimpse of these habits in Updike's 2004 poem "Death of a Computer," which describes his retaining an ailing machine that could still read his older disks, and that then "turned them into final printed versions, / dark marks on paper safer than electrons.") Moreover, the casual references to answering letters at the typewriter and to the word processor as the place where "everything gets typed up" speak to a blurring of boundaries between the author as a composer of texts and the author as a compositor of them. (Before long we would call this desktop publishing.) That contention is reinforced by what the innumerable proofs, galleys, and other papers at the Houghton tell us about Updike's relationship to his editors and proofreaders and the compositors who ultimately typeset his work professionally. He took an active interest in all of these matters, and was sensitive to the impact the word processor would have on what Adam Begley describes as his "literary production line."[48] In a May 1983 correspondence to one editor, he notes: "The trouble with a word processor is I haven't figured out a way to put the page numbers in the margins. . . . I hope the checkers will still find my manuscript useful."[49]

Updike was to continue writing for another two and a half decades, until his death in 2009. The Wangwriter stayed with him for over a decade before he made the transition to an IBM-compatible computer and a word processing package called Lotus Ami Pro. Eventually he moved on to

Microsoft Word. In 2010 Updike's Olivetti 65c electric typewriter, which he had passed on to his daughter sometime in the mid-1990s, was auctioned at Christies.[50] It was purchased by the collector Steve Soboroff, complete with an intact but jammed ribbon that dates from early 1983, as its various textual remainders, such as an introduction Updike was then writing to a collection of Kafka's short stories, testify. Included on it are also these three snippets:

"This ms. may be the last messy one you get I've bought a word processor and we're slowly coming to an understanding. It's quick as the devil, but has very little imagination, and no smalltalk." That was addressed "to Roger" (Angell, his editor at the *New Yorker*) and dated March 12—only a day prior to the date marked on the typescript of "INVALID.KEYSTROKE."

Next: "I'm having a mechanical crisis; this is an electric typewriter, I have a manual, and also a word processor, and in going back and forth between them I keep hitting wrong keys, mostly the return button here which sends the carriage flying. Back to goose quill perhaps." That one was to "Susan," April 19.

And finally, addressed to his typist, on May 23, 1983: "Why don't you charge me $1.25 per page? I have a word processor now and won't be needing too much more typing." And that is the last thing there is to read on the ribbon.[51]

FIVE

SIGNPOSTS

In introducing his 1981 articles for *Popular Computing*, Isaac Asimov was at pains to emphasize his status as a word processing novice despite his stature as a renowned scientist and author: "When someone asks me if I work with computers myself, I shudder and say, 'I am a signpost, sir. I point the way. I don't go there.'"[1]

And yet he did go there, and in relatively short order, too—becoming not only a signpost but also something of a sandwich board through his participation in a subsequent advertising campaign for Radio Shack.[2] As a group, science fiction (SF) authors accounted for more early converts to word processing than any other community or constituency within the literary field. In my estimation, SF authors were ahead of the popular adoption curve by three to four years; the community's most vocal proponents for the technology were active from roughly 1978 onward as opposed to from around 1981. Asimov, in other words, followed behind many in his cohort—as he himself acknowledged, claiming in that same year that "almost every writer of my acquaintance is using a word-processor, or is getting one, while I cling (more or less in terror) to my electric typewriter."[3] While a matter of three or four years may seem trivial, the landscape for personal computing was changing rapidly in that same time period. From the tactile feel of keyboards, the size of display screens, and the reliability of disk storage to the availability and capabilities of software, a writer working in 1978 or 1979 would likely have had a very different experience with word processing than a writer first coming to the technology even just a few years later, as Asimov did.

Why did science fiction—collectively—get to word processing first? There was the obvious reason: Nothing may be more natural to us now, but thirty or forty years ago, pressing a key on a keyboard and watching the corresponding letter, number, or symbol wink into existence on a glowing glass screen must have seemed like something out of a space opera. David Gerrold, author of well-regarded science fiction novels like *When HARLIE Was One* (1972) but perhaps best known as the writer of the classic *Star Trek* episode "The Trouble with Tribbles," had, as early as 1973, augmented his standard IBM Selectric with a $4,000 cassette-driven memory system called a Savin 900; by 1978 he had traded up to a North Star Horizon II, yet another early Z-80-based microcomputer; after that he moved on to a Kaypro. Gerrold was a frequent commentator on word processing in the press, and believed science fiction authors were predisposed to become early adopters: "I'm convinced that there is a specific connection here," he wrote in 1981. "These are the people who've been living with the idea of home computers ever since Asimov wrote his first robot story."[4]

Perhaps. But the computers that populated science fiction until well into the 1970s typically involved physically massive machines that were also sentient (and often poorly disposed toward their human creators). Asimov himself recognized how much he had failed to foresee, starting with miniaturization: in the 1950s, extrapolating from such vacuum-tubed behemoths as the UNIVAC, he had gone even bigger, giving us the MULTIVAC, "half a mile long and three stories high."[5] The epitome of this trend was Arthur C. Clarke's HAL 9000 in *2001: A Space Odyssey* (1968), but there are numerous other exemplars. "I, BEM," a 1964 short story by Walt and Leigh Richmond, features a sentient computer that begins life as an IBM typewriter; from these humble beginnings computers and robots take over the Earth, but by the end of the story they find themselves in danger of being replaced by the "biologicals" they have engineered to serve them.[6] *Colossus* (1966), by Dennis Feltham Jones, posits a pair of opposing Cold War supercomputers that become self-aware and threaten to destroy the world. In 1967 (still a year before Clarke and Kubrick's *2001*), Harlan Ellison published a short story called "I Have No Mouth and I Must Scream." The story is interspersed with messages resembling punched paper tape from "AM," the megalomaniac supercomputer that sadistically persecutes the five remaining human denizens of its postapocalyptic world (the computer's name has various meanings in the story, the most fundamental of which—it remains unspoken if its messages are not translated—is that of the cogito, "I think, therefore I am"). The communiques, which Ellison called "talkfields," are encoded in ITA2, the International Telegraph Alphabet No. 2, derived from Baudot Code and a precursor to the computer lingua franca of ASCII. The talkfields are thus likely the first attempt to represent computer speech in a work of fiction in a technologically realistic manner. However, the talkfields also posed a challenge to typesetters, and Ellison has stated they were not printed without being "garbled or inverted or mirror-imaged" until the story's inclusion in a collection in 1991.[7] Meanwhile Barbara Paul offered readers a benign alternative to HAL and AM in her short story "Answer 'Affirmative' or 'Negative,'" published in *Analog* magazine in 1972; it posited a supercomputer containing "the sum total of man's knowledge" called the WOMAC. When queried with some especially difficult problem, WOMAC would sometimes—frustratingly and inexplicably—return a line or two of poetry instead of one of the story's titular absolutes. Eventually the operator in charge of the machine finds out why, and considers reprogramming it to correct this behavior—but then has second thoughts: "The whole world depends on WOMAC. I think I might just wait and see what happens when the whole world has to learn poetry. Yessir, I might just do that."[8]

Artificial intelligence and giant, (mostly) malevolent supercomputers were to remain more appealing subjects for science fiction than word processing. But the day-to-day realities of early personal computing were also quite different from the kinds of colossi the genre was typically given to imagine. Consider the tangle of decisions and deliberations that were necessary to purchase a home computer in the late 1970s: even assuming one had made the decision to go with a microcomputer instead of a dedicated word processor, potential buyers still had to sort through whether they wanted one of the few comparatively weak off-the-shelf "integrated" systems then available or whether they wanted to pick and choose their own mix of components; whether they wanted an 8080- or Z-80-based chipset; 5¼- or 8-inch disks; letter-quality or dot matrix printing; what kind of keyboard; what kind of screen; what kind of software; and on and on and on. Clearly some found such immersion in technological minutiae vulgar and distasteful; Harlan Ellison was openly scornful, ridiculing what he perceived as "this whole lemming-like rush to pick up the latest toy."[9] (John Varley, as we have already seen, was also to have similar reservations.) Any number of science fiction authors would, like Asimov, declare themselves at least mildly technophobic or else simply ignorant when confronted by the prospect of buying and learning a computer or a word processor.

For all of their interest in computers, robotics, and sentient machines, then, science fiction writers no more predicted or successfully anticipated word processing—the seemingly suddenly commonplace experience of using something like a typewriter attached to a TV set to create and edit text—than any other genre or constituency.[10] Nonetheless, and despite some vocal refuseniks like Ellison, they were noticeably quicker to adopt it once the technology was within reach. By 1982 Gerrold posits that there were "probably several hundred" authors with their own computers, the majority of them in science fiction.[11] By way of context, in one editor's estimation there were around 400 individuals actively producing publishable science fiction in that same time period.[12] But if neither the powers of literary prognostication nor an innate affinity for technology are adequate explanations, then we must look to other causes and factors. Of these there are several. The first concerns the part played by several prominent personalities within this still rather small community; a second causal factor involves economic dictates and the marketplace for science fiction; and finally, there was interest from the computer manufacturers themselves, who actively promoted their products by placing them in the hands of the genre's luminaries (as happened with Asimov and others).[13]

Not every early literary adopter of word processing wrote science fiction or even genre fiction. But the rapid uptake of the technology among these constituencies is too conspicuous to ignore. Harold Bloom once said that as far as he was concerned, "computers have as much to do with literature as space travel, perhaps much less."[14] Of course he was wrong on both counts.

Dr. Jerry Eugene Pournelle was born in Shreveport, Louisiana, in 1933. Since 1969 he has written some three dozen science fiction novels—a number of them collaboratively, most often with Larry Niven—along with numerous short stories and essays. Some of the best-known include *The Mote in God's Eye* (1975), *Lucifer's Hammer* (1977), *The Mercenary* (1977), and *Oath of Fealty* (1981). (As the titles suggest, his storylines tend toward the martial.) He is also a prolific science writer, and has held such positions as science editor for the National Catholic Press. He saw active duty during the Korean War and later earned his doctorate in political science from the University of Washington, publishing a monograph about national defense. His interest in military matters was further nurtured by time spent in the aerospace industry at Boeing in the 1950s. Later he worked with NASA on the Mercury, Gemini, and Apollo programs. He briefly served in Los Angeles mayor Sam Yorty's administration, and then did a stint as a professor at Pepperdine before turning to writing full time. In 1980 he began contributing a monthly column to *Byte* magazine: so influential would his column become that at his insistence early versions of Microsoft Word included an option—the so-called "Pournelle feature"—to adjust the text and background to white on blue to emulate the color scheme of his then-preferred word processor, Symantec Q&A.[15] In 1984 he published *The User's Guide to Small Computers:* the cover was a mock-pulp illustration depicting a bespectacled Pournelle triumphant atop a conquered heap of computer hardware, a keyboard clutched like a broadsword (or space rifle) in his hands. In 1985 his mere mention of activities on the ARPANET (the still rather secretive nascent Internet) in the pages of *Byte* resulted in his account being terminated, an incident since enshrined in computer lore.[16] He has had a long-standing friendship with Newt Gingrich, and been an outspoken right-wing political commentator through his Chaos Manor website, a sort of blog *avant la lettre.*[17] He remains a recognized name in neoconservative politics, the science fiction genre, and technology journalism. He has worn a pencil-thin mustache for most of his life and speaks with a high-pitched voice that still bears traces of his Louisiana roots. To this colorful biography we can add that he has a strong claim to having been the first author to have written published fiction on a word processor in the manner in which we would conventionally envision it: composing and editing with a keyboard connected to a screen, using a commercially available personal computer system—a "microcomputer"—to create, revise, and store text before outputting it to a printer.

Pournelle himself is not shy about promoting this claim, though he is also unable to pinpoint the specific text that might have been the first (probably something written in 1978). A working writer's life can be messy, he will explain; there is always more than one project under way at a time, and work is not always published in the order in which it is finished.[18] Frustrating though this may be for the historian, it does not prevent us from reconstructing the circumstances that placed Pournelle in the position of certainly being *one* of the very first fiction writers in English to make use of a microcomputer (as they were then known) and one of the most widely read voices in computer journalism.

If he is not shy or reticent, Pournelle is resolutely unsentimental regarding his own work. And his writing is indeed his *work.* Throughout his career,

Pournelle has repeatedly laid emphasis on the actual labor of being an author. "Writing is hard work" is a mantra he has repeated over and over again.[19] There is an upside: almost anyone can learn to do it, he believed, and with enough persistence can enjoy some success.[20] But make no mistake: writing *was* work. So was correcting copy. Rewriting was even harder work than writing, because typing was part of rewriting. Managing correspondence and contacts and contracts were all work as well. And managing the financials of a writing career was work, tedious and very demanding at that—the IRS seemed always to be asking for more documentation. Throughout his career Pournelle has been uninhibited in addressing these matters. "Somebody's always getting me to come lecture to their writing class, and I don't talk about writing at all," he says. "I talk about the business of making a living at this racket. That's what I do, I make a living at it."[21]

By temperament, then, Pournelle was an author given to approaching his writing—his work—with an accompanying cost-benefit analysis. But the particular economics of science fiction as a publishing genre also forced such matters to the fore. In 1973, when he served as president of the Science Fiction Writers of America, Pournelle believed there were fewer than twenty SF authors in the country who could make a living from sales of their writing.[22] But the times were rapidly changing. Interest in the SF genre had been piqued by real-world events throughout the 1960s, a decade that began under the shadow of *Sputnik 1* and concluded with the *Apollo 11* moon landings in fulfillment of John F. Kennedy's pledge. The same decade saw the debut of *Star Trek* in 1966 and Kubrick's *2001: A Space Odyssey* in 1968; Frank Herbert's *Dune* (1965), meanwhile, proved that written SF could also produce blockbusters. Science fiction was increasingly in demand, but changes in the publishing landscape were also mitigating away from the dominance of SF's early serial publications and more toward novels as the basic economic unit of the genre.[23] Increased reader demand, coupled with greater expectations for long-form work: that meant opportunity, but it also "presented a problem," as Pournelle notes in a typical summation: "In order to make a living at writing I had to write a lot, and writing is hard work. Actually writing wasn't so bad: it was *rewriting*, particularly *retyping* an entire page in order to correct half a dozen sentences. Typing neatly involved correction fluid, carbon paper, fussing with margins; a lot of work, most of which I hated."[24]

Pournelle had been introduced to computers early in his career as tools for solving engineering problems, but it did not occur to him that they might also be of use for writing. That changed sometime in 1977. He had been given an opportunity to bring his first two novels, published under a pseudonym, back into print, but they required extensive revision: "Wouldn't it, I mused, be marvelous if those books were in some kind of electronically readable form so that I could do the scissors-and-paste job without so much retyping?"[25] It is important to emphasize that the technological landscape at this time was dramatically different from what it was around 1981. Dedicated word processing machines like the Wang were available, but a microcomputer was more versatile, something that appealed to Pournelle. There were as yet no truly viable "integrated" systems, like the IBM PC, Kaypro, or Tandy, that came tidily packaged in a box. (The Apple II, Commodore PET, and TRS-80 Model I all came on the market in 1977, but all initially had significant limitations—at the time the Apple couldn't even display lowercase letters, obviously a nonstarter for a writer, while the PET's diminutive name seemed to speak for itself.) Instead, it was understood that a user with any pretense toward serious applications pieced together their computer's components from individual hardware vendors—choosing the processor and boards, choosing the keyboard and the display, the method of disk storage, the printer, and of course the software. All of it was custom assembled. To help him navigate this morass, Pournelle turned to a "mad friend" who was a devout computer hobbyist.[26] The system the two settled on represented no small investment: $12,000, about half of it for the printer.[27] Pournelle had to borrow the money, and claims he subsequently battled the IRS for a write-off, having to explain why a computer could be a legitimate business expense for a writer.[28] The essential software program was Electric Pencil, which preceded WordStar as the first word processing program designed specifically to run on a CP/M microcomputer. Because of the Z-80 chip at its heart, Pournelle named the machine Ezekial *(sic)*, or Zeke for short.

Initially he first typed his drafts on the computer, then printed them out and amended the hard copy by hand. However, Pournelle soon found himself composing and revising as a single unified activity on the screen.[29] In 1979 he wrote an article for the inaugural issue of *onComputing* (the magazine that became *Popular Computing* and published Asimov's essays in 1981) in which he detailed the components of his system and offered advice to other writers. It is clear even at this early stage that Zeke had already fulfilled the hope of increased productivity—in the article Pournelle details revising a screenplay in an afternoon, and then composing and writing a 15,000-word story in three days. He estimated that it was saving him whole "months" of typing and retyping, and that it let him produce prose at "double" the usual rate. "It doesn't get in the way of writing: no paper to change, no erasures and strike-overs, no Sno-pake," he concluded, referring to the correction fluid. "Best of all, every draft is a clean draft—but it's so easy to produce another clean draft that there's no hesitation over a rewrite."[30] A few years later he elaborates: "Ezekial changed my life. He did most of the real work of writing. I never had to retype anything, and I could fiddle with the text until I had *exactly* what I wanted. Computers not only let you write faster, but by taking the mechanical work out of writing they let you write *better*."[31]

The 15,000-word story Pournelle discussed but did not name in *onComputing* is very likely "Spirals," which he cowrote with his frequent collaborator Larry Niven. It was published first in 1979 in *Destinies* magazine and reprinted that same year in an anthology entitled *The Endless Frontier;* in the introduction, Pournelle remarks on having used a computer to write the first draft: "I write everything into a computer and edit on a glass screen."[32] Was "Spirals," then, his first published fiction written with Zeke? It seems the most likely candidate.[33] Also around the time "Spirals" was published, he sat for an interview with TV journalist Tom Snyder, who quizzed him about his new writing system. Like Michael Crichton's exchange with Merv Griffin, the dialogue is worth lingering over, for it captures the colloquialisms with which word processing was then described to a general audience. "I have a little home computer," Pournelle begins.

> And I type; stuff goes up on a glass screen. If I don't like what I've typed, I type over it and it puts the new stuff in and I can spread the lines apart, move paragraphs around. . . . I can move blocks of text around, I write something, say, "Well that's pretty good but does it belong here," I can store it over yonder and later on insert it and that type of thing. When I get this thing all done and it's on a record—it looks just like a phonograph record, but it's a magnetic record. I push a button and an automatic typewriter types it out onto a piece of paper. I mail it to New York, where somebody hands it to a guy who sits there and types it all into a typesetting machine so that it goes back onto a little record that looks just like mine.[34]

It is telling to hear something as familiar to us as word processing described by way of such careful and sometimes strained language. (He loses Snyder halfway through.) But as awkward as Pournelle's description must seem to us now, the reality is that we might also find many of the functional particulars of word processing, as he experienced them, difficult to recognize. To write something with Zeke meant first booting the machine, which was done from a diskette, not a hard drive. There was no "desktop" as such, only a cursor prompt waiting for the appropriate command to be input. Launching the program displayed the message: "THE ELECTRIC PENCIL (C) 1977 MICHAEL SHRAYER." Nothing else happened until one started typing; this message would then disappear to be replaced by the user's text.[35] The Hitachi screen was 64 characters by 20, just about as big as a typescript page though other systems were smaller. Often chapters were spread across multiple disks, and one became accustomed to swapping them in and out to access different portions of the manuscript. Functions like spell-check and dictionaries were separate programs. Fonts as we know them did not exist—the keyboard simply translated the character input to ASCII code, which at least had the virtue of speed. Basic graphical effects like boldface, underlining, and italics were thus beyond the capabilities of Electric Pencil, which at that time was the state-of-the-art.

Here, meanwhile, is how the Electric Pencil manual described the procedure for moving a block of text:

> In order to move a block of text to another part of the file or to delete a block of text, it must first be marked. The character used to mark the boundaries of a block is [\]. This character is also called a "marker." To mark the boundaries of a block, the cursor is placed over the first character of the text desired and then the [\] key is depressed. The action will be the same as if the Insert Mode was entered. The text will shift right and the "marker" will be placed. The cursor should now be moved to one character beyond the end of the block of text desired and similarly marked. Exactly two markers must be used; otherwise, a MARKER ERROR message will appear on the video display screen when a move is attempted.[36]

There was no autosave feature, so documents had to be manually saved to the disks on a regular basis or else they could be obliterated by a power outage or program glitch. And Electric Pencil was notoriously glitch-ridden; for example, it had a habit of dropping the last character from every line. Making a backup copy of absolutely everything was essential. The disks themselves rotated continuously, a fact that, as Pournelle notes, generated white noise that some writers might find distracting.[37] Printing was in theory a "push of a button," but Pournelle also notes of most printers, "The darn things take just enough attention that I like mine in sight."[38] High-quality printing was also slow—a novel-length manuscript could take many hours. The paper, of course, came in the iconic fanfold format with tear-away perforated edges for a tractor feed. When a print job was finished, it was incumbent upon the writer to separate the sheets and remove their perforated edges, an unexpectedly satisfying chore Pournelle refers to as "busting out" the manuscript.[39] The conclusion is unmistakable: although the computer was undeniably a timesaver, it also inaugurated its own attendant labor regimen and introduced a fresh array of time-sinks, from the learning curve of operating the system itself to the routine maintenance and upkeep of each and every component.

Pournelle might have opted instead for a dedicated word processing system like the Wang, Lexitron, or Lanier. The cost would have been about the same. But he understood that a true microcomputer would be far more versatile. He was interested in using the system for his taxes and other financials, as a correspondence database, and, most intriguingly, as a simulation engine to assist with world-building in his science fiction.[40] He wrote what is known as "hard" science fiction, meaning that the stories were grounded in as much scientific fact or plausible scientific theory as was possible. What percentage of a planet's atmosphere consisted of oxygen? What force did gravity exert on an object of such-and-such a density and mass? What was a rocket's trajectory for achieving a low-Earth orbit? All of these details mattered, and many of his readers could tell the difference between the real thing and hokum. Pournelle thus immediately grasped the potential of the computer as a world-building tool, and learned enough BASIC and FORTRAN programming to transfer the formulas he worked out on a pocket calculator to a set of custom programs for his system.[41] (Despite his engineering background, he claims he found the prospect of actually learning programming "terrifying.")[42] Other writers grasped this potential as well: Frank Herbert, for example, imagined a seamless integration between a simulation engine (complete with visual renderings) and a text editor, whereby details from his planetary models could be made to populate his prose on demand: "You will *know* when it's spring on Planet X or when the tides rise four hundred feet on planet Y."[43] Such examples are tantalizing, and the close integration of simulation and modeling with word processing seems to me one of the richest roads-not-taken in thinking about the potential of computers as writing platforms, not just for science fiction but for all manner of genres demanding close correspondences between storyline and setting, words and worlds.[44]

The first beneficiary of Pournelle's proselytizing was his neighbor and longtime writing partner Larry Niven, also a highly regard science fiction writer in his own right (his 1970 *Ringworld* had won the Hugo Award). The story both tell is that not long after Pournelle purchased his system, Niven ordered a duplicate, his "magic typewriter" (two of them in fact, the second for his wife, who kept his accounts).[45] This marked the beginning of a pattern: Pournelle's residence in Studio City (a brick manse he refers to as Chaos Manor) was a frequent stopping place for science fiction writers passing through town, especially to visit the nearby Jet Propulsion Laboratory in Pasadena. (The JPL at the time was mission control for the *Voyager* and *Pioneer* spacecraft, and it would open its doors to writers and journalists as part of its Planetary Encounters series.) Pournelle would introduce his guests to Zeke, putting the word processor through its paces. Guests would leave suitably impressed, and find themselves thinking about getting some RAM and ROM of their own on the way home.

Later, Pournelle took to bringing portable computers to conventions, where he would show them off to the same effect. He quickly garnered a reputation as the go-to person in the community for advice and assistance as, one by one, writers began making the transition. And though he was the last person on whom it would have been lost that his friends were also his professional competition, he unhesitatingly obliged. Authors he advised in this capacity include Gregory Benford, Gordon R. Dickson, Joe Halderman, his onetime mentor Robert Heinlein, Frank Herbert, Robert Silverberg, and others. In a 1982 correspondence with Silverberg, he is able to rattle off the names of other writers and their systems or software as though reciting from a Rolodex: "Norman Spinrad recently bought a Kaypro 10 and he's very happy with it."[46] "The only major writer I know who uses WordMaster is Gordon Dickson, and he does so because he started with it back when there were very few [text] editors his equipment could run."[47] He knew that Gary Edmundson, Carolyn (C. J.) Cherryh, and Joe Halderman all preferred WordStar.[48] Such throwaway comments underscore the extent to which science fiction was a close-knit community of authors who regularly saw each other at conventions and other social gatherings, who corresponded and collaborated, and rarely missed an opportunity to talk shop. In 1981 Pournelle convinced his longtime editor at Ace and Tor, Jim Baen, to buy an IBM PC.[49] Baen, himself an influential personality in the genre, quickly became a convert and encouraged his other authors to make the switch.[50] In 1983 Baen left Tor to start his own imprint, and because Baen

Books was small and was not burdened with a base of existing legacy technology, he was able to champion now-standard industry practices like the electronic submission of manuscripts.

While Pournelle is hardly single-handedly responsible for every science fiction writer's move to computing, his evangelizing does suggest the extent to which a single individual was sometimes enough to exert broad influence on a genre—whose practitioners would be motivated not least by their perceived need to keep up with the competition. Whether or not he was truly the "first" author to produce fiction on a microcomputer, his influence is undeniable, not only among authors but in the computer industry as well. He built his popular *Byte* column around his identity as a user, an everyman. He ruthlessly took systems and software apart, stress-tested and benchmarked them, and badgered and harangued developers and manufacturers.

Zeke was eventually followed by another Z-80-based system, Zeke II, and Pournelle moved from Electric Pencil to a custom-built application simply called WRITE and then to Symantec Q&A. Eventually he moved on to Word, which is what he still uses today. Zeke remains the one computer about which he is unabashedly sentimental, and it would go on to have an interesting afterlife (as we will see in Chapter 10). Otherwise he did not romanticize the technology. He was a working writer—for him the word processor was simply, obviously, the better tool for the job. "Boy," he said at the end of his first article about word processing, "has this thing made it easier to write science fiction."[51]

Frank Herbert was what we might nowadays call a "maker." In 1977, the year he published *The Dosadi Experiment* (arguably his most important book outside of the *Dune* franchise), he filed a patent for a panemone, or vertical-axis windmill, which he hoped might one day supply power to his homestead in Port Townsend, Washington. By then Herbert had been writing science fiction for some twenty-five years. *Dune* (1965), his acknowledged masterpiece, and its sequels had cemented his stature as perhaps the foremost writer of the genre, for a time at least surpassing even the likes of Heinlein, Asimov, Bradbury, and Clarke in fan devotion as well as critical acclaim. By the late 1970s his advances were setting industry records. His financial success had been a long time coming, but Herbert eventually found himself able to function as a patron of sorts for the local technorati.

One such individual was Maxwell Barnard, a former Boeing engineer. It was Barnard who introduced Herbert to the emerging phenomenon of personal computers.

Computers had figured in Herbert's fiction for some time, of course, at least speculatively or notionally. In *Dune* there existed an intergalactic ban on the technology as a consequence of events 10,000 years before the story begins: "Thou shalt not make a machine in the likeness of a human mind" is one of that universe's core dictates. (*Dune* instead posits Mentats, individuals trained from birth to function as human computers.) His *Destination: Void,* a novel serialized in *Galaxy* magazine the same year that *Dune* appeared, essentially inverts this premise: it is set inside an enormous sentient spaceship whose central nervous system fails, leaving its thousands of human passengers to attempt to replicate its functions artificially, by constructing a synthetic intelligence. (In the book's sequel, *The Jesus Incident,* this entity declares itself a god.) Herbert's own early computing efforts were rather more mundane, however. Sometime in the late 1970s, likely as a result of seeing Jerry Pournelle's setup, he acquired a computer of his own, very likely a TRS-80 Model I (which had debuted in 1977); he may have used this machine for the composition of at least some of the book that became *God-Emperor of Dune* (1981), the fourth in the series.[52]

Regardless, by early 1979, driven by his own predilections and tantalized by Barnard's ideas, Herbert had set about designing a custom-built computer system to meet the needs of writers and other creative individuals.[53] He attached more urgency to this pet project than to some of his previous ones: Herbert believed that personal computers would be transformative, and that those without the wherewithal to use them would quickly become disadvantaged socially and economically. Though Herbert was also in the thick of writing *God-Emperor,* he and Barnard accepted a proposition from his editor, David Hartwell (who had just moved to Simon and Schuster), for a nonfiction book spelling out their ideas about personal computers. It would function as a guide to the technology for the uninitiated—one of the first books of its kind—and the custom-built machine that Herbert and Barnard were busily designing was to be its centerpiece.

As is the way of such things, that computer was never actually finished. It was simply too ambitious for its time, and Herbert had abandoned it by the end of 1980 (by one account he had spent some $40,000 on it by then).[54] But the book, at least, had gotten written. Entitled *Without Me You're Nothing* (1980), it is one of the more intriguing artifacts of its era, more in line with Ted Nelson's visionary *Computer Lib / Dream Machines* (1974) than the innumerable other home computer guidebooks soon to be on the market. The cryptic title was meant to be understood as words addressed to the computer itself: "Without our intervention they are useless junk," Herbert and Barnard repeated over and over again, a variation on the standard programming axiom "garbage in, garbage out."[55] Nonetheless, they begin with a dire-sounding warning: "You are already being taken advantage of by people with computers," they write. "You will not be able to meet that challenge or keep up with other changes unless you acquire a computer yourself."[56] They then raise the stakes even further: "Get your own computer. Learn how to use it. . . . If you don't do this, the Bill of Rights is dead and your individual liberties will go the way of the dodo."[57] Throughout the book this libertarian rhetoric is offset by the promise and potential of what computers can offer. "Computers also amplify creative imagination," we read just after the urgent prognostications about civil liberties.[58] Computers, they assert, will be used by writers and artists, by journalists and doctors, by account executives and their secretaries, and by architects and government clerks. We find early anticipations of assistive and adaptive applications for the disabled, such as screen readers and self-driving cars. Computers can be used to control the environment in one's house, and even (with "appropriate sensors") can be used for pest control. Machine translation, electronic voting, and sensors to detect spoiled food or contaminated water are also all on offer. Nonetheless, Herbert and Barnard are careful to emphasize, computers are only tools. More precisely they are assemblies of switches, on or off, wired in sequence. And without us they are nothing.

Other chapters explain principles of how computers work, their specialized vocabulary, and the components of a typical system. There are also, however, attempts to teach elements of the BASIC programming language in layperson's terms, and repeated encouragement for people to experiment with computers on their own. But given the limitations of the existing systems then on the market, the real heart of the book is its insight into what Herbert and Barnard desired in their self-designed computer, what they term a "Rolls Royce" (they never give it a proper name). They begin with alterations to the physical hardware components: The keyboard was particularly important—it must have the right "feel," they assert, and be glitch-free. It must also be silent. These might seem like odd or arbitrary stipulations, but the keyboards at the time were often buggy and awkward, and such considerations were important to Herbert even with his manual typewriters. "Dad always said the keys could be sprung if they weren't hit just right, with an even rhythm," recalls his son Brian Herbert. "He was extremely sensitive to the touch of his keyboard."[59] It was more than a casual affectation: "Frank Herbert believed his creative process was partly in his fingers. . . . He described it as a kinesthetic link, in which thoughts flowed from his brain through his body to his fingers and onto the paper."[60] Other elements of the system were equally important. The screen would display in vertical format, like a typewritten page, and it could be set to either black-and-white or color. The printer would be a "phototypesetter . . . giving you on paper an exact match of what you had on screen"—an obvious anticipation of the WYSIWYG principle. Indeed, they note, an editor will not be able to tell that a manuscript was prepared on a computer, again signaling that peculiar early anxiety—one perhaps heightened by Herbert's troubles submitting early dot matrix printouts to Berkley Books.[61] Storage would be both internal and external, on disks capable of holding up to "21 million bytes" (a mere 21 megabytes!) as well as tape for backup.

For writers, there would be additional features. Obviously the user would have the ability to instantly navigate to any point in the manuscript via a search function. Spell-checking and a dictionary would both be standard. Moreover, the computer would also function as a kind of literary minder: One of the more colorful scenarios they offer involves prompting the author with "ringing bells and flashing ERROR messages" if he or she unthinkingly attempted to make use of a character previously deceased in the story. Graphics—what the two termed "image writing"—could be seamlessly integrated into the text.[62] Similarly, the machine was to have the world-building and simulation capabilities already detailed earlier in this chapter.

The core of the system, though, was to be its integration with a revolutionary new programming method called PROGRAMAP. To use computers, Herbert and Barnard believed, people needed to be able to tell them how to do things, even things they didn't already do. BASIC was then the language most often recommended to beginners, and though it was designed to be easy and accessible, it still demanded rigid adherence to a syntax and the capacity to express one's wishes in the twisty loops and branches of programming logic. PROGRAMAP was to be different. As the name implies, it was a visual language, like a flowchart or a map. The idea was to furnish a set of graphical conventions to help individual users organize their thoughts—the objectives for the program—into a sequence of algorithmic steps. The resulting PROGRAMAP diagram consists of visual glyphs—

ideograms—arranged in four columns. The first indicates the actual system components that are to be involved—display, keyboard, printer, speakers, and so forth. The second presents the logical operations of the processor in schematic form. The third column, called the REROUTE, imposes conditionals or branching logic. The fourth and final column was for descriptive notes, human- rather than machine-readable text. Herbert and Barnard believed that a universal visual vocabulary for representing programming concepts would allow a larger number of people to not only use but in fact program computers. "If you can type a letter on a typewriter and tell a stranger how to get to your house, you can write your own programs," they promised.[63] Nonetheless, they understood at the time that the user would still have to manually transliterate the PROGRAMAP to an actual executable language like BASIC. What they foresaw, however, was a further iteration of the concept in which an actual keyboard outfitted with PROGRAMAP iconography would allow people to create and edit programs in a purely visual environment, with the computer automatically compiling its internal machine language from the input. Of course such ideas were not in and of themselves original—work with so-called visual programming languages had begun in the 1960s, and they remain an active area in both computer science and education research today. But PROGRAMAP itself appears to be largely forgotten in this history, although Herbert and Barnard filed a patent on the keyboard design in 1984.[64]

Without Me You're Nothing contains appendices that include sample programs in BASIC and PROGRAMAP for a car maintenance utility and a mortgage calculator; Simon and Schuster initially balked at printing them—it must have seemed like a waste of paper—but Herbert insisted.[65] But before these, there comes a confession. The book proper concludes with a postscript, and it exists to address a reader's inevitable question: "Did we write this book on a computer?" The answer was no. The "Rolls Royce" had remained a purely speculative design. Jerry Pournelle suggests that Herbert, under increasing pressure from Simon and Schuster, returned to his typewriter to finish the manuscript.[66] Herbert and Barnard offered a more considered rationale: "This book was partly a test project to set the design requirements for a writer's computer. This required that we use conventional methods while comparing those methods to available computer methods."[67] At face value this perhaps also explains why Herbert, if indeed he had a Tandy computer at the time, chose to forego its use. Or else perhaps that system had already been torn down to furnish the chassis for the Rolls.

Regardless, we are left with the irony that a book intended to evangelize for the importance of personal computers—with special attention to the particular needs of writers—was written on an Olympia typewriter.[68]

The book performed moderately well despite the cryptic title and was an alternate selection for the Book of the Month Club. A trade paperback followed the next year. It has now been largely forgotten, even by most of Herbert's fans, but we can conceive of *Without Me You're Nothing* as an unintentional exercise in speculative fiction, documenting in considerable detail an imaginary computer that can be explored alongside the many other systems and machines in his novels and those of his contemporaries. Herbert and Barnard were particularly prescient in their insistence that literary authors had needs that were not well accommodated by existing products on the market; as late as 1984, for example, an advice article in a science fiction periodical begins with the complaint that most word processing systems are tailored toward secretarial needs, not those of creative writers.[69] Herbert and Barnard went a step further, tapping into a debate that is still relevant today about the necessity of the individual end-user's knowledge of programing. "The machine must be fitted to you," they concluded. "Not the other way around."[70]

Not every science fiction writer would come to word processing by way of Chaos Manor. Barry B. Longyear was a contemporary, based on the other side of the country in Maine; by the end of the 1970s he had built a successful writing career that included a novel, *Enemy Mine* (1979), that won the Hugo, Nebula, and Locus Awards. Like many of his colleagues, he dreaded the chore of typing manuscripts, and though he considered himself something of a technophobe, he came to the conclusion that a word processor was the only solution.[71] Intimidated by microcomputers like Zeke, Longyear opted instead for a Wang System 5 Model 1, acquiring it in May 1979 (several years before a rather more famous Maine writer would buy a slightly later version of that same system).[72] A recent heart attack added impetus to the decision.

Initially the machine was an adjustment. Longyear reports that the Wang's CPU made so much noise he had to move it (and the daisy-wheel impact printer) to a separate room. It typically required three 8-inch disks to store a complete novel, and many more for his reference files; this meant that searching for a particular passage often meant repeatedly swapping the large, clumsy disks in and out of the drives until he found it. But "it didn't break down very often and took care of one thing I hated: retyping."[73] Indeed, he would soon claim that he was learning to savor the rewriting process, revising his prose in a way he never had before.[74] He also came to the realization that the Wang could alter his process in more fundamental ways: "Now that the burdens of typing, correcting, and retyping had been lifted from me by my Wang . . . there was no need to begin a story with a title and a perfect first sentence."[75] He describes the composition of a short story in which he simply began with the first scene that occurred to him, then wrote another scene, and another, "hopping" from one to the next and "sort[ing] it out electronically afterward. . . . It was like having a chronic pain, a debilitating brain disease, or insufferable stress banished forever. I could begin with what interested me, have fun with it, and continue having fun until I was finished."[76] Given Longyear's health concerns, the prevalence of pathological language in the preceding description is notable; the Wang was an anodyne, the agent of a new, stress-free lifestyle.

In 1981 he entered into a kind of amiable disputation with Pournelle in the pages of *Asimov's Science Fiction* magazine, in which the two debated the pros and cons of their respective computer systems. (It is illustrated with a Jack Gaughan pen-and-ink drawing depicting the two authors as ribbons of fanfold paper being fed into their respective machines even as they type athletically on them, the Möbius-strip-like logic posing the question of who is in control, the writer or the technology.) Pournelle argued forcefully for the advantages of a general-purpose computer like Zeke over a dedicated word processor. Longyear conceded Pournelle's technical acumen, but held that the Wang was the right choice for him precisely because it assumed little or no specialized knowledge. "A typewriter is nothing but an extension of the fingers, as is a pen or pencil," he concluded. "However, a word processor is an extension of the *mind*."[77]

At about this same time Robert L. Forward, a physicist working at Hughes Aircraft, the California-based aerospace company originally founded by Howard Hughes, began authoring science fiction novels; the best-known of them, eventually titled *Rocheworld*, began as a serial in *Analog* magazine (1982–1983) and was first published under the title *The Flight of the Dragonfly* in 1984. Forward's work at Hughes gave him access to a mainframe computer running the UNIX text editor called TECO, which he began using probably around 1980 or 1981.[78] TECO (Text Editor and COrrector) was originally developed at MIT in the 1960s, a successor to the Expensive

Typewriter; it was designed not as a "word processor" but as a "program editor" to let programmers enter and edit their source code directly on a terminal screen rather than encoding punched cards or paper tape. (In fact, TECO in its earliest incarnations did use paper tape—the "T" in its title originally stood for Tape.)

Though it did nothing in the way of graphics or formatting, TECO would have had several features conducive to Forward's needs as he reworked what eventually became a 160,000-word novel through multiple versions and publication venues, starting with its original *Analog* serialization of about a third the total length. There was an extremely sophisticated search function, for example, and the host computing system—with accompanying tape storage—would have been capable of holding the entire text in its memory. Forward eventually moved on to IBM PCs and painstakingly had his tape files migrated to diskette.[79] His experience serves to illustrate yet another way in which science fiction writers found their way to word processing, in this case through professional proximity to the kind of computing resources still largely unavailable outside of corporate, government, or university settings.[80]

Neither was Eileen Gunn within Pournelle's immediate orbit. She has had a noteworthy career writing short stories (including a 2004 Nebula Award). Gunn started out in 1969 as an advertising writer at Digital Equipment Corporation, where she produced copy for the PDP-8, among other systems; by the mid-1980s she had moved to Microsoft, where she eventually became director of advertising. Her experiences at DEC and Microsoft informed what is probably her best-known story, the biting corporate techno-fable "Stable Strategies for Middle Management." In 1981, while still freelancing for DEC, Gunn had accepted a contract with them for an ambitious series of in-house technical publications. The company had a dusty, aging PDP-8 delivered to her Seattle home. The unit, which had a built-in CRT display, sat atop a wheeled base and was about the size of a household washing machine. Gunn cleaned it up, followed the instructions in the manual, plugged it in, and flipped the power switch. It ran WPS-8, a dedicated word processing system. For practice, Gunn began typing in one of her current stories in progress, "Contact," which was eventually published in an anthology in 1983: "I typed it all in and I pushed the button and printed it out. And it was really an exciting moment. I was thrilled. . . . There was a point in the story, about 18 pages in on the printout, that if I made a change there it made the ending make so much more sense. So I went back to the computer and I made that change. . . . And I pressed the button and it printed it out again. I was an instant convert. . . . I can have a perfect manuscript with the press of a button. I should have known that, I'd been working in the computer industry for 15 years. But somehow just pressing that button and it became reality. And it was just enormously exciting to me."[81] DEC took its time asking for the old PDP-8 to be returned, so Gunn continued to use it for another three or four years, until she accepted the job at Microsoft. In the meantime, she wrote more stories: "It was like having someone give me a beat-up old Mercedes to drive around."[82] By her own description, she was never a fast or an accurate typist. "Word processing," Gunn says, "made that irrelevant."[83]

Across the Atlantic, Terry Pratchett is said to have bought a Sinclair ZX81 as soon as they became available (1981), and he maintained a houseful of computers until his death in 2015; the original ZX81 became the inspiration for the fictional "Hex" computer that figures prominently in the Discworld novels.[84] Then there was Anne McCaffrey, famous for her Pern books about dragons, who was another early adopter of the Kaypro. McCaffrey's neighbors near her rural Irish estate, doubtless imagining something very different from the self-contained Kaypro II, however ungainly its twenty-five-pound bulk might be, were reportedly on record as fearing it would consume all the electricity in County Wicklow.[85] But no British or Irish SF writer from this era is more heavily associated with computers than Douglas Adams. "I remember the first time I ever saw a personal computer," he writes. "It was at Lasky's . . . on the Tottenham Court Road, and it was called a Commodore PET. It was quite a large, pyramid shape, with a screen at the top about the size of a chocolate bar. I prowled around it for a while, fascinated. But it was no good. I couldn't for the life of me see any way in which a computer could be of any use in the life or work of a writer."[86]

Nonetheless, by 1982 Adams had purchased a dedicated word processor called a Nexus; a year or two later (the *Hitchhiker's* trilogy already complete) he bought the British computer known as an Apricot, which ran MS-DOS.[87] He was an enthusiast, in short order accumulating a plethora of models and brands, also including a BBC Micro and a Tandy 100.[88] Sometime in 1984 he found himself at a small company in Cambridge, Massachusetts, called Infocom, where they were adapting his breakthrough novel *The Hitchhiker's Guide to the Galaxy* (1979) as a piece of interactive fiction—a surprisingly popular form at the time that allowed players to read prose presented on their screen and type responses in something approximating natural language, thereby to advance the story through one of its branching plotlines.[89] (Future poet laureate Robert Pinksy would also dabble in the genre, producing an "electronic novel" called *Mindwheel* with the assistance of a team of programmers for Brøderbund Software in 1984.) While visiting Infocom, Adams saw one of the just-released Apple Macintosh computers and was instantly captivated by it; he subsequently claimed to own the first one in England, and was profiled in *MacWorld* magazine.[90] He would go on to acquire at least one of most Apple Macintosh products released thereafter, and the number of Douglas Adams Macs that thus ended up populating the world would eventually yield at least one memorable surprise, as we will see in Chapter 10. But it wasn't just Adams's very visible association with Apple products that makes his computing history noteworthy here. He also helped move science fiction toward a more realistic mode of portraying computing technology through the fictional device of the Hitchhiker's Guide itself, which—as portrayed in the novel of the same name—was in effect an e-book. As his biographer M. J. Simpson notes, "It didn't think, and it displayed text on a screen."[91] In other words, several years before acquiring one for himself, Adams bucked the then-dominant assumption that what was most interesting about computers was their capacity for artificial intelligence; instead, he imagined a personal media tablet like the ones we use today.[92]

M. J. Simpson credits Douglas Adams with fundamentally changing the public's perception of how computers worked as a result of their portrayal in (or as) the best-selling *Guide*. Individual writers' relationships to the technology, however, were to remain deeply idiosyncratic. Jack Vance, for example, had more reason than most of his colleagues to make the switch to word processing, and had done so by the time he was writing *The Green Pearl* (1985): his eyesight, never good, was progressively worsening, and so his computer became a platform for early experiments in what we would today call adaptive technology. Raised embossments were affixed to the keys, the font was enlarged, and a speech synthesizer was added to read text aloud from the screen.[93] All of this was accomplished with a custom DOS-based word processing package named BigEd written for him by Kim Kokkonen, then president of Turbopower, a major software vendor. Vance used BigEd in conjunction with a thirty-inch screen, and this set-up enabled him to continue writing even after he became legally blind.[94] His dependence on the computer was thus unusually strong, but it serves to dramatize what David Gerrold, writing in 1982, had already understood: "The computer has evolved into a partner, a tool, and an environment." Gerrold goes on to make cause and effect explicit: "Computers are no longer malevolent iron brains. . . . This shift in perception is a direct result of humans finally getting their hands on computers of their own."[95]

We might see that same effect in the experience of Samuel R. Delany, who has won four Nebulas and two Hugos even as he has bent or eschewed many of science fiction's conventions. Delany bought a Kaypro II with Perfect Writer in July 1983 and attended users' group meetings at New York University with the artist and essayist Richard Kostelanetz.[96] In his correspondence from that period he would often mention the purchase in conjunction with the private school tuition for his daughter, which suggests the perceived magnitude of the expenditure. Nonetheless, the computer must have meant something more to him than just another check to write. Always an early riser, he begins one letter by setting the scene thusly: "Good morning. It's moving on toward five-thirty, though outside my office window it's still dark. If I rise a bit in my seat and look down, over the back of my word processor and out the window, there are only four lights beside the back-alley doors of the tenement buildings to the south; if I look up, I can see the beacon on top of what I assume is the Beresford, over Central Park West."[97] At this quiet hour, the way in which the blocky geometry of the Kaypro seamlessly merges with the darkened shapes and silhouettes of the cityscape beyond (so important in Delany's fiction) to form a single unified perspectival field perhaps suggests the extent to which the computer had worked its way into his consciousness.

Still, some of the genre's most notable luminaries rejected the technology outright. Besides Harlan Ellison, Andre (Alice) Norton, best known for her *Witch World* series, was quoted in *Life* magazine in 1984 as saying she "wouldn't have a word processor on a bet."[98] Others came to it only later, sometimes much later: Ursula K. Le Guin did not yet have a word processor when she was working on *Always Coming Home* (1985), but she comments in an interview that she was "aware even back then that the computer might encourage certain complex movements of narrative—recursions, implications to be followed, forking paths, etc.—that were very much in my head as I wrote the book."[99] In interviews given in 1996 and 1997 Octavia Butler is still talking about finishing *Parable of the Talents* (1998) on one of the succession of manual typewriters she had been using since she started writing at the age of ten, though manuscript evidence suggests she subsequently completed the novel on a newly-acquired computer.[100]

Sometime in 1982, while Gerrold was tendering his preceding observations about personal computers evolving into partners, tools, and even whole environments, an American ex-patriate living in Vancouver who had published a handful of promising short stories sat down to write the novel that would popularize the word "cyberspace."[101] As is widely known, he wrote it not on a computer but on a Hermes 2000 manual typewriter.[102] It wasn't a romantic or ironic gesture: William Gibson simply couldn't afford his own personal computer. *Neuromancer* is indeed the novel that introduced readers to cyberspace when it was completed and published in 1984, but it is also a novel about a rogue artificial intelligence. It thus pays tribute to the already long tradition in science fiction of stories about computers breaking bad (or breaking free), even as it also helped usher in the cyberpunk genre, the near-future noir settings first glimpsed by many in Ridley Scott's film *Blade Runner* (1982). It also resonated with another film released that same year, Walt Disney's *Tron*: Like that garish neon fable of a hapless hero forced through the looking glass to do battle with a malevolent computer program on its own silicon turf, Gibson's novel imagined the world inside a computer, or—more precisely—an all-encompassing environment derived from the totality of our world's networked computers. "Cyberspace. A consensual hallucination experienced daily by billions of legitimate operators. . . . A graphic representation of data abstracted from banks of every computer in the human system. Unthinkable complexity," is the novel's most explicit definition, rendered as an anonymous voice-over culled from a database.[103]

It is widely known that Gibson took his inspiration for this vision from watching players stand transfixed in front of the very same arcade games that figured prominently in *Tron*. Less well known is that although he didn't yet own a computer himself, he nonetheless took some share of inspiration from them as well. Not "computers" in the abstract, but the actual consumer products then rapidly appearing on the market. Gibson recalls the moment in his *Paris Review* interview: "The only computers I'd ever seen in those days were things the size of the side of a barn. And then one day, I walked by a bus stop and there was an Apple poster. The poster was a photograph of a businessman's jacketed, neatly cuffed arm holding a life-size representation of a real-life computer that was not much bigger than a laptop is today. Everyone is going to have one of these, I thought, and everyone is going to want to live inside them. And somehow I knew that the notional space behind all of the computer screens would be one single universe."[104] The computer depicted in the ad was presumably an Apple IIc, their 1984 entry into the luggable genre inaugurated several years previously by the Osborne 1 and Kaypro II. (This same computer was to make a cameo in the movie version of *2010;* Roy Scheider's character is briefly seen pecking at one on a bright, sunbaked beach, the moment a stark counterpoint to the more dramatic renderings of the damaged HAL adrift in deep space near the end of the film.) Gibson has often stated in interviews that his ignorance about the actual workings of computers was an asset in conceiving the notional technologies of cyberspace. But the bus stop ad for the Apple—publicly visible but personally unattainable, at least for the time being—seems to have allowed him to see computers with just enough specificity to bring his imagination into focus. When he did finally get a computer of his own (the same Apple IIc, and a dot matrix printer, in 1986),[105] it was a revelation: "I'd been expecting an exotic crystalline thing, a cyberspace deck or something, and what I got was a little piece of a Victorian engine that made noises like a scratchy old record player. That noise took away some of the mystique for me; it made computers less sexy. My ignorance had allowed me to romanticize them."[106]

Gibson had already heard about word processing by then, from his friend and fellow cyberpunk author Bruce Sterling, who had gotten an Apple IIe in 1984 and, as Gibson tells it, called him up excitedly.[107] Once Gibson tried it for himself there was no going back. "It was hugely freeing," he recalls. "Typographical errors were now of no import, and text became literally plastic with cut-and-paste."[108] His third novel, *Mona Lisa Overdrive* (1988) became his first actually written on a computer; shortly after he and Sterling would collaborate on *The Difference Engine* (1990), a novel set in an alternative Victorian era. Just as *Neuromancer* is widely regarded as a touchstone for cyberpunk, *The Difference Engine* inaugurated the "steampunk" genre, which posits that the Victorians (typically in the person of Charles Babbage) had been successful in inventing digital computers. And like cyberpunk before it, steampunk offers an homage to the idea of the sentient machine. Gibson and Sterling's novel is in fact narrated by a computer, a device we have seen before from John Barth and Michael Frayn. Here the "Narratron" assembles the story from a bricolage of sources encoded among its vast libraries of punched cards. What makes *The Difference Engine* notable, however, is the way in which this conceit in fact originates with the circumstances of its composition: Each author had his Apple computer, and disks containing the latest draft were shipped back and forth between Vancouver and Austin or else text was sometimes faxed.[109]

From the outset the authors adopted a rule in which they were forbidden to copy and paste any of their own preexisting text into the working copy of the book—to restore a previously expunged passage it had to be re-created by being rewritten. Just as the enormous clacking engines of the story iterate through their bundles of perforated cards, sifting and sorting data to weave the text, Gibson and Sterling iterated through one another's prose, writing and rewriting until it felt like a fusion. In an interview given at the time Gibson comments, "This is a word-processed book in a way that I would suspect no book has ever been word-processed before."[110] He describes the way in which in he and Sterling liberally pilfered source material from Victorian novels and journals, often transcribing them verbatim. "Then we worked it," says Gibson, "we sort of air-brushed it with the word-processor, we bent it slightly, and brought our eerie blue notes which the original writers could not have."[111] Word processing, he concludes, became the "aesthetic engine" of the book, even as he also extends the musical analogy by comparing their technique to sampling. He and Sterling would make this point again—and also retain the striking "air-brush" metaphor—in the afterword to the novel's twentieth anniversary edition: "We could blur the edges of the collage, effortlessly blend disparate parts, render the juncture invisible," Gibson concluded. "A different level of magic."[112]

Sydney Coleman, a Harvard physicist, liked to spend such spare time as he could reviewing books for the *Magazine of Fantasy and Science Fiction.* He followed developments in computing closely as part of his day job. By the early 1980s he had come to a realization: "We all knew computers were coming," he'd say, "but what astonishes us is it's not the scientists but the word people who have taken them up first."[113] Kathryn Cramer, an SF editor and anthologist, likewise pegs the time frame as "the early 1980s," and recalls, "You would go to science fiction conventions and what writers would talk about was what word processors they used. . . . That was shop talk. They were comparing notes."[114] Longyear too recalls these years as the ones when he was routinely dragooned to appear on panels about computers at the conventions.[115] During this same time period, David Brin, who then worked on an Apple II and was serving as secretary for the Science Fiction Writers of America, started a computer newsletter called *Space Chips* for the organization.[116] The newsletter was next taken over by Sheila Finch (also an Apple user), who recalls it was filled with "the newfound wonders of computing and word processing and comparisons of various machines and their memory size."[117]

Doubtless there were also other early adopters—signposts—among authors in other genres and literary domains, including Robyn Carr in romance (with her Burroughs Redactor III word processor) and, for the thriller or techno-thriller, Richard Condon (an Olivetti TES 501), Michael Crichton (also a TES 501), Frank M. Robinson (an IBM Memory Typewriter), and Thomas N. Scortia (a North Star Horizon II like David Gerrold's). The British historical novelist Malcolm Ross-MacDonald recalls going to a computer exhibition in London in 1979, looking for a system geared toward a writer's needs. "I want to draw a line round a bit of text and say to the computer: 'Move that bit of text to . . . here!' and touch the screen where I want it to be moved to," he remembers having told a salesman, who replied, "The programs to do that probably exist somewhere, for a particular machine in a particular lab, for a particular application. But general computing isn't moving in that direction at all."[118] Stuart Woods, meanwhile, claims that he remembers visualizing the idea of an electronic screen attached to a keyboard while writing his first book in the mid-1970s.[119] Not long afterward he saw an ad for the Lanier word processing system in an airline magazine; upon landing he called the company and was quoted the price tag. It was too expensive, but he soon (sometime in late 1979) wandered into a computer shop in Atlanta and emerged with an S-100 microcomputer called a Polymorphic, with a nine-inch black-and-white screen, 5¼-inch disk drives, and a printer. He used it to write his breakthrough novel, *Chiefs* (1981), which was later serialized as a CBS television series starring Charlton Heston; Woods also recalls transmitting the text to the computer of his publisher W. W. Norton, an IBM 360—the book was supposedly typeset directly from Woods's electronic manuscript.

As a general rule then, "word people" working in various forms of genre fiction tended to adopt word processing earlier than writers who perceived themselves to be engaged in the craft of *belles lettres*. For some this may be taken as proof positive that word processing marked an irreducible rupture with the muse—that to use a word processor was to admit that one's words were mere commodities waiting to be properly milled and packaged. Certainly science fiction authors and their compatriots believed word processing was a way to write more books more quickly with less fuss and less tedium.

But it was also—hopefully—a way to write them better, as Pournelle, Longyear, and others would testify. Many writers professed a newfound commitment to the craft of writing after their sojourns in the electronic empyrean. Whether or not the prose really did get "better" (by whatever standard) is a question I leave for readers of particular authors to debate.

Similarly, it is a commonly held belief across a number of different genres that novels became noticeably longer after the advent of word processing. But it is important to remember that the marketplace for science fiction (and other genres) was changing rapidly during the 1980s, and in addition to the rise of chain bookstores and the newfound demand for best-sellers came the popularity of film and television tie-ins, especially in the *Star Wars* and *Star Trek* franchises.[120] The bibliography of Vonda N. McIntyre is noteworthy here: the prolific SF writer had gotten an Osborne 1 very soon after they first became available in 1981.[121] What followed was one of the most productive periods in her career, with a half-dozen *Star Trek* or *Star Wars* novelizations or tie-ins, and several other books besides, published in the next five years. Was all of this activity a function of the word processor? Was it a function of changes in the business of selling books? Or did it result from other, less obvious circumstances idiosyncratic to McIntyre's life and career?

Such questions seem to me impossible to adjudicate without focused biographical and critical scrutiny, and I am suspicious of any claims to account for broad shifts in literary trends solely through technological factors. Put another way, word processing doubtless played its part in the numbers and the girths of novels or other books—except for whenever it didn't, as high-profile examples like Ralph Ellison (another early Osborne user) or Harold Brodkey should remind us.[122] Ultimately what gets published is a function of what the marketplace will bear, as much as or more than it's a function of what writers do with their fingertips at their keyboards, whether typewriter or word processor. And some authors face challenges that no writing machine is going to resolve, no matter its pedigree. Scholars who have examined the manuscripts of Octavia Butler's never-completed final novel, *Parable of the Trickster* (which she had begun as early as 1989, almost a decade before acquiring a computer), report that the surviving materials consist of "dozens upon dozens of false starts." Butler worked and reworked her opening pages and scenes but never got very far before her death in 2006.[123] Intensive iteration could create "magic," as Gibson and Sterling had discovered; but it could also prove devastating. The literary history of word processing is marked by many different kinds of signposts.

SIX

TYPING ON GLASS

Writing with light, writing on glass: "The day may come," Wilfred A. Beeching declared on the closing page of his 1974 opus, *Century of the Typewriter*, "when the letter typed will appear in front of the operator on a television screen in any selected type face, and by pressing a button be transferred electronically to sensitized paper."[1]

In fact, however, such technologies were already well advanced even by 1974, both in academic and in organizational settings. In the commercial sector, companies like Lexitron, Linolex, Vydec, and Wang had all begun selling dedicated word processing systems featuring video display terminals in the early 1970s. (Lexitron's was the first in 1971; Vydec's in 1974 was the first to display a full page of text on the screen.) Beeching's fanciful scenario would thus have already been familiar—mundane—to any office secretary who had been trained on Lexitron or its competitors' equipment.[2] The creation of a working "TV Typewriter" (TVT) also soon became a rite of passage for the home computer hobbyist. It was a key stepping stone for Steve Wozniak on the way to the Apple computer, and it featured prominently in the announcement for the first meeting of the Homebrew Computer Club: "Are you building your own computer? Terminal? T V Typewriter? I O [input/output] device? Or some other digital black-magic box?" read the ad that was posted around Silicon Valley in February 1975.[3]

Computers themselves, of course, compute: which is to say they work by fundamentally arithmetical principles. An algorithm is a sequence of arithmetical steps. A program is a sequence of algorithms. The basic architecture of input, processing, storage, and output has been canonical in computer science since John von Neumann published his "draft" report on the EDVAC in 1945.[4] All of those different abstract components, however, require a corresponding instantiation in some physical medium. Computer storage, for example, has historically taken the form of everything from disks and tape to punched cards to magnetic drums, wires, mesh, and ringlets, even cathode ray tubes (before they were used as output devices). For much of the twentieth century the most common medium for receiving output from a computer was paper. Typewriters were first connected to computers as early as 1957.[5] The most versatile output device was the Friden Flexowriter, which could deliver keyboard input to a computer as well as render its output, *and* do so on punched cards, paper tape, or a roll of typing paper. ASCII, the international standard of binary numbers designating various alphanumeric signs, was originally designed for use with teletype machines, and so to this day includes seven-bit codes dedicated to vestigial hardware features such as the ringing of the terminal bell. (The initial choice of seven bits for the electronic alphabet was—at least according to one account—a concession to the tensile strength of paper.)[6] Today, with the advent of virtual keyboards on our touchscreen tablets and phones, the screen and keyboard have at last merged or converged, physically coming together to occupy the same visual and tactile space.

But the iconic conjunction of a cathode ray or video display and typewriter has a much more tangled history than many of us realize. When Peter Straub made out the check for his IBM Displaywriter, he remembers, "What excited me in the first place was the concept of being able to write on something like a television screen. I found that idea immensely appealing. I couldn't exactly tell you why. It certainly smacked of the future."[7] Word processing on a personal computer must have seemed like future shock incarnate to any number of writers, Alvin Toffler's widely read prognostications arriving humming, glowing, blinking on their desks in front of them. In this chapter we will look back to some of the first researchers and writers—working in close proximity to one another—who were in a position to glimpse that future, just beneath the surface of the glass.

Douglas Englebart's presentation of his oNLine System (NLS) on December 9, 1968, at a San Francisco computer science conference—known colloquially as "the mother of all demos"—is an obligatory touchstone in this as in so many other computing histories. The NLS demo famously debuted such innovations as the mouse, windows, multimedia, collaborative document editing, and remote videoconferencing. For much of this it relied on software for entering and editing text with a keyboard and rendering that text on what was then a five-inch television screen (for the dramatic public demonstration the screen was projected on a twenty-foot display, itself a notable feat).

Watching in the audience that day was Andy van Dam, a Brown University computer scientist who since 1967 had been independently working on his own screen editing systems, in partnership with fellow computer pioneer, Ted Nelson. Their collaboration was to prove fraught, with Nelson departing Brown the following year.[8] Van Dam had always emphasized the importance of being able to deliver print-ready documents to a user and steered the design of the system in that direction, whereas Nelson regarded the dictates of the printed page as far too limiting and an abdication of the full potential of what he preferred to call hypertext. "Hypertext" was a term coined by Nelson around 1965 or so, and it denoted not a particular system or technology so much as the idea (or ideal) of nonsequential reading and writing, with documents connected associatively by "links" or "trails" and fully integrated with images, video, and other media.[9] Today, of course, we take such things for granted on the Web, though even the Web falls short of what Nelson imagined. Regardless of their disagreements, however, the Hypertext Editing System (HES) that van Dam and Nelson produced (which was reworked and refined after Nelson's departure by van Dam and colleagues as the File Retrieval and Editing SyStem, or FRESS) was a landmark, one of several early efforts that had independently and at about the same time arrived at something recognizable as word processing, using metaphors like cut and paste to manipulate and edit text on a video display screen. FRESS in particular was intended for a possible commercial deployment, and van Dam recalls shopping demos to several prominent newspapers and publishing houses; he was told the technology was marvelous, but that it would take "decades" to catch on.[10] Nonetheless, HES and FRESS would have important legacies at Brown, eventually involving prominent writers associated with the campus, notably Robert Coover and Rosmarie Waldrop, as well as writing instruction for students.[11]

Meanwhile yet another young researcher back in the San Francisco Bay Area found himself thinking along similar lines. Like Douglas Englebart, Larry Tesler was affiliated with Stanford University, in his case through the pioneering Stanford Artificial Intelligence Laboratory (SAIL). At the time (to the best of his recollection either 1968 or 1969) he also volunteered at the Midpeninsula Free University, one of the community educational collectives that had become fashionable in the wake of the Berkeley Free Speech Movement. Tesler's responsibilities included assembling a quarterly catalog of courses; he vividly remembers working with a mathematics instructor named Jim Warren, fussing with scissors and X-Acto knives and glue and tape to finalize the galleys for printing. "Someday this will be done on a computer screen," Tesler commented. Warren didn't understand, so Tesler found himself struggling to explain. "I tried to paint in the air what it would be like. . . . You had the screen, you could see the page on it, you could see the little clippings, and you would take some kind of pointing device . . . and Doug Englebart has this thing called the mouse . . . and there were light pens and touchpads."[12] SAIL in fact had Englebart's NLS system installed, with display screens on individual researchers' desks. "They were pretty crude at the time," recalls Tesler. "White on green, all text."[13] Nonetheless, he was at least working from concrete experience as he haltingly explained the basics of the technology to Warren. But the more advanced features Tesler was imagining were not yet available in NLS or any other text editing system he had yet seen. "While I was describing it to him, I was trying to make it up as I went along. I said you'd have a cut command and you'd have a paste command that would do what we're doing here with scissors and tape."[14] Eventually he brought Warren to SAIL and showed him NLS. Warren was sufficiently impressed to go on to a significant career in computer science and publishing, founding a number of influential industry publications; and he would create the annual West Coast Computer Faire (1977–1991), where the Kaypro and other notable computers made their debut. And Tesler, for his part, would soon have a chance to put his ideas into practice, as we will see.

Not far from Stanford University lies the wooded expanse of Xerox's Palo Alto Research Center (PARC), the famously cloistered research campus that opened in 1970 and is now venerated as the place where Steve Jobs first saw a demonstration of the ideas and concepts that would help inspire the Macintosh. While the story about Xerox's failure to effectively commercialize those ideas is well known, the sequence of events behind the actual systems and their development is complex, involving multiple generations of hardware and software.[15] The computer that Jobs famously saw demonstrated at PARC in 1979 (by none other than Larry Tesler, who had by then been working there for the past six years) was called the Alto, which had been prototyped as early as 1973.[16]

The Xerox Alto is not to be confused with the Altair 8800, the machine that famously graced the cover of *Popular Electronics* in January 1975 and (so the story goes) motivated Bill Gates to leave Harvard. (The Altair communicated its output to users, not on a screen or even on paper, but through gnomic blinking LED lights, the epitome of the "digital black-magic box" mentioned by the Homebrew Computer Club.)[17] Unlike the Altair, then, the Alto was a fully functioning graphical computer system, years ahead of its time. Its first demo program displayed a high-resolution image of *Sesame Street*'s Cookie Monster. Like Englebart's NLS, the Alto adopted a series of visual metaphors based on windows, and it featured a mouse along with the keyboard as the basis of user interaction. Unlike the NLS, however, the Alto was built around a new technical approach to display technologies called bitmapping. Bitmapping treated the entire screen as a kind of grid whose individual units—pixels—could be turned on or off by the programmed logic of the computer. This in turn enabled far more sophisticated forms of graphical interaction, as well as the kind of visual fidelity portended by the Cookie Monster rendering. Michael A. Hiltzik compares the significance of that disarmingly playful image to Alexander Graham Bell's first throwaway words spoken over a telephone line: "He [Alto designer Chuck Thacker] knew he had done more than create a novelty. He and his colleagues had reduced the computer to human scale and recast its destiny forever."[18] The difference between the Alto, in other words, and the green-and-white display screens Tesler had brought Warren to see at SAIL—which were known as "calligraphic" displays for the way they rendered characters as individual lines or strokes—could not have been more pronounced. Moreover, the Alto's screen was designed to emulate paper, with dimensions of 8½ by 11 inches and a vertical orientation.[19]

Unsurprisingly, one of the first applications was a text editor. The job of creating it fell to Charles Simonyi, a Hungarian-born computer prodigy who by 1972 had found his way to the United States and to PARC. In 1981 he would leave Palo Alto to go to Microsoft, where he would become the lead architect for Microsoft Word and Excel. In between, however, the project with which he was most closely associated at PARC was called Bravo. (In key respects Bravo would become the model for Microsoft Word.) When Simonyi began work on it, there were no direct precedents to draw from. Like Tesler speaking to his fellow university instructor, he groped for metaphors and models: "I could see books in their entirety flowing in front of you, virtual books and everything," he comments. "In retrospect it all seems so obvious. Uh-uh, it wasn't obvious to anyone. This stuff was in the future then."[20] That sentiment is confirmed by Andy van Dam, who concludes, in a 1971 paper surveying the state-of-the-art in screen editing, "the imaginative use of computers for on-line composition and extensive manipulation of free-form text is still in the early stages of experimentation."[21] Van Dam, a distinguished computer scientist and certainly one of the leading authorities on the history of text processing systems, is in fact speaking with some precision here; two terms in his expert summation—"on-line" and "free-form"—are worth closer examination.

Today we mainly use "online" to mean the Web or social media, but the term once had a more specific connotation, as the slight dissonance of the vestigial hyphen in "on-line composition" reminds us. Because the status of systems or devices as being "on-line" can be important for legal or regulatory purposes, the use of the term is codified in a federal standard; its technical application is dependent upon one system or device being under the control of another system or device.[22] The emphasis on whether one device is being "controlled" by another or not makes sense when we remember the actual conditions of computing at the time: personal computers as such didn't exist—the Alto was arguably the first (actually, technically, its processor was housed in a so-called "minicomputer"). Because computing resources were so precious, the idea that a machine might be dedicated to the exclusive use of an individual was all but unthinkable. Computers thus worked through batch processing or time-sharing models, with users connected to the central processing unit by terminals. Terminals themselves took a variety of forms, differentiated as both "smart" and "dumb" dependent on their capabilities; initially a terminal was nothing more than a typewriter connected to the computer by a telephone line to receive input and output—a teletype. By the mid-1960s, there was rapidly increasing use of CRTs, sometimes called "glass teletypes" (though the less expensive and more reliable paper-based teletypes remained commonplace well into the 1970s).[23] Writing "free-form" prose—the second of van Dam's key distinctions—would never have been encouraged "on-line" because it was perceived to be such a mundane task. (Specifically, "free-form" text was taken to mean "insertions and substitutions of arbitrarily sized character strings at arbitrary points in the manuscript," perhaps as stark a definition of word processing as can be imagined.)[24] Even actual program code—the opposite of free-form text—was ideally meant to be edited offline, though in practice programmers often flouted that dictum in the interest of expediency. Investment in systems for editing "free-form" text while a terminal was in its "on-line" state was thus a considerable conceptual leap as well as a professional risk, and the foresight of key figures like Englebart, Simonyi, Tesler, Nelson, and van Dam—all of whom were coming to a more or less common set of ideas at about the same time—should not be underestimated. "With a CRT display," van Dam wrote in 1971, "the editor may think out and implement his changes simultaneously."[25] The elimination of any delay between "thinking out" and implementation comes very close to the essence of word processing, the simultaneous real-time manipulation of a representation of a document coupled with the suspension of its inscription in immutable form that we examined back in Chapter 2.

Those privileged enough to be close to these developments were quick to grasp their potential. One such story concerns a young researcher named Douglas Hofstadter. Hofstadter had recently earned his doctorate and was looking for a teaching position, but in the meantime, by late 1975, he was living at his family's home in Palo Alto—largely distracted from his academic career by ideas for a book that would somehow unify the contributions of key figures from mathematics, music, and the arts through a new theory of cognition. These ideas became *Gödel, Escher, Bach* (1979), his "metaphorical fugue on minds and machines in the spirit of Lewis Carroll." Hofstadter wrote its final version using TV-EDIT, yet another example of an early mainframe text editor, which he had access to at nearby Stanford University, where (like NLS) it had been developed under Englebart's supervision. Hofstadter has since stated he couldn't imagine having finished the book without it: the prose "just flowed out . . . ever so smoothly."[26]

Back at PARC, the specific technical innovations on which Simonyi built Bravo ended up saving enough in the way of system resources—memory—that rich graphical effects and embellishments could be used to augment the on-screen text, thus taking advantage of the bitmapped display.[27] Basic formatting features such as boldface, italics, and underlining, which we take for granted today, were unheard of at the time—but easily fell within Bravo's unique capabilities. Bravo was so good at duplicating the appearance of a printed page that Simonyi and his PARC colleagues created a canned demonstration that consisted of showing a business letter with complex formatting features (including the Xerox logo) on the screen and then superimposing a transparency that had been printed from the same electronic file on one of PARC's laser printers, another key technology that was being developed at the same time. As Simonyi tells it, at one such demo a representative from Citibank spoke up with the tagline used by comedian Flip Wilson when playing his character Geraldine on the TV show "Laugh-In": "What you see is what you get."[28]

WYSIWYG was to become one of the perennial grails of word processing, requiring ongoing innovation not only in software but also in hardware technology, notably laser printers, which would not be commercialized until the end of the decade. Alto and Bravo were never commercialized at all, though PARC eventually manufactured several thousand of the systems and farmed them out to various academic and government locales. At PARC, though, even in the early days, once word about Bravo began to spread, employees and their families would come in after hours to work with it. PhD dissertations (many for nearby Stanford) were one favorite use of the machines. Other people produced newsletters or fliers, or simply wrote letters. Simonyi suggests this is probably the first time computers as we would recognize them were used by what he describes as "average persons."[29] Perhaps a teenager would come in to write a school report. And likely as not, someone, enticed by the softly glowing screen that looked so much like a sheet of paper, was moved to write some poetry, or a piece of fiction. If so, however, these efforts are lost to us.

But at least two other literary episodes associated with PARC are worth recounting. One we will explore in the next section; the other is a direct outgrowth of further refinements to the Bravo interface that were undertaken by Larry Tesler, who had moved to PARC from SAIL in 1973. The basic features of Bravo were finished by the following year, but there was then a push to improve the user experience. In particular, Bravo relied heavily on modes, a concept we have already encountered in conjunction with WordStar. This created problems. If a user typed the four-character sequence E, D, I, T into Bravo while it was in the wrong "mode," rather than echoing the word "edit" on the screen the system would instead do the following: first, it selected the Entire document; second, it Deleted it; third, it would Insert whatever the next character to be entered happened to be—in this case, T.[30] Simonyi, cleverly, had implemented a Replay feature in Bravo which could be used to rewind and recover the document in such instances—but only if one timed things so that the "replay" stopped just short of the fatal character sequence.[31]

1. MT/ST prototype, 1957. Two separate devices were used to store and retrieve text: a keypunch (left) for input and an electric typewriter (right) for output. Note also the mount for the recording tape in the center.

2. Len Deighton at work on *Bomber* in his high-tech home office at 29 Merrick Square, London, in 1968. The MT/ST is visible in the foreground. Also visible are his Selectric typewriter with its roll of telex paper atop the improvised standing desk, residue from the telex paper littering the floor (and the telex machine itself to the right), and a map of North Central Europe (the setting for the book) on the far wall.

3. In a contemporaneous photograph taken from the same angle, a glimpse of Ellenor Handley at work on the MT/ST.

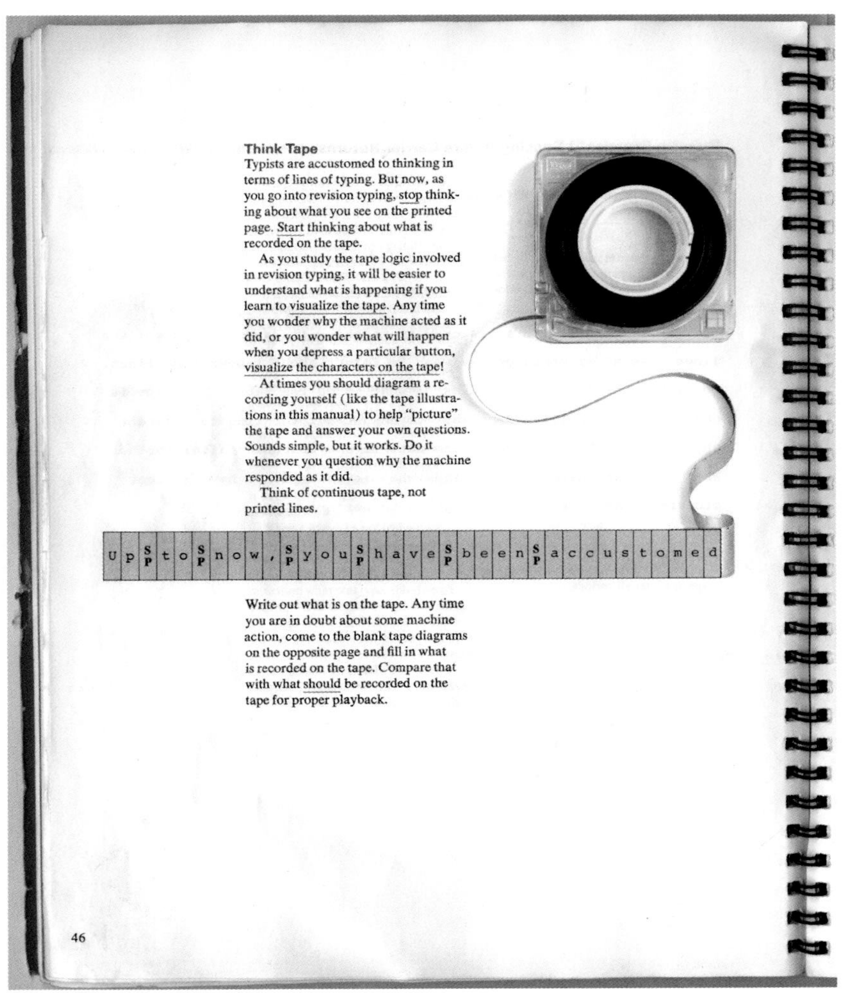

Think Tape
Typists are accustomed to thinking in terms of lines of typing. But now, as you go into revision typing, stop thinking about what you see on the printed page. Start thinking about what is recorded on the tape.

As you study the tape logic involved in revision typing, it will be easier to understand what is happening if you learn to visualize the tape. Any time you wonder why the machine acted as it did, or you wonder what will happen when you depress a particular button, visualize the characters on the tape!

At times you should diagram a recording yourself (like the tape illustrations in this manual) to help "picture" the tape and answer your own questions. Sounds simple, but it works. Do it whenever you question why the machine responded as it did.

Think of continuous tape, not printed lines.

U	p	SP	t	o	SP	n	o	w	,	SP	y	o	u	SP	h	a	v	e	SP	b	e	e	n	SP	a	c	c	u	s	t	o	m	e	d

Write out what is on the tape. Any time you are in doubt about some machine action, come to the blank tape diagrams on the opposite page and fill in what is recorded on the tape. Compare that with what should be recorded on the tape for proper playback.

46

4. "Think Tape." A page from an IBM MT/ST operator's manual. Note the way in which the diagrammatic illustration of how characters are stored is spliced to the photorealistic rendition of an actual tape reel.

The Death of The Dead-end Secretary

It's been long overdue.

For 45 years women like you have dominated the secretarial field.

Many young women first entered the field because they saw secretarial work as a steppingstone to a career in business — a chance to compete with men on their own turf.

After all, until the feminist, Katherine Gibbs, founded her secretarial school in 1911, women were as welcome in a business office as mongooses in a cobra pit.

As more and more women entered the field, the secretary's status diminished. And because of the staggering increase in the volume of office paperwork, the job became a monotonous routine of type and retype.

Chained to their typing desks and virtually forgotten, secretaries had reached a dead-end.

Today, the dead-end secretary is dead.

For two big reasons.

One reason is the change in you.

Unlike your complacent sisters of the previous generation, you want the freedom to get into more interesting and challenging work — work that'll give you the chance to move into a staff, administrative, or managerial position — where you can call some of the shots.

The other reason is the change in the tools of your trade.

With the development of automatic editing typewriters and word processing systems, the conditions of dead-end typing have been practically eliminated. Unlike standard typewriters, word processing systems handle the paperwork avalanche quickly, more efficiently, and more accurately.

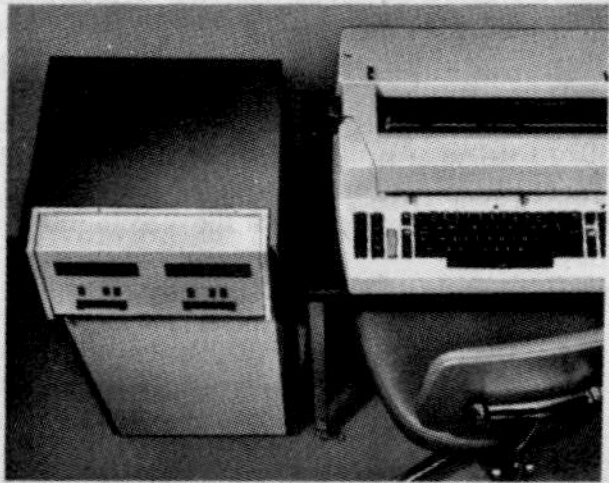

Here, at Redactron, we manufacture and sell word processing systems that were developed with the latest "state of the art" technology and the new you in mind.

We realized that the retyping routine was leading you to a dead-end.

That's why we made our Data Secretary the easiest to operate, most versatile, and most time-saving unit on the market.

And because we're a new company that's been in a position to learn from the mistakes of others, we've been able to add unique features and capabilities that correspond to your natural typing habits and eliminate the time-wasting, complicated formatting and logging procedures that are necessary with other systems.

Ask your boss about the Data Secretary.

We've already told him about it.

To show him that your changes and ambitions are for real, start your own campaign.

We're offering FREE memo pads, buttons, and stickers that say — "Free the Secretary". Just write our president:

Ms. Evelyn Berezin
Redactron Corporation
Dept. 100
100 Parkway Drive South
Hauppauge, L.I., N.Y. 11787
516 / 543-8700

It's the least you can do for the dearly departed.

5. The Redactron Corporation's "Death of the Dead-End Secretary" ad for their Data Secretary product (a direct competitor to the MT/ST). It appeared in the December 1971 preview issue of Gloria Steinem's *Ms.* magazine. The connections between the supposedly emancipatory promise of word processing and the burgeoning women's movement were made explicit, complete with the promise of what we would nowadays call swag. The ad is signed by the company's president, Evelyn Berezin.

June 16, 1975

Larry Tesler.... I have gotten hooked on this system! I would love to write a novel on this machine.

A MESSAGE TO LARRY ....IT IS FIVE A.M., AND I AM FEELING REGRET AT HAVING TO GIVE UP WORKING ON THIS SYSTEM. It <u>is</u> an unusual experience, and one I never expected. I appreciate your kindness more than you may realize. Have a joyful summer.

Nilo

Nilo

Monday 3:00 a.m.

Larry —

It's been a great experience, and I'm sorry to be leaving Gypsy behind. See you in the Fall . . . unless you come East again — if you do, come have lobsters in Rowayton —

Nilo

6. Note to Larry Tesler from Nilo Lindgren, a technical writer hired by Xerox PARC to produce documentation for Bravo / Gypsy. Lindgren had stayed up all night experimenting with the system. His note tells Tesler, the lead designer, that he felt like using the software to write a novel. The note was typed on Gypsy, and printed from one of PARC's then cutting-edge laser printers.

CA[31,353]...............22-May-73..........17:44...............Page 1

26 CA, CB
24 CD, CE
29 CF, CG, CH
28 CI, CJ
22 CK

This morning I have some hope of reaching the petition-windows.

The tight-packed column of citizens four abreast stretches back along Church Street to the corner of Elm and around toward Orange and out of sight but not imagination. I have been on this line since deep dark, since before five.

I have now come within a couple of hundred yards of the bureau building. There is less than an hour till I must start for work. This is the sixth day in a row I have tried to make it to the windows.

I felt a flash of line-fear just then, but I am all right now.

As usual at this hour, downtown streets are glutted with busses and cargo conveyors and people shuffling on foot to their jobs. Every square inch of concrete and asphalt is taken up. Wheeled traffic worms along at the stipulated pace. On this sidewalk, at the outer edge of our waitline to the right, one solid mass of pedestrians, facing us, inches toward Elm, and another, beyond, toward Chapel. It takes them fifteen or twenty minutes to move a single block. This is the familiar suffocating physical crush of the morning hours: breast touches shoulder blade, hip rubs hip, one's shoes are scuffed by others' shoes.

24
50

In the street, the busses and cargo vans creep along so close together they almost touch each other. The vans, uniform in design, squarish and chunky, white and immaculate, with no writing on them

7. Hard-copy draft of John Hersey's *My Petition for More Space,* date-stamped May 22, 1973 and printed from the PDP-10 at Yale University.

HMPFMS

Monitor Commands
Day.
DIR.
DTA (DECTAPE
R LIST
$ C1/T/D
title/double space
Stop Control C

LOG IN:
LOG space 31, 353 Ret

LOG OUT
EXIT> K/F Ret

To make file ENTER (ESC) (DSKA:) Give file name. CONTROL V
DSKA: XX <177>

To set or re-enter a file ENTER " " " " B

To go from monitor to EDITOR R#E - RET

~~Then CONTROL Z (EXIT)~~

INSERT SPACE CONTR A eg. ENT. 6. Cont. A OR

LINE CONTR D

DELETE SPACE CONTR S DELETE WORD: cursor under initial of word to be deleted <ENT> move cursor to initial of word after <CONT S>.

LINE CONTR F DELETE SEVERAL SPACES ENT (Repeat enter button until end of space to be %) CONTR S

PICK CONTR K

PUT CONTR G

MOVE SCREEN UP L. CONTR T ONE PAGE = 20 LINES.

DOWN L. CONTR W

UP-P. CONTR Q eg. ENTER-5-+PAGES.

DOWN P. CONTR. Y

GO TO CONTROL P eg. ENTER - 50 (meaning %) · = CLEAR PAGE

SEARCH +(Below) CONTROL R

-(Above) CONTROL E eg. ENT "WORD" SEARCH.

INSERT WORD ENT "WORD" PUT Also works when substituting longer word.

EXIT CONT Z.

IF SYSTEM STOPS. YALE REAL-TIME TEST means system restarted. Then must LOG-IN again

PROTECT: PRO C2<177> (Monitor command)

ITALICS:

COPY TEXT COP Destination title = Source title. RET

DECTAPE ASSIGN: AS#DTA | Response DTA0 (or 1, 2, 3.) DIR # DTA0

8. John Hersey's spiral notebook where he kept a crib sheet of commands for the Yale Editor.

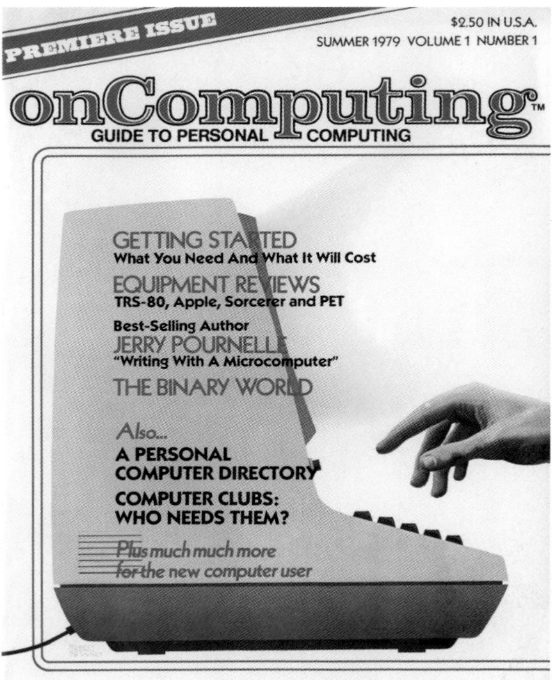

9. Cover of the 1979 debut issue of *onComputing* magazine, a spin-off of the better-known *Byte.* Jerry Pournelle's "Writing with a Microcomputer," in which he first details coming to grips with his Cromemco Z-80, "Zeke," is a feature story. *onComputing* would become *Popular Computing* in 1981, the year Isaac Asimov would write about his own word processing experiences for the magazine. Cover design by Robert Tinney.

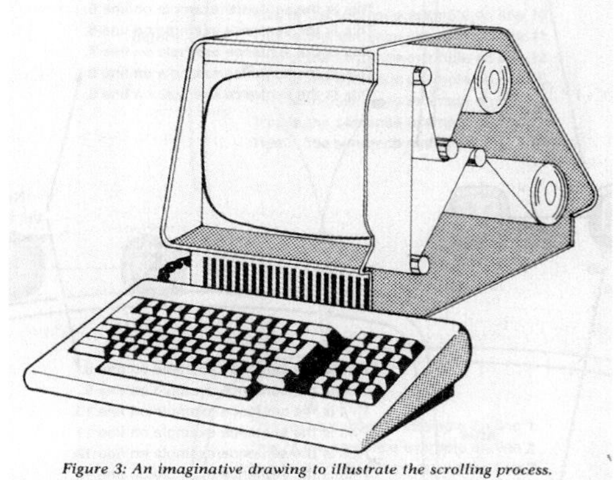

10. In this fanciful drawing from a 1982 Perfect Writer software manual, the concept of "scrolling" is illustrated for the novice by means of an imaginary roll of paper depicted as if it were moving inside the monitor. Perfect Writer was a CP/M-based word processor that shipped with the Kaypro II, and a competitor to WordStar. New users tended to worry about what happened to text when it scrolled off the top or bottom of the screen.

11. Stanley Elkin at the "Bubble Machine," his affectionate name for the Lexitron word processor that Washington University purchased for him in 1979.

12. This caricature drawing by Jack Gaughan, which accompanied the debate between Barry Longyear (left) and Jerry Pournelle (right) in the August 1981 issue of *Asimov's* magazine, captures much of the anxiety around early word processing. The two writers get swallowed up by their machines to become both input and output in a never-ending paper loop.

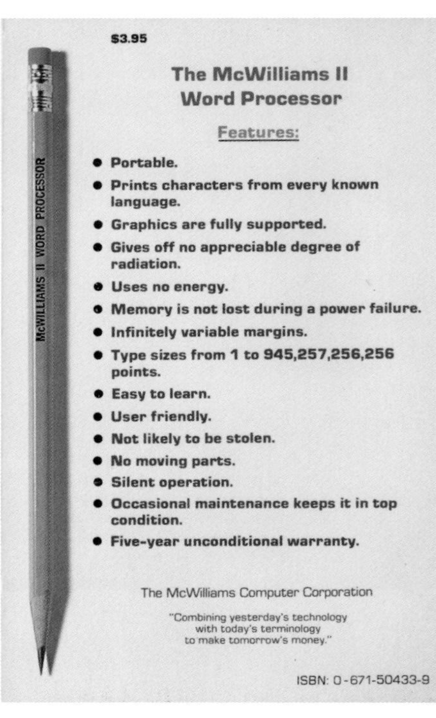

$3.95
McWILLIAMS II WORD PROCESSOR
The McWilliams II
Word Processor
Features:
• Portable.
• Prints characters from every known language.
• Graphics are fully supported.
• Gives off no appreciable degree of radiation.
• Uses no energy.
• Memory is not lost during a power failure.
• Infinitely variable margins.
• Type sizes from 1 to 945,257,256,256 points.
• Easy to learn.
• User friendly.
• Not likely to be stolen.
• No moving parts.
• Silent operation.
• Occasional maintenance keeps it in top condition.
• Five-year unconditional warranty.
The McWilliams Computer Corporation
"Combining yesterday's technology
with today's terminology
to make tomorrow's money."
ISBN: 0-671-50433-9

13. Gag pieces positing graphite pencils or fountain pens as feature-laden "word processors" were a staple of the 1980s computer press. Peter A. McWilliams stretched the joke to a full-length book in *The McWilliams II Word Processor Instruction Manual* (1983), product of his faux "McWilliams Computer Corporation," spoofing the kind of titles that were suddenly ubiquitous on bookstore shelves.

14. Stephen King with his Wang System 5 Model 3 word processor in 1982, as seen in the short film "I Sleep with the Lights On," made that year for the Public Information Office, University of Maine at Orono, by Henry Nevison in association with WABI-TV, Bangor.

15. "The Smartest Way to Write." Trading on Dr. Asimov's public image as a brainy writer, Tandy Radio Shack featured him in a series of advertisements for the TRS-80 and the Scripsit word processing program, the software and computer Asimov used for over a decade. The ads were a common sight in newsstand computer magazines. This one is from *Byte,* January 1983.

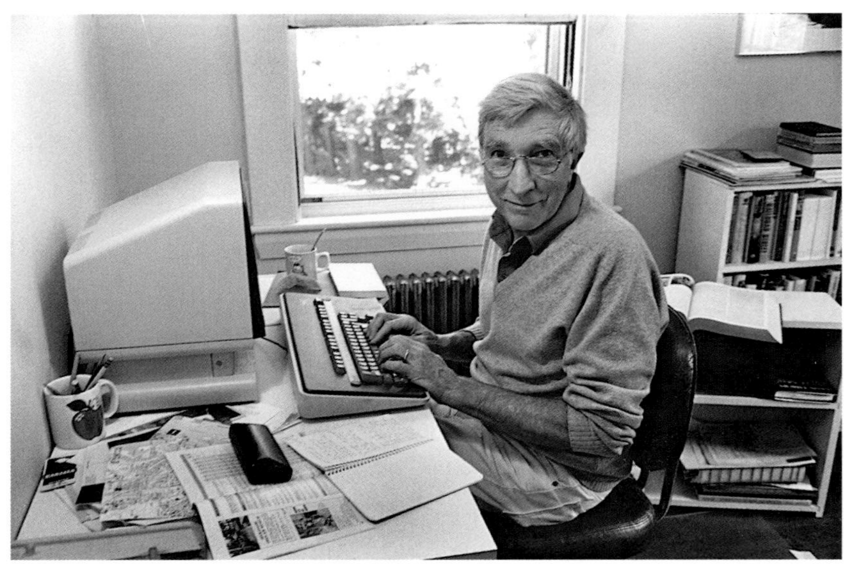

16. John Updike in 1987 with the Wangwriter II he used for much of the decade. The word processor never replaced Updike's typewriters or pencil—both of which he continued to favor for drafts—but it did replace his typist as he became accustomed to editing and revising directly on the screen.

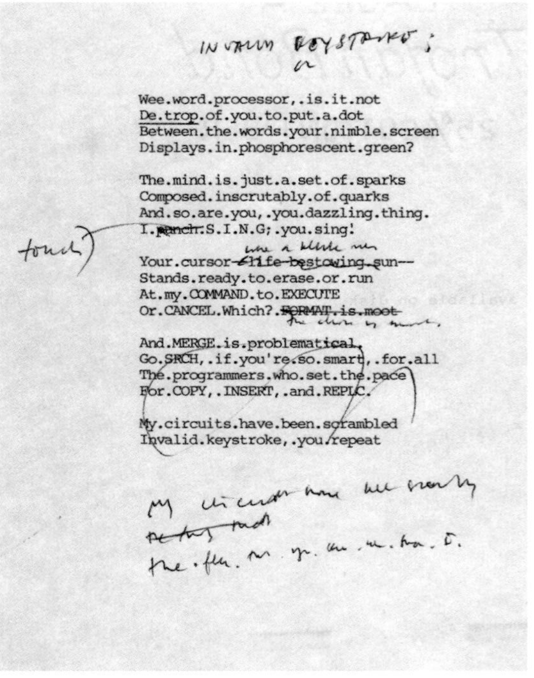

Wee.word.processor,.is.it.not
De.trop.of.you.to.put.a.dot
Between.the.words.your.nimble.screen
Displays.in.phosphorescent.green?

The.mind.is.just.a.set.of.sparks
Composed.inscrutably.of.quarks
And.so.are.you,.you.dazzling.thing.
I.S.I.N.G;.you.sing!

Your.cursor--life-bestowing.sun--
Stands.ready.to.erase.or.run
At.my.COMMAND.to.EXECUTE
Or.CANCEL.Which?.FORMAT.is.moot

And.MERGE.is.problematical.
Go.SRCH,.if.you're.so.smart,.for.all
The.programmers.who.set.the.pace
For.COPY,.INSERT,.and.REPLC.

My.circuits.have.been.scrambled
Invalid.keystroke,.you.repeat

17. Hard-copy draft of John Updike's "INVALID. KEYSTROKE," with handwritten revisions (including the title), as printed from his Wangwriter II in March 1983. In a departure from his usual habit of composing poetry with a pencil, Updike had produced these lines directly at the computer shortly after its arrival in his home.

18. Ralph Ellison in 1986 with the Osborne 1 computer he had purchased four years earlier. Note the external monitor used in lieu of the diminutive built-in screen. Sold bundled with WordStar and nominally portable, the Osborne was a popular first computer for many writers, including also Michael Chabon, Vonda N. McIntyre, and Anne Rice.

19. Eve Kosofsky Sedgwick at work on *Between Men* ca. 1982, with her Osborne 1 computer. Like the laptops and tablets that would eventually follow, first-generation "luggables" like the Osborne and Kaypro changed the array of potential postures and locales for writing with a computer.

20. "Old writer . . . leers into computer." By 1995 when R. Crumb produced this drawing of Charles Bukowski, the computer was fully on its way to becoming assimilated into the stock of cultural imagery around writing and authorship. Bukowski had been using a Macintosh IIsi since Christmas 1990.

21. Over a dozen of John Updike's "lost" Wang computer diskettes were fished from his curbside trash between 2006 and 2009 by a private collector. Any of these disks—whose contents may still be readable using forensic computing technology—might hold electronic drafts of works or versions of works that don't otherwise survive. This one was apparently once used to save portions of *The Witches of Eastwick*, and then later for reviews.

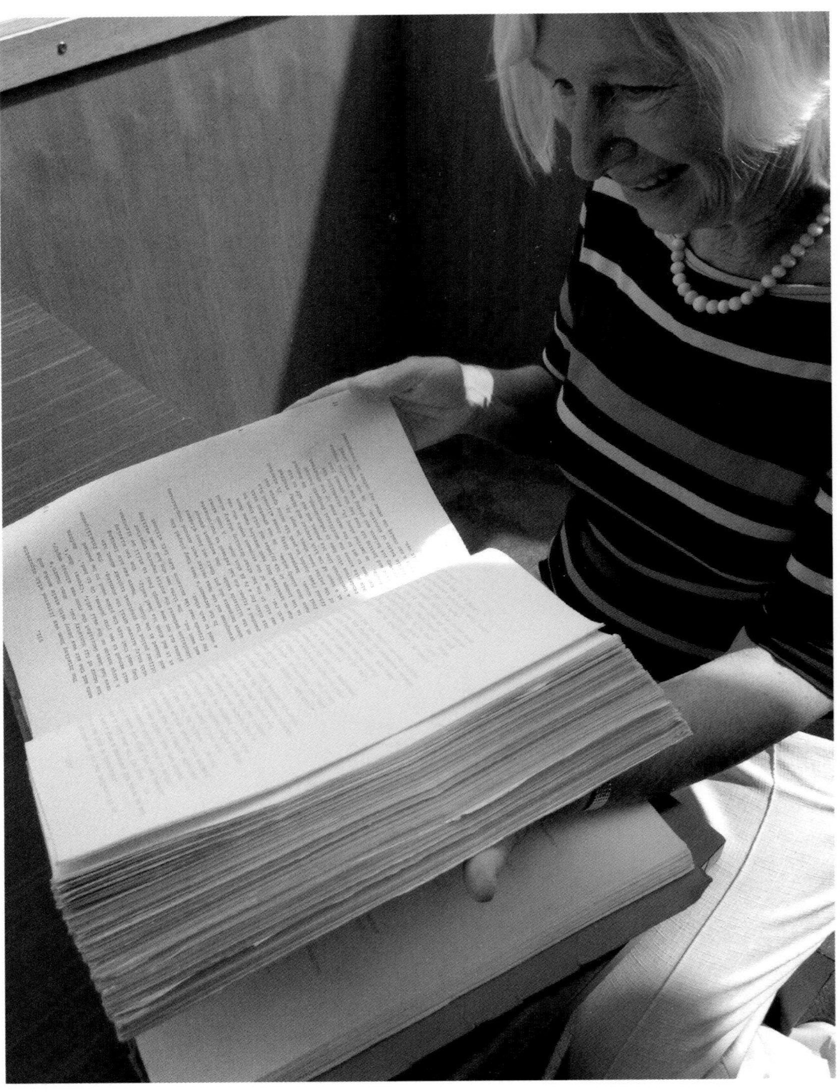

22. Ellenor Handley with the "master" manuscript of *Bomber* in 2013. Note the bite marks along the edge of the page where it was gripped by the "unseen hands" of the MT/ST's tractor feed.

Around 1974 Tesler and a few colleagues commandeered one of the Altos and set to work implementing a modeless interface for Bravo, which they named Gypsy. "This was the first editor, the first word processor, that worked like the ones today," he claims. He narrates a typical workflow: "Point with a mouse, drag, double-click, cut, copy, paste, undo—that combination did not exist before Gypsy."[32] Xerox was sufficiently impressed with Tesler's efforts that they hired a science writer, Nilo Lindgren—who was also then researching what proved to be an unfinished book about Douglas Englebart—to document the new editing system.[33]

As Tesler tells it, he sat Lindgren down to demonstrate how to operate the Gypsy interface.[34] He then went home for the night. The next morning—June 16, 1975—Tesler found a note that began: "Larry Tesler . . . I have gotten hooked on this system! *I would love to write a novel on this machine.*"[35] Lindgren had added: "It is 5 A.M. and I am feeling regret at having to give up working on this system. It is an unusual experience and one I never expected. I appreciate your kindness more than you may realize."[36] It was signed "Nilo," but the note itself had been typed and printed from Gypsy, which, with its menus and icons, was remarkably contemporary in its appearance and behaviors.

Lindgren never did write a novel, and died in 1992. However that plainly joyous note, composed after an apparently sleepless night spent exploring the system, testifies to just how powerful such a prospect must have seemed.

The Altos remained in use at Xerox PARC throughout the 1970s. Steve Jobs's famous pilgrimage to PARC wasn't to come until December 1979. It would be another half decade before the Alto's most important design features, including the bitmapped display, windows, and the mouse, were brought to market as Apple's first Macintosh computer in early 1984. (Xerox did attempt to introduce a graphical computer system incorporating some of Alto's capabilities in 1981; called the Xerox Star, it was marketed to businesses rather than to personal users, was expensive, and had to compete with dedicated word processing systems; thus it was never a commercial success.) Though the Jobs visit is the one that will remain inscribed in legend, there were many outsiders who came through PARC, and the Alto, Bravo, and Gypsy teams did many demos. A year or so before the group from Apple, a young Hollywood screenwriter named Bonnie MacBird showed up. She had an unusual background: she had read about PARC in Ted Nelson's book

Computer Lib / Dream Machines (1974) and had studied with Donald Knuth, one of the world's foremost computer scientists, at Stanford. She was also working on a rather unusual script.

MacBird had recently left a writing position at Universal Studios for a new production company run by Steven Lisberger specializing in animated films and computer graphics. Lisberger had an idea for a movie about video games, which were quickly becoming the rage in coin-operated arcades (*Space Invaders* had just been released in the United States by Midway). The film Lisberger envisioned would feature a computer warrior character named Tron (as in "electron"), and take advantage of new special effects techniques. Lisberger hired MacBird, and they went out on a listening tour, which brought them to PARC, where they were received by Alan Kay.

Alan Kay is one of the truly legendary names in the computer industry: he coined the term "object-oriented programming" and foresaw tablet computing with a prototype device called the DynaBook.[37] His career would eventually take him to both Apple and Walt Disney. Like Andy van Dam and Larry Tesler, he had been present for Englebart's famous NLS demo in 1968. And like Tesler and Simonyi, he had spent the better part of the decade at Xerox PARC. One of his projects there was called SmallTalk, an early so-called object-oriented language. Kay was interested in democratizing knowledge of computer programming, and conceived of SmallTalk as a kind of creative complement to the Alto, allowing an untrained user to write programs by pointing and clicking through a graphical interface (Frank Herbert's ambitions for PROGRAMAP were thus squarely in line with it, and Kay would have grasped the purpose of Herbert's flowcharts and keyboard design immediately). MacBird vividly recalls her visit in late 1978:

> The last interview of the day was Alan Kay, then still at Xerox PARC. . . . It was Alan's group that inspired Steve Jobs to envision what was to become the Mac. Alan's group is credited for developing object-oriented programming and the first implementation of the graphical user interface that is the face of personal computing today. Alan even coined the term "personal computer." So this visit was really to the mother lode. And way ahead of the curve. People had not even heard of this stuff yet. It was a big deal for all of us. Alan talked for close to four hours about computers, education, music, early childhood, theatre, storytelling, science, psychology, learning, artificial intelligence, programming, science fiction, biology, humanism, evolution,

> Bach, Buckminster Fuller, philosophy, neurology, aesthetics, and the future, the future, and the future. Steven and I had our minds blown.[38]

MacBird had already been thinking that the film would hinge upon the contrast between the world inside and outside of the computer; the three developed a rapport, and she and Lisberger hired Kay as their technical consultant. MacBird also decided that the film needed an Alan Kay–like character, and wrote him into the script as a master programmer named Alan Bradley, or "Alan-1." Kay, meanwhile, had an idea. He knew that MacBird was still in the thick of writing and editing the script, which would eventually go through eight major rewrites. At the time she was working on legal pads and an electric typewriter. Kay invited her to input the text into the Alto and finish writing and editing it on site at PARC (much as Xerox employees and their relations had been doing after hours for years).

MacBird first used a standard teletype to connect to PARC remotely from Los Angeles and inputted the existing draft of the screenplay so it would be waiting for her when she arrived. "While the data entry was kludgy for getting *Tron* into the Alto from afar," she recalls, "once up in Palo Alto, working on the Alto was gold and I wrote and rewrote, cutting and pasting, drafting like the wind in writer's heaven."[39] The Gypsy interface, with its sophisticated formatting capabilities and WYSIWYG printing capabilities, was ideally suited for the task. Kay showed off the variety of different fonts the system could handle, but MacBird insisted on Courier. "Why? Because only something that looked like it was hot off a writer's typewriter would be received with interest by the studios. I know that from my years as a development exec at Universal prior to becoming a writer."[40] (As we have seen, this attitude would persist well into the next decade: Publishers sometimes shunned manuscripts that were self-evidently prepared on a word processor.) Much like Nilo Lindgren before her, MacBird quickly grasped the potential of this new mode of on-screen writing. Her recollections capture that enthusiasm: "There was very little ramp up. It was truly WYSIWYG, and though I had never seen anything like it, it was instant love. Wow, rewriting on the fly! Fantastic! Seeing a whole page at a time, amazing! Printing almost instantly, wow!"[41] Indeed, the availability of laser printer technology at PARC proved vital to her experience with the Alto. Unlike authors like Pournelle, who could spend hours printing and then "busting out" their manuscript pages on large, loud, and temperamental impact printers, MacBird recalls that "the printer down the hall spit out pages at less than one per second."[42] She elaborates: "Familiar as I was with scripts due to my experience at Universal, I could sense the flow of a scene by how it looked on the page—and being able to see a whole page was critical. Unfortunately it was a long time before I had that again!"[43]

Tron was released in 1982, the same year *Time* magazine named the computer "Machine of the Year." (The issue featured conspicuously *Tron*-like imagery on its cover.) The film's box office fell short of expectations, though reviews were generally positive. It has since, of course, become a touchstone of popular culture. Like so much science fiction before it, *Tron* revolved around the machinations of a malevolent supercomputer. What was different about the film, though, was that *Tron* created a visual identity for the computer's internal dimensions. Not rows of blinking lights and spitting teletype printers, not hulking room- or building- or planet-size conglomerations, but something very much like "cyberspace," the word we already know that William Gibson (himself inspired by arcade machines) would coin the following year, limning something very close to the luminescent ray-traced imagery of *Tron* in his prose. *Tron*'s significance, however, lies not just in its advancement of science fiction as a genre, and not just in its special effects and innovations, but also in the symbiosis between the technical specifics of those innovations and the film's subject. Its computer-aided graphics were placed in the service of the first cinematic and one of the very first widespread visual depictions of a computer's internal operations, the world that existed on the other side of the glass screen.[44] Ironically, despite multiple Oscar nominations, the Motion Picture Academy refused to consider the film in the special effects category because the use of computers was seen as cheating, an attitude that we have also seen manifest itself in relation to writing.

Looking straight through the glass screen that more and more people by that time were just becoming accustomed to looking *at* on its surface, *Tron* took advantage of computers to imagine (and depict) the interior space of computing, opening up a new landscape for science fiction. But MacBird's story of how the final drafts of the script were written on an Alto computer at PARC (running Gypsy) in 1979 is not nearly as widely known. This is all the more striking given that the very same concepts of graphics and interactivity widely acknowledged as essential to the aesthetic of the film were, as we have seen, the conceptual foundation of the Alto and Gypsy, and thus the essence of her experience using it as a working writer. Indeed, though MacBird instantly became a convert to word processing, she had to wait until the Macintosh before finding a computer that could again approach the Alto's capabilities. She and Kay are unequivocal, though, in their description of the central role Alto played in finishing the script (which was, in the normal Hollywood way, to be rewritten extensively once again, following its acquisition by Disney—her screenwriting credit was eventually resolved by arbitration).[45] As for Kay and MacBird's relationship, it too flourished. In 1983 they married.

The walk from Pierson College to the Becton Center on the campus of Yale University is a short one. Pierson, Yale's largest residential college—all red brick and white columns in the Georgian style—is three blocks from New Haven Green. In the early 1970s it was home to some 350 undergraduates and had a reputation as one of Yale's more rambunctious addresses. Within the building housing Pierson Tower (the tall spire that is its most visible landmark) was the Pierson Press, founded in 1948, part of the tradition of letterpress printing and the book arts preeminent in the Yale residential colleges (one source describes these as, collectively, "an infrastructure unmatched in the United States, and quite possibly the world").[46] The Pierson Press's centerpiece was a Colts Armory platen press, in the estimation of many the finest instrument of its kind.[47] In those days before desktop publishing it was used by students for chapbooks, little magazines, broadsides and advertisements for campus activities, ephemera, and small jobs of all kinds.[48]

The Becton Center, by contrast, presented a rather different milieu. A 133,000 square-foot modernist concrete blockhouse on Prospect Street, designed by Marcel Breuer, it had opened just recently (in 1970) to house Yale's School of Engineering and Applied Science. To get from the one place to the other, one might follow Pierson Walk out of Pierson Tower and out onto busy York Street, then proceed past Trumbull College, Sterling Library, and the Law School until reaching the unexpected expanse of the Grove Street Cemetery. There, to the left looms the gothic façade of the Payne-Whitney Gymnasium, and to the right, and thence onto Prospect, lies the Becton. No more than a ten-minute stroll, even if one paused for a casual word with an acquaintance.

In the spring of 1973 John Hersey would have occasion to make that walk often. Born to missionary parents in 1914 in China, Hersey had been educated at Yale (Skull and Bones 1935) and so had long-standing ties to the campus. In 1965 he had returned to take up a five-year residence as master of Pierson.[49] Before doing so, however, he had managed to see a good bit of the world and had become one of the most recognized writers of fiction and nonfiction in America. After a brief stint working for Sinclair Lewis, he had begun his career as a war correspondent, reporting from throughout Europe and the Pacific during World War II. By the time the armistice with Japan was signed he was a regular contributor to *Time*, *Life*, and the *New Yorker.* Hersey found himself in Hiroshima during the winter of 1945–1946 and began the research for what would become a groundbreaking 30,000-word story about the bombing and its aftermath, centering on the experiences of six survivors. The piece consumed the entirety of the August 31, 1946, issue of the *New Yorker*—there were no other articles printed, nor were the magazine's signature cartoons present. The editors took this unusual step, as they wrote, "in the conviction that few of us have yet comprehended the all but incredible destructive power of this weapon, and that everyone might well take time to consider the terrible implications of its use."[50] *Hiroshima* was published in book form by Knopf later that same year, and it has remained in print ever since, widely lauded as one of the most significant nonfiction books of the twentieth century. Notable for its unsparing descriptions of the blast effects and the wounds suffered by the bomb's victims, it has also frequently been identified with the rise of the New Journalism, a label Hersey himself sought to resist. Hersey was also by then an accomplished fiction writer, and his 1944 novel *A Bell for Adano* (set in Italy during the war) had won him a Pulitzer Prize.

By 1965, however, after two decades of global travel, writing, and reporting, he was ready to put down roots with his second wife, Barbara, and so accepted the unexpected offer of the position of master at Pierson (the first nonacademic ever to hold the post). Besides his collegiate duties he maintained an active writing regimen every morning, publishing two additional novels and a nonfiction book about an incidence of police violence during Detroit's so-called 12th Street Riot. Hersey remained the college's master for five tumultuous years—his term for them was "explosive"[51]—that saw him through campus activism related to the Vietnam War and civil rights, culminating in the dramatic Mayday protests of 1970 when Bobby Seale and Erikka Huggins of the Black Panther movement were on trial in New Haven. (He wrote about these events in a volume entitled *A Letter to Alumni* published at the end of his term.) After a year abroad at the American Academy in Rome, Hersey again returned to Yale and a more relaxed routine at Pierson to begin teaching writing seminars there. He also soon found himself at work on another novel—not an unusual circumstance for him, in the least, but the manner in which he would end up writing it would prove very unusual indeed.

Hersey was also an avid and accomplished letterpress printer. During his tenure at Yale, he had overseen the relocation of the Pierson Press from its original basement room to its then-current locale in a converted two-story squash court adjacent to Pierson College Library, where he had made it a showcase among the college presses. This work had resulted in a friendship and collaboration with Alvin Eisenman, head of the graphic design department in the Yale School of Art. Sometime in late 1972 or early 1973, not long after his return to campus, Eisenman came to Hersey with a proposal.[52] A faculty member named Peter Weiner had recently arrived from Princeton to establish Yale's Department of Computer Science, the discipline still being very new at the time. Weiner had a particular vision for computing at Yale: not wishing to see it sequestered in engineering and the sciences, he sought to make a campus-wide push to integrate computing into as many subjects as possible.[53] With the department's cofounder, Edgar ("Ned") Irons, Weiner had been able to acquire a half-million-dollar DEC PDP-10 mainframe computer (a descendent of the PDP-1, which ran Expensive Typewriter at MIT). The PDP-10 was a time-sharing system, meaning many different users could share it—interact with it—at the same time. This capability alone suggested radical new possibilities and was fully compatible with Weiner's vision. He and Eisenman had already worked together to connect a Mergenthaler Super-Quick phototypesetting machine to the PDP-10, writing custom control software for it with the assistance of colleagues and students. Crucially, Weiner also had a way of inputting text, something called an "editor"; it was like a typewriter, Eisenman would have explained, but the words appeared on a glass screen instead—they could be edited and revised at will before being printed. With the Mergenthaler, it would even be possible to typeset the book and deliver camera-ready copy to Hersey's longtime publisher, Knopf. It seemed like an interesting experiment, a way of validating Weiner's idea that computing could be applicable far beyond the usual narrow range of disciplines. Was Hersey interested?

He was. By then Hersey had completed a longhand draft of the current book in progress, a work of dystopian speculative fiction entitled *My Petition for More Space.*[54] Given his experience as a printer, the idea of having complete creative control over the book's appearance was irresistible. And the "editor" seemed like a good way to save some time. So Hersey agreed to give it a try. The Yale Editor as it was known (it was also called simply "E") was the work of Ned Irons, who had accompanied Weiner to New Haven.[55] He had built a previous incarnation of it at the Institute for Defense Analysis in Princeton, sometime around 1967; it was certainly one of the very earliest CRT-based text editors, more or less contemporaneous with the initial experiments of Andy van Dam, Ted Nelson, and Douglas Engelbart. It would go on to spawn a number of derivatives, notably the so-called RAND editor, and NED (New Editor, also eponymous of its creator).

What was most distinctive about the Yale Editor, however, was not just its reliance upon a CRT but the actual computational model of text that it instantiated. Other editors that were available at the time treated the text as essentially one long alphanumeric statement, arrayed in an endless horizontal row (in practice, constrained only by the limits of system memory and resources).[56] Line breaks were a function of special characters or the properties of the associated display system. Irons's editors were different. They introduced what became known in technical parlance as the "quarter plane" model, meaning that the text became something like a canvas. The user worked through a rectangular window onto that canvas—the top left-hand corner (or quadrant) by default—with potentially unlimited horizontal and vertical extensibility. In other words, the editor treated the writing space as a vast two-dimensional sheet of paper rather than as a one-dimensional string.[57] Irons himself used the analogy of working with the window of a microfilm reader.[58] The consequences were significant in terms of both the features that could be enabled and the technical specifics of how the terminal managed information vis-à-vis the mainframe. Hersey, of course, would not have known any of this, and doubtless would have cared less. What he knew was that he had something new and potentially very powerful on his hands.

The PDP-10 was housed in the Becton Center in its own air-conditioned room with raised floors for the wiring.[59] The terminals, however, some ten of them, were in nearby Dunham Laboratory. Student users, at least, were expected to reserve time in one-hour blocks and bring with them ¾-inch magnetic DECtape reels (available at the bookstore for ten dollars apiece) for storing their data. Working on the book would have meant walking over to the lab and initiating a session on the PDP-10.[60] Hersey would have been seated in front of a display known as a Sugarman Laboratories CRT, one if the first of its kind commercially available, which accommodated 80 characters of text across the slightly convex glass screen, and 20 lines down. The text was presented in plain white phosphor. By his side was a spiral notebook in which he kept a careful shorthand list of instructions and reminders. Just a few pages covered all of the basics needed to create and open files, perform the typing and revision functions, print a hard copy, and save the results to the tape reel. At least initially, he would have had to refer to it often.

We can reconstruct in some detail what would have been a typical session with the Editor for Hersey. Just as with many computers today, first he had to login. He typed LOG 31,353 and pressed a black key on the right-hand side of the keyboard labeled ESC for Escape (the seemingly random numerals were in fact a "project" and "programmer" number, in other words his unique account ID). Logging onto the PDP-10 also automatically placed one into the Editor, so Hersey was now ready to start working (the assumption was that most users would immediately wish to write a program to run on the machine, thus the Editor was the default environment). Most of the Editor's commands were initiated by pressing the ESC key (also referred to as the Enter key in the documentation) and then inputting a parameter of some sort, followed by the desired command to execute, itself specified with the blue Control key on the left-hand side of the keyboard and a seemingly arbitrary letter in chorded combination with one another. Opening a file thus meant pressing ESC, entering the file name, and then chording the Control-B combination. Hersey maintained separate files for each chapter. (He named them simply ONE, TWO, THREE, and so on.) Inserting and deleting text similarly required Control-key combinations, and Hersey needed to learn separate ones for inserting (or deleting) a line versus inserting (or deleting) a character. Moreover, unlike today's word processors, the Editor stipulated that the requisite *number* of spaces first be inserted *into* the document *before* one could type the desired addition.

Let's take an actual example. The second line of the book begins: "The tight-packed column of citizens four abreast stretches back along Church Street to the corner of Elm." Hersey made no changes from his handwritten draft except for the addition of the word "back," which was performed with the Editor. To accomplish this, Hersey would first have had to maneuver the cursor (more Control-combinations!) to the blank space between "stretches" and "along." He would have pressed ESC, then entered "5," then chorded the Control and the A keys. This would have inserted five blank spaces into the document, one for each letter of the word plus one additional.

Only *then* could he proceed to type "back" into the spaces he had thus petitioned for. To "pick" (or copy) four lines of text, Hersey would similarly situate the cursor at the beginning of the first line, press ESC (in other words, changing modes), enter 4, and then execute the Control-K combination. To "put" (paste) this same text, he would follow the same procedure, only with Control-G. Moving upward or downward in the file? More Control-combinations. Nonetheless, he caught on soon enough. "It took about a month to get used to looking at words on a screen, almost as if in a new language," he recalled. "But once that was past, it seemed just like using a typewriter."[61] He found the Search function (initiated by pressing ESC, entering the desired search string, and then either Control-R or Control-E to search forward or backward from the current location) of particular benefit: "If I used an out-of-the-way word and had a dim memory of having used it a hundred pages earlier, I could simply type the word and ask the machine to find it, and there it would be, in its context, right away, instead of my having to riffle through a hundred pages and spend two or three hours looking for it. It was simply a time-saver."[62]

Hersey printed out his chapters as he worked, one every couple of days, carrying the spindled, fanfold sheets back to Pierson, where he would sometimes mark changes with a pencil to be input at the next session. Though he revised numerous words and phrasings, he did not appear to add or remove substantial portions of the text once it had been input into the Editor. In this he conforms to what Hannah Sullivan has termed a substitutive revision practice, which she sees as characteristic of screen-based authorial editing.[63] But revising the text was only part of the allure of the computer for Hersey. From the outset, as evidenced by a hand-drawn colored design on the manuscript title page, he intended that the visual layout would reinforce the book's core themes. Here the capabilities of the PDP-10, via the attached Mergenthaler equipment, would be of paramount importance.

The Mergenthaler Super-Quick was state-of-the-art in phototypesetting at the time. It had four photographic glass plates (they were sometimes said to resemble a windmill), and it was typically fed its text input and formatting codes by paper (TTS) tape.[64] By the early 1970s, however, it was becoming increasingly commonplace to control the codes by computers, most often an IBM machine or a member of the DEC PDP family.[65] Weiner's colleagues (Bob Tuttle, Walt Bilofsky, and Jim Sustman) had written their own custom software to accomplish this feat, which they dubbed LINTRN. The documentation they provided was not particularly encouraging. The software, they admitted, would make the equipment "if not easy to use, at least usable."[66] Nonetheless, Hersey, the trained printer, threw himself into the task of learning the arcane cant of LINTRN with gusto—it was far more complex than just coming up to speed on the Editor—filling pages with notes about formatting codes and his mockups for how the layout should look. By June 1973 he was fully invested in the task. No detail was too small to escape his attention. "The theme of the book is crowdedness and the pressure of numbers," he wrote to his publisher, "and the folios [that is, page numbers] are huge in order to give the reader, if only subliminally, a push toward that theme."[67] Similarly, "the big margins, and especially the very big one at the foot of the page, are intended to support the theme of a need for more space."[68] Using the LINTRN program in conjunction with the Mergenthaler, Hersey and his collaborators were able to provide Knopf with camera-ready copy that reflected all of these intentions.[69]

My Petition for More Space was published on October 11, 1974. It is set in a near-future New Haven that is part of a grotesquely overpopulated planet (it can take twenty minutes to traverse a single city block). The first-person protagonist, Sam Poynter, waits on an endless line of fellow citizens to reach a government service window where he can submit a petition for more living space—currently he lives in a seven-by-eleven-foot cell. With its ideas about bureaucratization and population control, the novel has some significance for students of environmental literature and dystopian science fiction, though at least one early reviewer judged it unfavorably next to Orwell and Huxley.[70] It is tempting to try to spin a reading of the novel as an allegory of Hersey's experiences working with the Yale Editor: the whole of it takes place on a line, after all, which is to say *on-line.* Lines of listless people shuffle toward a remote bureaucratic window even as lines of luminescent text shuffled up and down the screen, itself a "window" onto some vast, inscrutable space; and then there is the novel's overt theme of the oppressive computerization of society and all that it portends. But such readings are not licensed by the circumstances of the book's conception and composition. With his longhand draft Hersey had fully articulated the core themes of the novel—its imagery and devices, as well as its title—before ever hearing the first mention of the Yale Editor.

Yet *My Petition for More Space* doesn't need the conceit of any thematic premeditation to justify its historical significance to us. Hersey's use of the Editor and the PDP-10 predates Jerry Pournelle's move to microcomputing by some four or five years. Hersey did not initially compose the novel on the PDP-10, but he revised it there extensively, using a terminal setup—essentially a typewriter attached to a glass screen—that would be clearly recognizable to us today as "word processing." He inserted and deleted, copied and pasted, searched and replaced. Unlike Pournelle, he did not own the computer himself, and could not work out of his own home; he was completely dependent on the computing infrastructure at Yale. Nonetheless, he took pride in the achievement and remained a devotee of word processing for the rest of his life. His interest in computers also took him in other directions: in 1975 he was appointed to a Library of Congress commission on the copyright status of software. He wrote an insightful dissenting opinion in which he argued that software should not be copyrighted, on the grounds that while undeniably written out in the form of instructions or code, it was first and foremost an idea, not an inscription—thus anticipating a debate still not fully reconciled today. In 1986, pressed by his *Paris Review* interviewer Jonathan Dee on the apparently still-contentious subject of word processors, Hersey retorted: "I think there's a great deal of nonsense about computers and writers; the machine corrupts the writer, unless you write with a pencil you haven't chosen the words, and so on."[71] He made it clear that he rejected such views.

In fact, the specter of corruption that Hersey conjured was an anxiety he had experienced firsthand. Turn the pages of *My Petition for More Space* and you will admire the heavy weight of the paper stock and the rough cut edges (both specified by Hersey); you will immediately take note of the clean, distinctive layout, the generous margins and white space, the outsize page numbers and wide-spaced running head, the oversized initials at the start of each chapter. But within the pages of this book—so invested in its own presentation and layout—there is no mention of the role computers played in its production. There is no "note on the text" to impress the reader with the arcane intricacies of the process, no colophon with cryptic references to LINTRN or an "Editor," no acknowledgments naming a phalanx of computer engineers without whom not. This was all quite deliberate on Hersey's part. It had nothing to do with any lack of generosity toward his colleagues. No, he was afraid his readers would think a machine had written the book.[72]

UNSEEN HANDS

When the *New England Review / Bread Loaf Quarterly* asked him why he wasn't going to buy a computer, Wendell Berry didn't hesitate to give reasons. In fact he made a list. Most of the items on it were a direct outgrowth of the particular brand of agrarian conservationism that has defined his poetry and other writing: the increased dependency on the energy grid and fossil-based fuels a computer would entail, the unseemly price tag and immodest size of the equipment, Berry's inability to maintain and repair it himself, and the lack of small, locally owned businesses that were then capable of doing the same (then as now, he lived on a farm in rural Kentucky). Berry's airing of these concerns as early as 1987 is prescient—we are only now coming to terms with the impact of computers on the environment, for example. He also, however, took recourse in his long-standing workflow: "My wife types my work on a Royal standard typewriter bought new in 1956 and as good now as it was then. As she types, she sees things that are wrong and marks them with small checks in the margins. She is my best critic because she is the one most familiar with my habitual errors and weaknesses. She also understands, sometimes better than I do, what *ought* to be said. We have, I think, a literary cottage industry that works well and pleasantly. I do not see anything wrong with it."[1]

The essay was subsequently republished in *Harper's Magazine,* where this tranquil domestic vignette did not pass without comment. "Wendell Berry provides writers enslaved by the computer with a handy alternative: Wife—a low-tech energy-saving device," read one of the letters to the editor that followed. It continued: "Drop a pile of handwritten notes on Wife and you get back a finished manuscript, edited while it was typed. What computer can do that?"[2] Berry was sharp in reply, scolding his correspondents and branding them "gossips." He also made the point that in their emancipatory zeal they proceeded by presuming his wife, Tanya Amyx, to be torpid and subservient, unable to alter circumstances for herself if she found them objectionable. But the exchange—Berry's casual depiction of his dependence on his spouse and his apparent surprise in finding it called into question—also encapsulates a much longer history of keyboards and hands with very distinctive patterns of gender and labor that are essential to understanding word processing in its original contexts.

"As a farmer," Berry had said, "I do almost all of my work with horses. As a writer, I work with a pencil or a pen and a piece of paper."[3] The work of the page is thus perceived as different in its particulars but not its homespun values from the work of the field. Both require a firm and guiding hand, whether on the pen or on the bridle and traces. But while Berry may have had a clear sense of *his* work, the work of writing—writing as a vocation—entails more than the act of composition; it is also, as we have seen, the work of typing and retyping, to say nothing of the work of correspondence and copying, of filing and bookkeeping, and so forth. Whose work was that?

There were two killer apps of the early personal computer era. Not games, not at first: in the late 1970s games were still mainly the province of coin-operated cabinets in the arcades or television consoles like the Atari 2600. No, most people who brought home a home computer did so for one of two reasons: spreadsheets or word processing. VisiCalc, the spreadsheet program which debuted in 1979 for the Apple II, quickly sold several hundred thousand copies; indeed, its availability helped drive sales of the Apple computer itself.[4] Its conception was influenced by the kind of early computer-generated imagery then on the big screen in films like *Star Wars*, particularly the "heads-up" combat displays: "Like Luke Skywalker jumping into the turret of the *Millennium Falcon*, [Dan] Bricklin saw himself blasting out financials, locking onto profit and loss numbers that would appear suspended in space before him," wrote one commentator, tongue not entirely in cheek. "It was to be a business tool cum video game, a Saturday Night Special for M.B.A.s."[5] Journalist Steven Levy described the competitive culture that would arise around VisiCalc hacks and tricks, the quest for the "perfect" spreadsheet: "Spreadsheet hackers lose themselves in the world of what-if," he wrote.[6] Spreadsheets indeed lent themselves to speculation and scenario-spinning, to a future-oriented fugue state induced by the rows and columns scrolling past, figures rippling across the screen as fingertips adjusted a variable in a hidden formula. Not incidentally perhaps, all of the spreadsheet "cowboys" (his word) Levy describes in his article are male.

By contrast, the word processor was imagined from the outset as an instrument of what labor and technology historian Juliet Webster has termed "women's work."[7] Of course, the typewriter too had been feminized upon its inception by Christopher Latham Sholes with his stated belief that he had done something "important" for women heretofore condemned to menial, domestic labor.[8] Numerous critics have documented the typewriter's prismatic refraction of female identity and imagery ever since, including the popular vogue for novels featuring the eponymous "typewriter girl" as archetype.[9] But the stock of cultural imagery around the typewriter has since diverged: on the one hand the hardboiled noir writer, inevitably and inveterately male, the typewriter an extension of his intellectual firmament / fermented intellect as the words get punched out of his fingertips in an urgent rhythm; on the other hand, the secretary, a product of pink-collar office culture, the silent and passive conduit for the words of others. She listens but does not speak, she transcribes but does not compose, and she types but never reads. Or at least she's not supposed to.

Unlike the typewriter, then—which at least had its alternative reservoir of masculine imagery to draw upon—when writers began buying word processors, they were buying a product originally designed for a very specific usage and environment. "My model for this was a lady in her late fifties who had been in publishing all her life and still used a Royal typewriter," said an interface designer who worked with Larry Tesler on the Gypsy system.[10] Everything about word processing—starting with the advertising, where up until the advent of figures like Asimov in the 1980s the operators were always portrayed as female—would have conspired to reinforce this perception. For a writer beginning with a word processor in the late 1970s or early 1980s, then, the salient fact might not have been that the machine was a *machine* so much as that it was a machine so clearly intended for the girl at the office.[11] Sometimes, as in the case of science fiction writer Barry Longyear, this was a virtue; in his public disputation with Jerry Pournelle he made exactly that point: "The machine I have," he wrote, referring to his Wang System 5 Model 1, "was designed for secretarial work by someone who understood that not everyone in this world needs or wants to become a computer programmer."[12]

But the implications of the gendered status of word processing also went beyond whatever perceptions or reservations individual authors may have had. In an important study Jeanette Hoffman, another scholar of labor and technology, reconstructs the ways in which the specific design features of dedicated word processing systems reflected the industry's assumptions and expectations about their users. The logic of word processing itself is such that, as she contends, the act of writing must be "broken down into unambiguous operational units," thereby "subordinating writing practices to a formal, algorithmic order."[13] Similarly, the behaviors of users must be anticipated and likewise broken down into modular, predictable patterns. In the case of the Wang products like those eventually adopted by Longyear, King, and Updike (one of the specific systems Hoffman studies), this took the form of menu- and prompt-driven routines, in effect ensuring that the female office worker was always explicitly situated somewhere in a hierarchically ordered space within which no wrong choice or wrong turn was possible—there could be no invalid keystrokes. Moreover, certain actions—including deleting, copying, and renaming a file—required administrative access to the system. "In the digital world, producing text can be separated from physical disposition over the outcome," Hoffman concludes.[14]

Word processing software for personal computers reflected a different approach, or so it must have seemed. First WordStar and then WordPerfect (as well as Perfect Writer and many others) placed more freedom in the hands of users; menus and prompts, Hoffman notes, were replaced by a mostly blank screen. This resulted in a learning curve that was markedly steeper but deemed acceptable because this was the only piece of software the operator would ever need to learn. In other words, the word processor assumed that the user herself was a "word processor," a dedicated professional for whom the requisite training regimen was an acceptable trade-off for the leap in productivity to follow. The mostly blank screens of these products were not any true harbinger of workplace emancipation. Hoffman contends that it was only the widespread advent of graphical user interfaces and "direct manipulation" that finally broke this trend, as prefigured in the Xerox Star and Apple Lisa and finally popularized by the Macintosh. But the entrenched order of things died hard. As late as 1990, an independent vendor produced a menu-based interface called (what else) Perfectly Simple that could be installed as an add-on to WordPerfect.

For some literary authors, the allure of a personal computer with all its teeming intricacies may have been enough to offset word processing's pink-collar associations. Mastering a "microcomputer" was a proposition very different from operating a piece of office equipment by rote instruction. Thus the predilection for tinkering shared by Pournelle, Brodkey, Fallows, Tan, Straub, Clarke (but not Asimov), and many others. Can we recast some of the apprehensions and anxieties about word processing from male and female authors alike as a resistance to being superimposed onto the figure of that Royal typewriter lady? At stake would have been not only the stigmatization of "women's work," but the fear of assimilation into just the kind of modularized workflow—alienated from their own labor—that, as we will see in the next section, characterized the experience of word processing for most female secretaries, typists, and "word processors."[15]

> in an ideal world
> I would sit by a clear
> lake an occasional
> sailboat would
> flutter by an
> occasional butterfly fan
> my face in the
> hot sun in an ideal
> world the stones
> would sing . . .

These lines were written on a Wang word processor by Patricia Freed Ackerman, a San Francisco–based poet, while at her place of employment when what she really wanted to do was to leave the office that afternoon. We know this (or at least are encouraged to presume it) because the poem is entitled "Poem Written at Work on a Wang Word Processor Sometime in the Afternoon Wanting to Leave."[16] It continues for another sixteen lines, rocking back and forth in a loose free-verse rhythm of anapests and dactyls, a casual idyll confessing the author's desire to be sublimated by the natural world without. There is nothing extraneous or ornamental to the language; the first-person singular aside, there is no interior capitalization, nor is there is any interior punctuation. The words of the text are plain and, well, easy to process.

Conspicuously, the Wang word processor and the uninspired (and thus inspiring) work environment that attends it are never given entrée into the body of the poem—they serve only to name its occasion and instrument. Yet the very specificity of Ackerman's title encourages us to conjure more of the circumstances of its composition. We can assume that the act of writing the poem was at least marginally illicit: the word processor would have belonged not to her but to her employer, and it would have been intended for writing rather different sorts of things (correspondence, memos, reports and the like) that originated with people other than herself. She wants to leave, but she cannot—she is bound by her job, compelled to remain at the keyboard even if there is no work to be done. Like the word processor, she must be constantly ready, available, on call—on-line. In fact the word processor is a prosthetic, an extension of her identity—Ackerman is herself a "word processor," just as in an earlier epoch she would have instead been a "typewriter" or indeed a "computer"—and there is a kind of intimacy between her own identity and the machine that she perhaps thinks betrayed by the mostly generic texts she is called upon to produce with it: "in this ideal / everything I write / would be beautiful and true" she tells us later in the poem.

The unnamed office that Ackerman worked in when the poem was written was, we may surmise, a product of what was once widely known as the "office of the future"—the name the business administration literature of the late 1970s and early 1980s had given to the nexus of information technology, scientific management theory, and labor practices that was supposed to maximize workplace productivity in the face of the ever-increasing amounts of data, text, and information swirling through a modern organization. And word processing was envisioned as its cornerstone. "The electronic text exists independently of space and time," wrote Shoshana Zuboff of just such an office as Ackerman's (and at about the same time). "[It] can infuse an entire organization, instead of being bundled into discrete objects, like books or pieces of paper."[17] Yet word processing itself originally entailed something rather different from what we know today. It was understood as something far more than just a passive verb: it would have connoted a full-fledged system or paradigm for automating the flow of textual production in the office of the future. (In this sense, as computer historian Thomas Haigh notes, it was analogous to the term "data processing.")[18] To fully appreciate what it meant for a literary author to obtain a word processor—to choose to bring one into his or her home for purposes of writing fiction or poetry or drama—we first need to spend some more time with Ackerman at the office.

"Word processing, as a *concept* is the *single most revolutionary event in written communication since the invention of the typewriter,*" proclaimed an American Management Association publication in 1974.[19] All revolutions have their exigencies, and at the time that this was written, the Western business executive or office manager perceived himself (gender intentional) as operating in a climate of crisis. This had less to do with the global recession that was then at its peak than with the anxieties of the nascent information age, the "paperwork explosion" that we will look at in more detail in Chapter 8. As we will see, the proliferation of paperwork—repeatedly described via recourse to sublime naturalistic imagery such as oceans and mountains—and the costs in time and personnel associated with drafting, revising, refining, distributing, and filing it were the immediate context and catalyst for the widespread promotion of a new organizational model known as word processing in the business literature of the day.

But there was a second and even more urgent symptom of the executive's crisis, one that manifested itself explicitly in terms of pathology and contagion. This was, in the words of George R. Simpson, CEO of the Word Processing Institute (a subsidiary of the Office Management Systems Corporation), the "illness" or "social disease" of the "social office."[20] According to one source, six out of ten office workers were female, and the "vast majority" of those were employed in secretarial positions.[21] The position of secretary itself, however, was multifaceted: a secretary might be called upon to take a letter—in practice, a task that would require shorthand notation in the presence of the executive—and then she would "type it up" at her own desk. In between would be filing or telephone reception or appointment books and calendars or travel arrangements or opening the mail or making coffee. On the one hand, this meant that typing and allied skills—shorthand, filing, working with carbons and correction fluid—were only one aspect of a well-rounded secretary's portfolio. If a "girl's" typing skills were not all they should be, she might make up for it with competencies in other areas; nor was the typing ever likely to much improve so long as it remained a relatively ad hoc and unstructured part of a harried nine-to-five workday. This diversity of tasks also meant that the secretary would circulate freely about the office, interacting (promiscuously, some thought) with different members of the organization. This was the "social disease" Simpson had diagnosed, a terminal condition characterized by ceaseless circuitous movements through various arteries and corridors, impromptu meetings and interactions (obvious vectors for gossip and rumor), and other wasteful movements. (Another AMA consultant, Walter Kleinschrod, cites an unspecified study claiming that 17 percent of the time, "the average office employee is doing nothing but walking around and talking.")[22]

It is hard to overstate the force with which the case against the social office was made. Everything must change: salary scales, career paths, job descriptions, "attitudes," and perhaps above all, procedures—or processes.[23] A *total* division of work must take place," thundered the AMA.[24] Only this would "cut through" the accumulated "barnacles" and get to "the heart of the matter."[25] Here again, the conspicuously surgical figures of speech are overtly evocative of pathologies and bodily encrustations and intrusions. Where the rhetoric wasn't pathologized, it was overtly matrimonial—secretaries were routinely referred to as "office wives."[26] A key problem it was believed was that women, unlike men, were not accountable. Secretaries were "charming little nobodies," as one consultant charmingly put it, invisible on the organizational chart (this last was indeed true, in accordance with standard management practice).[27] "They lack supervision," lamented the AMA. "And their productivity is thereby in the main beyond accurate measurement and control."[28]

There is a linguistic irony here: "executive" derives from the verb "to execute" whose Latin root, *exsequi,* means to follow up, to carry out, and even to punish—all active measures that, as we have seen, would soon be literalized as an actual key on the word processor's keyboard. Yet it is the secretary who "executes" on her boss's decisions, transmuting his ideas and dictates into the tangible end-products of modern knowledge work. "The work done by the secretary is often more *visible* than that done by her boss," notes one contemporary commentator, *pace* the AMA. "She at least produces a pile of neatly typed papers to be signed at the end of the day; there is often some question about just what he has produced."[29]

Women in the office were all too visible in certain ways, even as they managed to stay invisible—and supposedly unaccountable—in all the ways that ostensibly mattered. This was the climate that created word processing as an organizational (as opposed to strictly technological) concept, the word processing that AMA authors hailed as the most "revolutionary" event in the history of written communication since the typewriter. Word processing would quickly make the social office "obsolete," they assured anxious readers.[30] In such a climate it is not too much of an exaggeration to say that the managerial regimen of word processing was about the regulation of female bodies, both their tasking and their situation.

Above all, the introduction of word processing into the office was about making female bodies accountable, and it did so by modularizing the anatomical functions of hand, eye, and ear. The AMA's consultants from the early 1970s carefully outlined how this prescription was to take hold:

> The term word processing becomes applicable when a simple (as with most truly revolutionary ideas) but genuinely profound change in perception is focused upon that good ol' office routine. This perception begins with the realization that such a "routine" should not be regarded as a series of related but discrete tasks performed in sequential but isolated segments of time and energy, but rather as a process—more precisely a system in which all the procedural and personal vagaries which, until now, separated each component in the process are seen as unnecessary and undesirable. The random, the haphazard, the unpredictable, the largely unmeasurable can thus—when understood in this new conceptual light—be replaced by the controlled, the organized, the automatable.[31]

This is a powerfully direct statement. In practice what it portended was a radical compartmentalization of office work in the form of dedicated word processing centers—physically distinct offices-within-an-office (sometimes even remotely located), staffed by personnel assigned exclusively to keyboarding tasks. Gone would be the "gal Friday" model of the secretary devoted to a single executive, catering to all his needs; in its place would be dedicated typing specialists who would move from task to task without regard for (or excessive loyalty toward) the identity of the originator. The word processing center was, in effect, a doubling-down on the typing pool concept of the earlier twentieth century. Once word processing was implemented technologically and organizationally, typing efficiency was expected to increase exponentially—figures on the order of "1,000%" were airily bandied about in the management literature.[32] Gone would be interpersonal interaction between the typist and the executive. Dictation for "processing" or draft copy for revision would be delivered to the "center" in the form of internal telephone transmissions (PBX) or else by a messenger bearing a tape cassette. Copy would be returned in the same way. Unseen hands would do the paperwork.

The word processing center of the seventies, both physically and organizationally, placed the secretary in a position not unlike the new technologies of electromagnetic storage that supported what we earlier termed suspended inscription. Just as magnetic tape, cartridges, and disks quickly rendered the friction of the eraser obsolete in correcting errors, other kinds of intra-office friction—the "personal vagaries" alluded to by the AMA above—would be eliminated by making the typing secretary into a kind of removable media herself. Like the word-processed text in its seemingly immaterial state, the typing secretary would be out of sight but always on-line (or on call). The argument I am making is thus that word processing duplicated its technical logic in the human "systems" (the word is used repeatedly throughout the management literature) who were tasked with operating all of that expensive hardware.[33] Recall, too, Hoffman's point about recasting the act of writing in terms of discrete, operational units subject to formal, algorithmic manipulation; words that were fully atomized and fungible demanded bodies that were equally modularized and interchangeable.

Organizationally, then, word processing was about restructuring work, reclassifying roles in the workforce, relocating people in the workplace, and reconfiguring the individual workstation, all under the guise of centralizing and automating the composition, transcription, reproduction, and distribution—the processing—of the printed word on paper. It was about procedure and control, accountability and measurement. In this, word processing was of a piece with the rising tide of office automation brought on by the computerization of successive subdivisions of what we nowadays call "knowledge work." Alan Liu, commenting on the same era of automation as Zuboff at the start of this section, lays particular emphasis on the way in which this computerization fostered the "systematization" of information in the office environment—the linkages, assemblages, and connective tissue of information that contributed to management's sense that all was under rationale control. "Computerization was a way to sense the overall systematicity of work," is how Liu puts it.[34] More than computerization at large, however, word processing was also about discipline and regulation, the fixing of dallying (or dangerous) bodies in space and the affixing of idle (or wandering) female hands to keyboards. Wang and Dictaphone, among other manufacturers, even included specific product features, with names like Mastermind and Timemaster, for tracking employee efficiency.[35] (Today's keystroke loggers are simply the culmination of that logic, supervising and surveilling word work in the smallest possible unit of productivity.)

Of course, as Christopher Latham Sholes had been the first to anticipate, female bodies were admitted into the office precisely *because* of the typewriter and its keyboard. Word processing thus promised to correct the typewriter's mistakes in more ways than one.

In late 1969 a new corporation consisting of nine employees set up shop in a 20,000-square foot manufacturing space in Hauppauge, Long Island. Hauppauge was part of Islip, a suburban bedroom community for New York City, a one-hour and thirty-minute train ride away.[36] It may have seemed an unlikely place for what we would today describe as a tech start-up, but in fact the Hauppauge Industrial Park was the largest such complex in the nation at the time, part of the postwar development of Long Island that also included planned communities such as nearby Levittown and Robert Moses's highway system. The company's name was the Redactron Corporation, it had $750,000 in operating capital, and its founder was Evelyn Berezin.

Born in the Bronx in 1925, Berezin had grown up reading *Astounding Stories* magazine in her bedroom under the El.[37] World War II allowed her to attend NYU on scholarship as a physics major; she graduated in 1945 and subsequently did graduate work there under the aegis of the Atomic Energy Commission.[38] But her career path quickly took her instead into the emerging world of computers, where she started out in 1951 at the Brooklyn-based Electronic Computer Corporation. The only woman in the shop, she became its lead logic designer. "All my life since I had been interested in physics and things like that I *always* was in with mostly all boys," she recalls. "The business of working with men was just part of my life, I never thought of it."[39] At the time, before transistors, computer design relied on vacuum tubes and primitive forms of magnetic memory. Berezin quickly learned how to work within the limits of available technology to design purpose-built systems for industrial and office applications, picking up a clutch of patents in the process. Eventually she found herself at a company called Teleregister, based in Stamford, Connecticut. With Berezin's assistance, Teleregister implemented the first automated airlines reservation system in the world (for United Airlines), predating the better-known SABRE system by at least a year.[40] When she tired of the commute to Stamford, she interviewed for a job that today we would call chief technology officer, at the New York Stock Exchange. Berezin recalls having been unofficially offered the position, before receiving an apologetic phone call several days later informing her that the Board had refused to confirm her appointment on the grounds that it would be "unbecoming for a woman" to be seen on the trading floor.[41] This and other experiences confirmed her belief that she would never be able to move beyond a backroom design position, so she began doing market research to launch her own company.

By this time it was the late 1960s, and Berezin had heard of a new office product from IBM, the MT/ST, which combined the capabilities of their Selectric typewriter with a tape storage unit. The machine was being touted as a revolution in secretarial labor—according to Berezin's research at the Bureau of Labor Statistics, 6 percent of the U.S. workforce was then employed as secretaries—but it also had the reputation of being "big, clumsy, expensive, difficult, unreliable."[42] Berezin, with her background in computers, thought she could do better. In particular, the MT/ST's transistor technology was on the verge of being replaced by the first microprocessors from Intel. "Word processing" seemed like an obvious growth area, though Berezin claims she had never heard that phrase at the time.

With the newly launched Redactron—the company's name telegraphing its promise of a kind of automated editing—she set about designing and building what would be called the Data Secretary, a product that would compete directly with the MT/ST. As it happens, Intel wasn't yet capable of producing the microprocessors that Berezin needed, so she contracted with the nearby firm of General Instruments instead. "The whole story is really determined essentially technically, by what was available, what *could* be done at the time," she recalls.[43] Like the MT/ST, there was no screen—a Selectric was still the input and output device, and keystrokes were stored on a pair of magnetic cassette tapes. But unlike the MT/ST, which was exclusively an electromechanical device, the Data Secretary would be a true computer, with thirteen on-board semiconductor chips and programmable logic driving its word processing functions. The complete unit was some forty inches tall. Its design and functionality were then tested on working secretaries. One signature feature allowed names and addresses stored on one tape to be merged with boilerplate text on another. The Data Secretary was thus "designed for a real problem in a real office."[44] At least one early UK user remembers it more prosaically, however: "We were based in Fitzroy Square at the time, in a Grade I listed building. When we turned the Redactron on, lights would be dimmed across Fitzrovia. When we ran it, conversation ceased and flakes of plaster would drift down from the ceiling. The room thrummed. We purchased earplugs in bulk."[45] Hyperbole, of course, but it conveys the essence of the experience.

Berezin remains particularly proud of the fact that despite its name the Data Secretary was not "just" a hardwired word processor. It was a real computer, Turing-complete as the jargon goes: "The only thing that made it a word processor was the program."[46] Its integrated circuits could, in theory, be repurposed for other applications, albeit not necessarily efficiently. By September 1971 Redactron was ready to begin shipping. Next came advertising and promotion, and this presented its own challenges; it was a "very new idea," as Berezin recalls.[47] She enlisted the aid of Roslyn Willett, an old friend who now ran a public relations agency—noteworthy in itself in an era when such agencies were dominated by (mad) men. In December 1971, Willett placed an ad in the preview issue of a new magazine intended for women.

The first issue of Gloria Steinem's *Ms.* was distributed as an insert in the December 20, 1971, issue of *New York* magazine. Its cover was a McLuhanesque *mise-en-scène* with a pair of female hands reaching into the frame to grasp the new publication, its red-orange cover-within-a-cover featuring a striking Vishnu-like drawing of a multiarmed multitasking woman (one arm brandishes a typewriter) by Miriam Wosk. The lead article was Steinem's "Sisterhood": "If it weren't for the Women's Movement I might still be dissembling away," its text declared early on. "But the ideas of this great sea-change in women's view of ourselves are contagious and irresistible. They hit women like a revelation, as if we had left a small dark room and walked into the sun."[48]

This and the other articles in the preview issue established the context in which Willett's ad for the Redactron appeared. Its headline, printed in a sixties-style bellbottom font: "The Death of the Dead-End Secretary."[49] This, however, is perhaps the least of its notable features. Its roughly 400 words of copy were a carefully prepared, multi-tiered pitch that positions the company's high-tech product within the ferment of larger social forces. It begins with a second-person address to "women like you." The copy then briefly recaps the history of the secretarial profession, even making mention of the founding of Katherine Gibbs's secretarial school in 1911; but nowadays, we are told, the secretary's prospects for professional advancement have been stymied by the mounting tide of paperwork and their sphere of influence has dwindled to the keyboard, a dull and monotonous routine of "typing and retyping."

Such is the state of affairs that has persisted for decades—but no more. "Today," proclaims the copy, "the dead-end secretary is dead." Again the copy appeals directly to its implied female reader, casting her in comparison to her "complacent sisters." "You want the freedom to get into more interesting and more challenging work—work that'll give you a chance to move into a staff, administrative, or managerial position—where you can call some of the shots." Enter the Redactron Data Secretary. Here the name of the device takes on its full significance: the secretary now has a secretary of her own, a device to absorb the drudgery of all that typing and retyping. She will be primed for management because she already has an automaton to do her bidding. The ad copy then dares its decisive maneuver: "Ask your boss about the Data Secretary," it enjoins. "We've already told him about it." This ploy is striking on several levels. First, of course, is the presumption that the secretary would be in a position to petition her (male) boss for the product at all. This seemingly empowered posture, however, is undercut by delivery of the news that Redactron has already communicated with him through other, unnamed channels. (Just because she was reading *Ms.* didn't mean the woman was actually out in front.) Finally, of course, there is the reality that in all likelihood the boss probably *hadn't* heard about Redactron before—the secretary was thus being actively enlisted as a conveyor for what ad men (and women) knew was the very best kind of advertising: word of mouth.

The ad concludes with the promise of what we would call swag—memo pads, buttons, and stickers—branded with Redactron's logo, all of which could be had upon written request to the president, Ms. Evelyn Berezin. This closing gesture, the name of a female CEO prominently displayed in the ad copy with an implicit promise of correspondence, would have confirmed the image of a company that was on the secretary's side, offering an emancipatory product allied with her own personal interests and the women's movement more generally.[50]

So it happens that at just about the same time as Roland Barthes was famously attending to funerary proceedings for the author (which is to say authorship) in France, in Hauppauge, Long Island, and in the pages of *Ms.*, Willett, Berezin, and Redactron were proclaiming the death of the secretary. Or at least the death of the dead-end variety. While the Redactron Data Secretary chattered out its typing jobs, the secretary would be free to take on other responsibilities, eventually breaking into management. Of course, it was not the case that tiresome typing tasks were the only thing standing between women and an executive's corner office, as Berezin well knew. Any promise of upward mobility for women in the workplace would have to contend with the gender politics then on full display in the AMA's publications and reports. The ad was not without other contradictions as well: a company founded and led by a woman was knowingly playing on working women's identification with the emerging feminist movement, selling them on a product that, while no doubt genuinely a boon for some, would also doubtless cost others their positions, or else relegate them to the very "dark rooms" (as full-time word processors themselves) from which Steinem was celebrating their collective emergence.

But Berezin herself was never much concerned about whether the technologies she helped pioneer and promote would take away jobs: "I didn't think of this as a problem for women. I always assumed that women would just move ahead," she says.[51] She subsequently worked with the American Women's Economic Development Corporation and other boards and organizations promoting the efforts of female business leaders and entrepreneurs. In 2011 she was inducted into the Women in Technology International Hall of Fame. Still, she remains a relatively unknown and underappreciated figure, with nowhere near the stature of other women who played significant roles in computer science and the computer industry and have since been recognized by historians.

By the end of the 1970s word processing as an organizational concept had been supplanted by so-called office automation, the beginnings of the vision that would eventually lead (a decade or so later) to personal computers on every desk in the office. Executives would increasingly be doing their own typing. Even at the height of its organizational vogue, however, word processing was never implemented as monolithically as the AMA's official publications might have desired. As one researcher puts it, its impact was "much less straightforward, and much more closely tied up with the context of the work organization in which it was introduced."[52]

But the fundamentals of the concept proved enduring—the monolithic tasking, the regimented workday, the limited range of intra-office motion. Condemned to such uninspiring surroundings, Patricia Freed Ackerman was not the only female writer to use the machine for her own devices.[53] Marina Endicott, as we have previously seen, first learned to use her IBM Displaywriter in a Canadian government office. She describes a "spacious room" in which she worked alone but for one other operator. "As almost always in government offices, there was hardly any real work to do. We weren't allowed to read, of course; we had to maintain the fiction that the administration of government requires constant effort. Learning the

Displaywriter took up the hours very usefully, but once I'd mastered the system, I still had to look busy. At first I wrote letters; then, running out of real life to tell, I started writing stories to amuse myself."[54] Terry McMillan, meanwhile, was working as a "word processor" at a law office in New York City as she was writing her first novel, *Mama* (1987); she used the firm's system after hours to print and send literally thousands of promotional letters to bookstores, reading groups, and review outlets, especially those with a record of supporting black writers—precisely the sort of task the technology was designed to expedite. The book quickly sold out of its first hardcover printing.[55]

Instances of such positive redirection of personal agency, though, were rare.[56] And the reality of the job was often far removed from the crisp flowcharts and workstation diagrams, and from the stock photos of trim, fashionably dressed young (white) women operating the machinery in the corporate promotional literature. Machines like the Redactron and MT/ST were loud. They also threw off a lot of heat. Most word processing centers that were centralized within the office were confined in a fully enclosed space, without windows. (One industry advocate suggested they were ideal workplaces for the blind.)[57] The work itself was monotonous; there was no variety in the tasks (this was, after all, the whole point); and perhaps even more importantly, there were only extremely limited opportunities for social interaction (the idea that the "social office," with all of the informal and spontaneous networks thus implied, was actually a virtue from the standpoint of workplace efficiency was a concept alien to most business minds of the day). Women, as a result, left the jobs in droves.[58] In 1981—a year we have gravitated to before—a new underground magazine began appearing on the streets of San Francisco. The cover of its first issue depicted an office worker whose head had been replaced by a computer. The magazine's name was *Processed World*—"If you hate your job you'll love this magazine!" it promised. *Processed World* ran essays, commentaries, cartoons, and other content critical of word processing, data processing, office culture, and the casualization of knowledge work in all its forms. It would remain in publication for the next two and a half decades.[59]

As for Redactron, high interest rates had made it difficult to borrow the money needed to grow the company, and toward the end of 1975 the decision was made to sell to the Burroughs Corporation. The subsequent rapid availability of home computers is something Berezin admits to not having anticipated—surprising in some respects, given her understanding of how much small, reliable microprocessors would change the industry back at the start of the decade. Nonetheless, she remained at the Burroughs Corporation until 1980, working with further iterations of the Redactron line, one of which, we have seen, was eventually used by the romance novelist Robyn Carr. "As far as I'm concerned that was the end of the word processing business," Berezin says. "It was only ten years long."[60]

The word processing organizational concept that took hold in the early 1970s wasn't just about bodies and hands and keyboards. It was also about the voice and the ear, speaking and listening, dictation and transcription. Before words could be typed, after all, they had to originate somewhere. This origination was assumed to be oral, from the mouth of an executive. To the extent that word processing described a particular category of hardware and office equipment, it fully included dictation devices. In theory, at least, an executive speaking into a microphone was "word processing" no less than his secretary—he was just participating in a different part of the document's life cycle. As early as 1959 IBM had signed an agreement to acquire Pierce, a manufacturer of dictation machines, its first external acquisition since the Electromatic Typewriter Company in 1933.[61] As we will see in Chapter 8, this decision was to have direct implications for the genesis of the word processing concept as such. And by the early 1970s, as Thomas Haigh notes, the inclusion of dictation equipment alongside the MT/ST, the Magnetic Card Selectric Typewriter, and even copiers, as part of its word processing line was commonplace: "The concept of word processing did not refer exclusively, or even primarily, to the use of full-screen video text editing."[62]

In the greater word processing scheme, dictation equipment was an essential element of a rationalized workflow and functioned to combat the ills of the social office. Secretaries, for their part, would no longer be called upon to linger in the executive's presence while "taking a letter." With the firm's investment in all that high-end word processing equipment, the notion that the keyboard was dormant while the secretary was off somewhere else, scribbling in shorthand, was unthinkable. Moreover, it was also deemed more productive for the executive to dictate without the distraction of a secretary in the same room. Indeed, once the monogamous boss-secretary relationship was severed, work could be distributed and directed much more efficiently.[63] The technology itself quickly achieved corresponding levels of sophistication. Recordings were stored on cassettes and cartridges, on magnetic reels and magnetic belts. Telephone handsets were frequently used as input devices, allowing for remote tape storage. More advanced products were outfitted with an automatic changer so that a busy executive would not be bothered with the need to physically handle and replace the media. The state of the art was a looping system known as a "tank type": because the recording tape revolved continuously in an endless loop, the boss could begin dictating (using the telephone), and the secretary could begin playback in something close to real-time. All of this activity could be monitored and controlled from a centralized location. One account describes the technology thus: "A supervisor's console or monitoring panel contains a row of dials which indicate which tank is in use and which is idle, as well as the amount of backlog that remains to be transcribed."[64] Like a word-processed text, the word-processed voice was rendered modularized and controllable, subject to mechanical repetition or acceleration, isolated from its source, detached and stored on physically remote media.

Just like Stephen King once placed a Wang-like machine at the center of a story about a writer who undertakes to redact and edit his own life, dictation equipment has also figured in fictional plots about word processing. In 1960 the final episode of the critically acclaimed first season of *The Twilight Zone* was entitled "World of His Own" (written by Richard Matheson).[65] After the show's familiar opening sequence, the story is introduced through a series of shots and voice-overs: "The home of Mr. Gregory West, one of America's most noted playwrights," host Rod Serling intones as we view the suburban manse. "The office of Mr. Gregory West" (the shot is centered on his desk). "Mr. Gregory West, shy, quiet, and at the moment, very happy" (a close-up of the successful author, obviously in his element). He is also very happy, presumably, because he is being attended by Mary, "warm" and "affectionate," who, we rapidly learn, is a figment of his literary imagination brought to life by speaking her character description into the Dictaphone machine that sits prominently on his desk. Mary mixes Gregory a martini, smiling benevolently. "Is it dry enough?" she asks. "Perfect, as always," he says.

As the episode unfolds, we learn that Gregory is trapped in a hostile relationship with his wife, Victoria, who berates and distrusts him and threatens to have him committed so she can take control of their assets. Returning home unexpectedly, Victoria glimpses Mary through the window of Gregory's study and accuses him of an affair, which he all but admits to. Only Mary is suddenly nowhere to be found, and the study has but one usable exit, the door through which Victoria herself has entered. We soon learn that Gregory "uncreates" (a word he uses twice—reminiscent of the Undo function in today's word processing software) his characters by taking a pair of scissors (Atropos's?) and snipping off the length of magnetic recording tape containing his spoken description of them, which he balls up and tosses in the fire. The characters then blink out of existence with the aid of some crude special effects magic. We also learn that Gregory has stored tapes in his safe containing character descriptions, not only of Victoria (who conveniently immolates herself when she throws it into the fire in a fit of pique), but also of Serling himself, who, appearing in Gregory's study for his traditional closing commentary, angers the playwright with the pronouncement that the story we have just seen was "purely fictional" and that in "real life" such "ridiculous nonsense" could never transpire. Notably, at this point in the life of the series, "World of His Own" is the only episode where a character becomes aware of Serling's presence as narrator. Gregory fetches Rod's envelope with its length of tape from the safe and into the fire it goes. He disappears, leaving Gregory West with the ever-affectionate Mary (now in fact Mrs. Mary West), his martini, and his Dictaphone: "Still shy, quiet, very happy, and apparently in complete control [fade] of *The Twilight Zone*." The show's first season thus concludes with the metaleptic dethroning of Serling, who ends up a character in one of his own stories.

Whether or not Stephen King had ever seen this particular piece of television when he wrote "The Word Processor" need not be our concern. The point is the pattern by which a new technological device seizes the literary imagination and spurs the author to play with the boundaries between fiction and the real (or reel), a distinction overtly challenged, as we have seen, by the very twilight zone in which this story takes place. Surface details aside, the two tales are even more alike than they might at first appear. Gregory West's mistake, for example, is that he creates his "regal" and "flawless" wife Victoria to be *too* perfect: "You're just the sort of wife I always used to think I wanted more than anything in the world," he says. Mary, by contrast is plainer, more wholesome, but possessed of that "inner quality of loveliness that makes a woman truly beautiful." Still, we suspect that Richard Hagstrom's *belle* Belinda is a lot more like Mary than like the "impeccable" but cold and calculating Victoria.

Likewise, it may seem that Gregory West's story reverses King's emphasis by focusing on a technology of the voice rather than a technology of the letter. Yet it would be a mistake to fully dissociate the Dictaphone from writing. The school of media criticism inaugurated by Friedrich Kittler reminds us that phonographic wax cylinders and then magnetic recording, together with photographic emulsion and the staccato mechanism of the typewriter, were all engines of inscription. Indeed, the foundation of Kittler's media theory rests on the premise that the near-simultaneous emergence in the nineteenth century of technologies for the recording of sound and images (followed soon after by the projection of moving images) challenged the premier place of writing as the means of "archiving" and communicating human experience. "Why simulate acoustic and optical data with words if they can be directly recorded, stored, and transmitted without any recourse to symbolic mediation," asks Kittler's most prominent English-language critic, Geoffrey Winthrop-Young.[66] If, as Walter J. Ong famously suggested, writing and speech are incommensurate ways of structuring our consciousness, Kittler, by contrast, gives us the tools for seeing them as part of a closed loop—office productivity technology conjoining spoken and written language in ever more efficient workflows. A typist transcribing dictation, headphones enclosing her ears, fingers moving over a keyboard, foot pumping a pedal to start and stop the recorder—here the body has become an interchangeable component in an integrated circuit, wired into the media apparatus from head to toe. For our part, we might simply note that the one thing Mr. Gregory West, "one of America's most noted playwrights," never ever does—not once in the twenty-five-minute episode that features him—is write.

Today dictation is once again rising in importance. Voice recognition software, once just barely usable, is now standard on tablets and mobile devices. A microphone icon is prominently displayed on the miniature onscreen keyboards in the hands of millions of users, inviting us to avail ourselves of vocal input as part of an increasingly heterogeneous spectrum of writing behaviors, which besides typing involves swiping, tapping, thumbing, and, once again, speaking. Richard Powers is perhaps the best-known advocate of voice recognition technology, describing his writing practice in the *New York Times:* "I write these words from bed, under the covers with my knees up, my head propped and my three-pound tablet PC—just a shade heavier than a hardcover—resting in my lap, almost forgettable," he begins. "I speak untethered, without a headset, into the slate's microphone array. The words appear as fast as I can speak, or they wait out my long pauses. I touch them up with a stylus, scribbling or re-speaking as needed. Whole phrases die and revive, as quickly as I could have hit the backspace. I hear every sentence as it's made, testing what it will sound like, inside the mind's ear."[67] Powers, who remembers programming what he describes as "interactive concrete poetry" on the University of Illinois's PLATO mainframe computer while still a student there in the late 1970s—and who bought a TRS-80 when they first came out—has long been drawn to various kinds of computer technology.[68] He views "word processing" as fundamentally multimodal, involving speech, pen or stylus, touch, and the keyboard.[69] Voice recognition first became an integral part of his practice in 2001, while he was writing *The Time of Our Singing* (2003).[70] But while Powers is clearly self-conscious of even the (minimal) weight of the tablet and takes care to mention the role of the stylus as well, the conceit of untethering is what seems paramount; in the *Paris Review* he amplified this same scene: "Since the computer is on the other side . . . of the room, I don't feel as if I'm tethered to it; the machine is a distant, nonintrusive recipient of the words that I'm thinking across the room. Later I go and collect them and note where they've stuck."[71]

Both conceptually and technologically, word processing was closely aligned with automation. Indeed, early word processing systems (the actual hardware as opposed to the organizational structures) were just as often described as automatic typewriters. And as both Darren Wershler-Henry and Lisa Gitelman have described, the typewriter was itself closely linked to automation (it was also a conduit for automatic writing, the supposed transcription of voices from the spirit world).[72] Word processing inherited that association with automation, which, as we saw in Chapter 2, drove authors like Arthur C. Clarke and Italo Calvino to write cautionary or satirical tales (even as others had to be persuaded that the machines weren't actually writing the books). That word processing devices also visibly performed in autonomous modes—most conspicuously when printing out a document—only furthered the notion that the act of writing was being automated in heretofore unprecedented ways. One common trope in writers writing about their first word processor was to linger over the free time they were suddenly afforded as the device did the hard work of churning out copy. Thus the journalist James Fallows: "If I am printing a draft of an article, I can hook up my tractor feed, push the print button, and go out for a beer."[73] Or Barry Longyear: "While the typing is going on, I can watch TV, read a book, or write another story."[74] Such winsome scenarios were proffered over and over again, seeming confirmation of the utopian promise that technology would inevitably lead to more leisure time. The tractor feed mentioned in passing by Fallows is as much a prosthetic for his hands as the word processor itself, alleviating him even of the burden of feeding individual sheets to the printer.[75]

Hands, both seen and unseen, have long been a trope in critical writing about technology. Kittler, by way of Heidegger, uses the relationship between hand and keyboard to structure his own idiosyncratic genealogy of typewriting. Kittler follows Heidegger in presuming that the typewriter "tears writing from the essential realm of the hand," by which he means the gap between the place where the hand and fingers do their work and the place where writing actually appears, in a physically and visually distinct locale.[76] Indeed, as has often been pointed out, the earliest typewriters (such as Twain's Remington No. 1) didn't allow the typist to see their text as it was being imprinted on the page.[77] This was in sharp contrast to the pen, where letterforms literally flow from the tip of a writing instrument that is consciously designed to be an extension of the hand. As Daniel Chandler has documented, many writers found typewriter and computer keyboards equally suspect, seeing writing and their creative muse as dependent on the manual labor of pressing pen or pencil to paper. "The word processor doesn't take as much time as actually forming the letters with your hand at the end of your arm which is attached to your body," says Denise Levertov. "It's a different kind of thing. They don't realize that this laborious process is part of the creative process."[78]

If the keyboard displaced hand from page, it also altered the relationship between an author and the basic building blocks of language. Letters, numbers, and punctuation were visibly arrayed in front of the writer, splayed beneath her (or his) fingers, a countable and renewable resource, as Scott Bukatman (following Kittler) likewise emphasizes.[79] Certainly this experience of letterforms as a constantly available inventory contributed to the typewriter's association with industrialization, as well as to the sense of language itself as a commodity—stored, warehoused, waiting to be milled and processed. (This same sensibility contributed to the avant-garde experimentation in visual poetry created out of typewritten letterforms.)[80] Nowadays word processing software can tell us the precise number of characters in a document as easily as its page length or the number of words, and the conventions of social media like Twitter and text messaging have made character counts the most vital metric for assessing the viability of a text.

The depth or granularity of this atomization could also be scaled up or down at will, from individual characters (correcting a typo) to selecting complete words, sentences, or entire paragraphs ("blocks") of text for editing and revision. Writers quickly began to grasp the ease with which sentences and paragraphs could be thus manipulated. "What happens if you write a chapter and then you're well into the second chapter and you find a couple of pages that really belong in the first chapter?" an interviewer would ask Anne Rice in 1985. Her reply: "You can just mark those pages off and copy them over to another disk, then take the later chapter put that in the computer and then copy those pages into the later chapter, you can move them anywhere you want, you can swap them around. That's what I mean by once you work this way, nothing stands between you and what you want to do."[81] The casual reference to "marking pages off" coupled with the tacit knowledge necessary for swapping around the disks to effect the transfer speaks volumes about the casual mastery of a new way of working. It also testifies to a new way of writing that nonetheless has its roots in office technologies of the previous decade, which placed a premium on precisely such a modularized view of text—merging it, flowing it, retrieving it, and outputting it. Thus Ray Hammond, proselytizing word processing to writers in 1984, unknowingly but eerily echoes the prescription of AMA consultants when he asserts, "Writing is broken down into its component parts; composition, editing, and printing."[82] As we have seen, the printer need not even be in the same room as the word processor, a far more jarring displacement of hand from page than Heidegger could ever have imagined.

Why not take the writer's hands out the equation altogether? In 2007 Margaret Atwood filed a patent on the LongPen, which allows an author to sign books anywhere in the world without having to be physically present.[83] An actual pen grasped in the servo-driven talons of a robotic arm faithfully reproduces the weight and direction of written strokes captured from a remotely linked input station, where the author is grasping a stylus to make their signature on a tablet computer. Although inspired by her experience as an author in high demand for signings at far-flung locations, the technology has been commercialized by the Toronto-based Syngrafii, whose target markets include commerce, government, and the legal sector. (One fan clutching a book newly signed with the technology described it as creepy. Quoth Atwood: "There are far creepier things in the world.")[84]

Today some believe that the grail of human-computer interaction is to operate machines directly by neural impulse. Writing can unfold at the speed of thought, without regard for fumbling fingers, inflammation of the carpal tunnel, or other bodily frailties.[85] Indeed, one of the key areas of application for this research is in adaptive computing technologies for the disabled. Stephen Hawking, whose first speech synthesizer was designed by Intel cofounder Gordon Moore and ran on an Apple II, is the most famous beneficiary of ongoing advances in adaptive character input; his current system uses a corpus of his published writing to predict upcoming words based on context.[86] But the aspiration of total disembodiment is part of a teleology that has historically been bound up with the nexus of gender, labor, and inscription that we have been exploring in this chapter.

Vannevar Bush's 1945 "As We May Think" essay for the *Atlantic* is best remembered for its description of the Memex, a hypothetical document processing station that anticipated key aspects of electronic hypertext.[87] But the Memex occupies only a portion of Bush's text, which also foresaw technologies as diverse as digital photography and voice recognition. Perhaps its most remarkable prognostication comes at the very end, where Bush imagines a machine that would capture sensory input directly from the optic nerve and route it to an external textual record, thereby eliminating the middleware of manual transcription—and the need for a secretary, or "girl" (he uses the term elsewhere in the essay, whose pronouns are rigidly gendered throughout). His thinking is as follows: "The impulses which flow in the arm nerves of a typist convey to her fingers the translated information which reaches her eye or ear, in order that the fingers may be caused to strike the proper keys. Might not these currents be intercepted, either in the original form in which information is conveyed to the brain, or in the marvelously metamorphosed form in which they then proceed to the hand?"[88]

Bush had been the wartime director of the Office of Scientific Research and Development, whose commissions included initiating the Manhattan Project. His essay, written expressly with the aim of transitioning the nationalized scientific establishment to a postwar footing, was to prove enormously influential. That this vision of what we might call "word processing" in arguably its ideologically (and biologically) purist incarnation is manifested so concretely in the closing paragraphs speaks volumes as to its centrality in the twentieth-century imagination of both writing and information. Frank Herbert and Max Barnard echoed the same scenario in *Without Me You're Nothing:* "Someday we will attach a computer directly to the human nervous system," they assured (or frightened) their readers.[89]

And of course the science is now well under way. Scientists in at least two separate labs have reported on results in which the neural activity associated with distinct letterforms is isolated and recorded, to be reproduced at will. At Mayo Clinic, patients "were asked to look at a computer screen containing a 6-by-6 matrix with a single alphanumeric character inside each square. Every time the square with a certain letter flashed, the patient focused on it and a computer application recorded the brain's response to the flashing letter."[90] As a result, according to the lead researcher, "We were able to consistently predict the desired letters for our patients at or near 100 percent accuracy."[91] Of note here are not only the results per se, but also the fact that what enables the isolation of the relevant brain activity from the background noise is the presentation of the individual letters on the screen in a matrix, very much like the spatialization and commodification of language Kittler and Bukatman discern in the typewriter keyboard. Another study at the University of Wisconsin–Madison produced results in which participants were able to tweet using EEG sensors alone. It's slow—no more than eight characters a minute—but speed and accuracy are said to improve with practice.[92] If and when they are reproducible at scale, such advances will mark a decisive break with an arc of technological development that began whenever the first people used sticks and charcoal to scratch marks onto the surfaces of their surroundings—the relationship between the work of the hand, an implement that it controls, and the inscriptions thus made defining the act of writing for millennia to come.

In the spring of 1979 novelist Stanley Elkin was in a bed in Barnes Memorial Hospital in St. Louis after yet another round of complications from his lifelong struggle with heart disease.[93] His first heart attack had been a decade earlier, when he was only 38. Another—his third or fourth—would kill him in 1995. His heart wasn't his only worry, however. Elkin, who along with William Gass and the poet Howard Nemerov formed the cornerstone of the thriving Parkview literary scene near Washington University, had also been diagnosed with multiple sclerosis seven years earlier. By 1979 it was taking its toll on the fine motor control in his hands. You can see the evidence in the powder-blue Washington University exam booklets he used to draft his fifth novel, *George Mills* (1982), which he was writing at the time and which many consider his masterpiece (it won him one of his two National Book Critics Circle Awards). Their pages are slashed and lashed by his pen. Whole sentences and paragraphs are scored by cross-outs. The handwriting leans to the right, all strong verticals and horizontals and acute angles. Lower-case a's become triangles, r's become inverted v's. The pen plows right through the red margins on either side of the pages to their edge, giving the eye no respite in white space. There are twenty-eight bluebooks in all, and the writing degenerates progressively across them.[94]

Elkin handed the booklets off for typing—depressing the keys on his Remington manual caused him too much pain in his fingertips.[95] But it was clear that this system was untenable; the pages were difficult for even a determined typist to decipher.[96] Somehow, in the course of a bedside conversation with his department chair (Elkin held a professorship in writing at Washington University), the idea of a word processor came up.[97] The keys would be physically easier to depress than those of the typewriter, and obviously the handwriting problem would be solved for good. His department chair, Dan Shea, convinced the university to invest in a Lexitron VT 1303 word processor for Elkin, which would have cost somewhere upward of $10,000. It arrived on June 6, 1979, and Elkin later told Shea that it was the most important day of his literary life.[98] He quickly set about composing the second half of *George Mills* at the keyboard and eventually revised the novel in its entirety.[99] The book was finished by early 1982, and his editor at Dutton was intrigued by the possibility it might be typeset directly from the computer diskettes.[100]

At a minimum the Lexitron would have provided a solution for Elkin's day-to-day challenges getting his writing done. But it proved to be much more than that, becoming an integral part of not only of his creative practice but also his colorful personal identity. At some level Elkin would had to have been self-conscious that the word processor was a product whose primary consumers were intended to be women working in an office (the training manuals would have driven home the point). If this troubled him at all, he made no mention. He spoke about the word processor often and affectionately in interviews, including a 1981 appearance on the *Dick Cavett Show* when it would still have seemed very much a novelty to the TV audience.[101] Elkin had a penchant for giving things quirky names and nicknames, and the dry, generic labels "Lexitron" and "word processor" were soon dispensed with. It became instead the Bubble Machine, and thus it would remain ever after in his personal lexicon. The name may have come from an episode in his previous novel, *The Franchiser* (1976), which features a scene with a Wurlitzer jukebox—"'Wait, before you go—the sound system.

Turn on the bubble machine. Hit the lights, please, Cisco.'"[102] (Jukeboxes were known as bubble machines for the visual effects produced by methylene dichloride boiled in colored glass tubes as the music played.)

In effect Elkin was domesticating the technology, dissociating it from the office and making it part of the house he shared with his wife, the painter Joan Elkin, which was filled with gadgetry and objects, including a Betamax video deck and a hotel soap collection stocked by their and their friends' travels.[103] Elkin was also quick to note how it was changing his writing:

> Plots have become very interesting to me since I started working with the bubble machine. The word processor has facilitated not just the mechanics of one's writing but has actually facilitated plot, and I'll tell you why. You write on a word processor: you open the store in the morning, you do what you do, and then it occurs to you, gee, wasn't there some kind of reference to that earlier on? So you put the machine into the search mode, and you find what the reference was earlier, and you can begin to use these things as tools, or nails, in putting the plot together. The word processor facilitates the plot—I really mean that. When I wrote "The MacGuffin," I was constantly looking back for general references to certain thematic key words, and that really tightens the nuts and bolts. Therefore, my plots have gotten better since I went to work with the word processor, which I did with "George Mills."[104]

Like a jukebox, the Bubble Machine allowed him to mix and match pieces of his prose, hopping back and forth around the book in progress the way the eponymous device shuffled its stacks of records (which perhaps for Elkin the eight-inch floppy disks resembled). And like a jukebox, it even lit up.[105] The Bubble Machine was not a cure for his MS, of course; but there is no question it was indispensable in allowing him to keep on with producing his MSS, which is to say his manuscripts—the work of his hands.[106]

EIGHT

THINK TAPE

What is very likely the first novel written with a word processor wasn't written on a word processor with a screen and its words weren't "processed" by the novelist who wrote it.

The novel in question is *Bomber,* Len Deighton's fictional account of the nighttime air war over German-occupied Europe during World War II. Published by Jonathan Cape in 1970 to enthusiastic reviews, *Bomber* follows the course of a single raid by the Royal Air Force (taking place on the deliberately fictive date June 31, 1943) through the eyes of dozens of different characters—British and German, combatants and civilians, in the air and on the ground. Deighton's preparation for writing the novel entailed thousands of hours of research, including site visits to many of the locations depicted in the book, forays into military archives, scores of interviews, and a cross-Channel flight in a restored German Heinkel He 111. He kept meticulous notes, all of them color-coded and cross-referenced. The walls of his South London townhome where he worked were papered with maps and air charts of North Central Europe to storyboard the unfolding action, tape and tags marking the positions of different aircraft over the course of the book—an uncanny re-creation of the big-board displays used by wartime air controllers to maintain situational awareness during the actual bombing raids. A chattering telex machine added to the atmosphere.[1]

By the time he began researching and writing *Bomber,* Deighton was already a cultural celebrity. He had published a half-dozen successful espionage thrillers, beginning with *The IPCRESS File* (1962), which was subsequently made into a blockbuster movie starring Michael Caine. Spies were then very much in vogue—John le Carré's career had been launched at the same time as Deighton's. In *Life* magazine, Conrad Knickerbocker declared spies (at least the ones then skulking through the pages of popular fiction) to be "hip, committed, *engagé* and morally relevant."[2] Riding this literary wave in swinging London, Deighton also wrote an illustrated culinary column for the *Observer* (spinning off a series of best-selling cookbooks) and was travel editor for *Playboy* magazine (hence the telex machine). But *Bomber* was to be a darker, more serious, and altogether more ambitious work, its origins lying in the author's childhood in London during the Blitz and his experiences as a photographer in the Royal Air Force shortly after the conclusion of the war.

Sometime in 1968, when writing for the novel was already under way, an IBM representative made a house-call to service the typewriters being used by Deighton and his assistant, an Australian woman named Ellenor Handley who had been working with him for several years. As we have already seen, a successful author can rarely indulge the sort of ennui and isolation affected by, say, the eccentric Mr. Earbrass in Edward Gorey's ode to the writing life, *The Unstrung Harp* (1953); instead he or she is likely to be something like the manager of a small office, with daily needs for typing, filing, accounting, correspondence, and other such tasks. The circumstances of the office secretary that we examined so closely in Chapter 7 thus have bearing on numerous professional authors and their typists or assistants. Individual situations might vary, as in the case of Wendell Berry, who testified that Tanya Berry had a conjugal backchannel to his muse. But while Updike dismissed his typist once he got a word processor, many others, like Stephen King, did not. Both before and after the advent of word processing, as Leah Price and Pamela Thurschwell have compellingly demonstrated, the literary secretary occupied a unique place, "iconic and yet invisible," at the intersection of labor, gender, and inscription.[3] And Ellenor Handley, it turns out, would occupy a more interesting place than most: "When I started," she recalls, "Len was using an IBM Golfball machine [Selectric] to type his drafts. . . . He would then hand write changes on the hard copy which I would then update as pages or chapters as necessary by retyping—time consuming perhaps but I quite liked it as I felt a real part of the process and grew with the book."[4]

As it turned out, the IBM man was to inform Deighton that Ms. Handley had been retyping some of his chapter drafts as many as two dozen times. Though it was designed for high-volume business environments such as the nearby Shell Centre, IBM had a new machine he thought might help. Would Mr. Deighton care to take a walk and see it in action?

He would. As Deighton's biographer Edward Milward-Oliver describes him, this writer was a "habitual early adopter of *practical* technology."[5] The machine in question was known as the MT 72, which was the European market's name for IBM's Magnetic Tape Selectric Typewriter (MT/ST). By all accounts the arrival of one soon after at Deighton's Georgian terraced home was an event: a window was removed, and the MT 72's two-hundred-pound bulk was lifted in with a crane. "Standing in the leafy square in which I lived, watching all this activity, I had a moment of doubt," Deighton says. "I was beginning to think that I had chosen a rather unusual way to write books."[6]

When the MT/ST was first announced on June 30, 1964, it was accompanied by the kind of fanfare that today is reserved for Apple product launches. Coverage appeared in the *New York Times,* the *Wall Street Journal,* the *New York Herald Tribune,* and the *Lexington Leader* (the local paper for the Kentucky bluegrass town where it had been developed and manufactured). It was clear from the headlines that this was something different: the *Journal* termed it a "brainier" typewriter, while the *Leader* lauded IBM for "revealing" a new kind of typewriting. IBM's president, Gordon M. Moodie, hailed it as "a radically new concept in typewriting which will change the traditional approach to typing jobs at thousands of typing stations across the country."[7]

As its bifurcated name implies, the MT/ST was a compound device: an IBM Selectric typewriter (still cutting-edge in its own right, having debuted only three years earlier) that was connected to a magnetic tape storage unit. At the heart of the hoopla around the new product was this simple fact: the MT/ST was the first mass-market general-purpose typewriting technology to implement something we can identify as suspended inscription—"stored typing through changeable, erasable magnetic media" was how Moodie put it at the time, capturing the essentials.[8] The basic principle was straightforward: at the same instant it was imprinted on the page, each individual keystroke was also recorded as data on a magnetic tape cartridge (each cartridge held approximately 24,000 characters), which could then be played back to have the machine go about the task of automatically printing (and reprinting and reprinting . . .) a page of text at the rate of some 170 words per minute. Backspacing to correct an error resulted in the usual blemishes on whatever piece of paper was in the Selectric's rollers at the time, but the revised sequence of characters was what got stored on the tape: clean texts could then be produced literally at the push of a button, without the need for time-consuming and imperfect erasures. Sentence spacing, line lengths, even hyphenated words were all automatically adjusted as revisions were introduced, one of the more technically impressive features of the product. Cumbersome carbons were eliminated because multiple copies of any document were so easily had. Even more significantly, the typing mechanism could be halted while in "playback" mode to allow for the manual insertion of additional text; this made it ideal for forms and form letters of all types. With dual tape reels in the storage unit (and Deighton would opt for such a model) a skilled operator could retain two different bodies of text at the ready "on-line," and blend them with one another in the course of producing hard copy—what we would today call a mail merge. Finally, and perhaps most tantalizingly, reference codes could be invisibly inserted into the stored copy of the text to act as markers or flags for later search and retrieval. (For a project such as *Bomber,* which involved continuous cross-referencing between the different narrative episodes, this was to prove a particular asset.)[9]

Development on what was to become the MT/ST had begun as early as 1956 at IBM's main offices in Poughkeepsie, New York—some four or five years before the Expensive Typewriter program was written for the TX-0 at MIT. By 1958 the design team had moved to IBM's newly opened typewriting manufacturing plant in Lexington. Leon Cooper was the lead engineer and recalls the early ambitions for the project, which were initially limited to storing keystrokes on tape or some other correctable medium and then playing them back to print out copies on demand.[10] Even in this nascent form, however, the notion of "suspended inscription"—a principle so convenient and compelling in theory—presented a host of practical problems. Because line lengths could shift as words were corrected or inserted, for example, the machine had to tell the difference between hard hyphens and soft hyphens: If a word was hyphenated merely because it broke at the end of the line, its soft hyphen could otherwise disappear; but hard hyphens had to remain no matter what. Initially, introducing two separate hyphenation keys, which the typist would have been expected to discriminate between, was what solved the problem.

More significantly, when the project first began, there was no IBM typewriter capable of functioning as both an input and an output device, which is to say capable of both creating and receiving character codes. The earliest prototypes therefore used two separate keyboards; one was from an IBM keypunch machine (which encoded the character, which was then stored in a relay); a typewriter then received the stored code and printed it on a sheet of paper. The result was thus an "amalgamation of what IBM used to do in keypunch and what we were headed for in digital recording on tape," as Cooper recalls.[11] The existence of this variety of equipment for keying, punching, and encoding words, and the place of such devices in the standard office environment, testifies to the extent to which the textual landscape the MT/ST was joining was already a complex skein of human-machine modulations: Text migrated from one medium to another, shape-shifted in and out of human legibility, flowed from reel to page and page to reel, oscillated between manual and "automatic" means of inscription. Different keyboards could punch cards, perforate tape, imprint ink, and encode messages to a computer. Tape could be either paper or magnetic, and could store programmed instructions, documents, or the human voice. Even celluloid could store the printed page in the form of microfilm.

The advent of the IBM Selectric eventually solved the character encoding problem, allowing a single typewriter to be used as both an input and output device. Other problems, however, presented themselves, seemingly at every turn. A special sprocket-driven tape had to be developed in-house at IBM, for example. It had to be sealed within its plastic cartridge with a novel technique utilizing ultrasonic vibrations; and it needed a special chemical solvent to strip the magnetic oxide coding in its lead; and so on.[12] The development team, which initially consisted of Cooper and a handful of others, grew to hundreds. As the work accelerated, further possibilities suggested themselves to the designers, most notably adding a separate track on the recording medium for the so-called reference codes (which enabled search) and then adding a second tape drive. A user could create and define a block of writing and store it separately from the main body of their document: suddenly text was not just recordable and replayable, it was modularly addressable and accessible.[13]

Meanwhile, IBM's marketing division was also getting ready. When it debuted in 1961, an IBM Selectric retailed for $500. The MT/ST would be priced a whole order of magnitude higher—at just under $10,000, an "expensive typewriter" if ever there was one. How to sell a product at such a staggering price point, especially when the same sales force had just been in and out of the same offices pitching the Selectric, itself heralded as revolutionary for its single-element interchangeable typing ball?

The first decision within IBM was to utilize a separate sales force, specifically the one charged with handling the Office Products Division's dictation equipment as opposed to the Selectric itself.[14] This decision was to have far-reaching consequences for what was to become known as "word processing," as it directly yoked the executive's microphone to the secretary's keyboard, thus ensuring that a single sales representative now handled a complete system for the origination, revision, and duplication of text. The next challenge for marketing lay in finding the best way to exploit the nature of the technology itself. The MT/ST was oriented toward revision-heavy typing tasks—so-called volume typing, or, in the phrase the marketing department soon hit upon, "power typing." IBM quickly realized that the correctable nature of the magnetic storage medium made it particularly relevant to scenarios wherein a line of prose might be worked over multiple times, which was precisely what slowed down even the most efficient typist and forced the endless dilemmas between stopgap solutions like erasers and correcting fluid versus retyping pages in their entirety. IBM produced detailed time management studies for the sales force to draw on, demonstrating the cost-effectiveness of the MT/ST even with its stratospheric price tag.[15]

There was another consequence of the MT/ST's price point: one didn't purchase such an expensive piece of equipment only to have it sitting dormant for large swaths of the day as a secretary attended to the multitude of other routine tasks with which she was charged.[16] As a consequence the MT/ST contributed directly to the word processing center as an organizational construct. An IBM sales representative would come through the door, and, armed with the latest data and statistics, make a pitch encouraging the business to radically change its internal system of text production. The sales rep would have examples and precedents from other businesses, right down to floor plans. He could furnish the complete suite of hardware necessary to carry a word or a sentence from an executive's lips to a typewritten page, as crisp and clean and blemish-free as a starched white shirt. While all of this seems an unlikely milieu for a novelist to step into, the fact that the MT/ST was marketed for revision-heavy office work would serve to make Len Deighton—committed, as he was, to "practical" technologies—and his assistant, Ellenor Handley, the ultimate power users.

When Deighton visited the ultramodern high-rise Shell Centre on the South Bank of the Thames to see the MT/ST in action, he was encountering it in exactly the setting it had been designed for. In much the same way that the perceived excess of information in digital form is now a common focalizer for doubts and fears about the online world, the overabundance of paper was the central anxiety of this earlier moment. Vannevar Bush was among the first to sound the alarm: he warned that the scientific research establishment that had been so crucial to the Allied war effort was increasingly in danger of being "bogged down" under a "mountain of research."[17] For Bush this was a generalized anxiety about the sheer quantity of information, and its division or compartmentalization into areas of increasing specialization. The scientist (or citizen) could not hope to keep up with it all without the advent of new tools for organizing information, the most important of which was the Memex, Bush's interactive workstation for manipulating what was then another cutting-edge document technology: microfilm. But while the Memex is widely cited as a precursor to the personal computer, it was never actually built; more than two decades later by the time of Deighton's visit to the Shell Centre, the technology of choice for business and science remained paper.

And not without reason. In their book *The Myth of the Paperless Office,* Abigail J. Sellen and Richard H. R. Harper summarize the manifold advantages of paper: it is "thin, light, porous, opaque, [and] flexible," and this in turn allows "many different human actions, such as grasping, carrying, manipulating, folding, and in combination with a marking tool, writing on."[18] Even today, in an era of touch-sensitive tablet computers with E-Ink, flexible screens, and ubiquitous computing just around the corner, these characteristics can still only be emulated and approximated by our digital devices. (To wit: the popularity of Moleskine notebooks with tech-savvy millennials.) But what's arguably the single most important property of paper—integral to its identity ever since the advent of wood pulping in the latter half of the nineteenth century—was also to prove its bane. For whatever else it was, paper was also, of course, cheap. This lent it a remarkable capacity to proliferate: "Paper tends, always, to generate more paper," declared two specialists writing on behalf of the American Management Association in 1974.[19]

It was plain to contemporary observers that the efficiencies or else the ineptitude of an office were externalized in the procedures it developed for handling paper: its routes and trails, its revision and duplication, its signing and mailing and filing in triplicate. The twentieth century's gradual compression of the term "paper work"—first two separate words, then conjoined by a hyphen, then a single compound term—as much as anything signals the term's inexorable encroachment. Paper begat not only paper, but also papers—formal academic studies. Statistics abounded on every conceivable aspect of office productivity: the cost of typing a letter (about $4, or perhaps closer to $15 if multiple rewrites are required); the cost of filing it (about 4 cents); the number of cubic feet (1.5 to 2) of stored paper a typical employee generates per annum; the cost per cubic foot of filing space; the cost per annum to maintain it; and so on.[20] The overall picture that emerges is of a corporate culture obsessed with the economics of managing and controlling the "mountain" of paper (the metaphor, the same one initially employed by Bush, appears repeatedly in the literature) and converting excesses to productivity.

No single artifact better encapsulates the anxieties of the era than a 1967 five-minute short film "The Paperwork Explosion," produced by IBM and directed by Jim Henson (who had previously designed a mascot character for the Electric Typewriter Division named Rowlf).[21] Its presentation and pacing—an avant-garde mixture of montage and electronic soundscapes composed by musician Raymond Scott—is deliberately jarring. For the first 30 seconds there is no dialog, just a rapid succession of technological and industrial images. The film then shifts to a pattern of juxtaposing brief sound bites from a variety of different speakers, each of which builds on the core theme: "At IBM, our work is related to the paperwork explosion." The montage continues, but paper and machines for handling it now dominate the imagery: copiers, feeders, shredders, sorters, mailers. "There's always been a lot of paperwork in an office." On cue, the film's title is literalized with a boom and burst of smoke, followed by pages fluttering through the blue sky.

The unifying figure in the film, appearing at intervals throughout, is a gaffer we initially see leaning on a farmyard fence and firing a corncob pipe, a stark contrast to the frenetic pace otherwise established. "Seems to me, we could use some help," he declares as the wind furls his white hair. Enter IBM, and specifically its Office Products Division, which had been the Electric Typewriter Division up until 1964, the same year the MT/ST was introduced. "IBM can help you with the time it takes to do the paperwork." Voice dictation recorders, along with the MT/ST, are featured at this point in the film with close-ups, accompanied by testimonials and statistics. "Used systematically throughout an office, these IBM machines can increase people's productivity by 50 percent." The film then enters its final movement. The pace quickens, the cuts come faster and faster. "IBM machines can do the work so that people have time to think." "Machines should do the work; that's what they're best at. People should do the thinking; that's what they're best at." "Machines should work; people should think." Cut back to the gaffer, who after a contemplative pull on his pipe, delivers one final line: "So I don't do much work anymore; I'm too busy thinking." Closure thus comes with a touch of comic relief, the realization that the gaffer—the only individual not dressed in business attire and situated in the only locale that is not an office interior or an industrial site—is perhaps the figure we can all aspire to be with the aid of IBM.[22]

Cinematically, the sort of imagery one finds in the Henson film—close-ups of machines for sorting and stacking and typing and copying reams of paper—would become the stock visual signature of its era, much like the cascades of luminescent ones and zeroes that would follow a decade and a half later. There would have been no question for any member of the professional managerial class that they were then living in an information age. "Information," declared one authority, "is the end product of paperwork."[23] This is what Shoshana Zuboff meant by what she termed "informatting," the way in which information was capable of autonomously generating more information.[24] And computerization notwithstanding, information was still largely made out of paper: "No data come out of the computer without having been somewhere, somehow, part of a paperwork operation," as the same industry authority put it.[25]

Though the words are never spoken in the film, the solution to the paperwork explosion—the means to harness all that undirected energy—was of course word processing. The proliferation of that term—again referring to an organizational concept rather than any one specific technology—can be traced through a variety of publications in the professional business management literature. Probably the term first appeared in print in the United States in 1970. Its definition at the time, given in a trade journal called *American Management*, includes the concepts of both dictation *and* typing, utilizes the language of centralization, efficiency, and flow, and compares word processing to what Henry Ford's assembly line did for automobiles.[26] The actual origin of the term, however, appears to have involved an interaction between one of IBM's German sales executives, Ulrich Steinhilper, and an American counterpart, Samuel J. Kalow.

The problem that initially brought the two together was IBM's dictation technology; German companies were proving unwilling to invest in the machines on behalf of secretaries who could take shorthand just as easily.

Kalow suggested that the dictation machines belonged instead in the hands of executives; the secretary would then type up a transcript of the tape. The dictation machine, insisted Kalow, should be the only device used by an executive besides the telephone.[27] Steinhilper, working from concepts he had been pondering for a number of years, then went a step further, formally yoking both dictation and typing into a complete workflow—or wordflow—that he termed *textverarbeitung*, the literal translation of which is "text processing."[28]

Likely as early as 1955 Steinhilper had sketched a flowchart that explicitly placed "text processing" in parallel to data processing: the first cell in the chart, unifying both in a common action, was labeled "Think"; thinking then diverged into the activities of either writing or calculation, which led in turn to the material realization of thoughts in the form of either text or data, respectively. Word processing or data processing (presumably implying actual products and services) were then applied to the text and data, leading the viewer to the final cell, reuniting both concepts under the aegis of the IBM brand.[29] (This diagram is an exemplary artifact of what Paul Erickson et al. have termed "Cold War rationality.")[30] To test his concepts, Steinhilper initiated a study at the Mercedes-Benz headquarters in Stuttgart to calculate the typical office costs involved in producing written textual communications. Text, he soon confirmed, was expensive. So expensive, in fact, that an outlay of a few thousand deutsche marks for an integrated text processing system would quickly be offset, saving time and money once an office committed to the IBM way.

The phrase "text processing" was in fact already in use within IBM's Data Processing Division, the great internal rival to the Electric Typewriter Division—which was itself renamed Office Products in 1964, the year of the MT/ST's launch. Kalow and others had lobbied for "Word Processing Division" instead, but they were unsuccessful. The novel construction, however, with its clear delineation in relation to Data Processing, nonetheless stuck and became widely associated with a variety of IBM products. Thomas Haigh cites a 1974 report noting that the Office Products Division was "calling virtually everything it makes a piece of word processing equipment—from a dictating machine on up to an office copier."[31] From there, word processing—as we have seen—took hold as a more general concept in office management, thereby recapturing some of Steinhilper's original ambitions for the term. (And in 1971 IBM had duly acknowledged him for having coined *textverarbeitung* for them.)[32]

Word processing was the fruit of business and managerial science's best attempts to come to grips with the paperwork explosion by automating and centralizing resources, maximizing the labor and productivity of executives and secretaries alike, aggressively advancing a variety of new technologies, and generally bringing to document production the same rigor and ruthless desire for efficiency that were characteristic of other areas of business operations, such as inventory control and accounting. Word processing, in short, was a means to an end. And that end was not pixels, but paper.

The early press coverage around the MT/ST yielded two very different kinds of stories: the revolutionary potential of magnetic tape storage in conjunction with typewriting, which promised to usher in an era of "perfect" final copy at a fraction of the previous effort, and a second phenomenon, hinted at in IBM president Moodie's comments about changing the traditional approach to typing tasks. The *New York Times* and the *Wall Street Journal* were both unabashedly more explicit: "Another giant step towards reducing secretaries to a purely decorative role was demonstrated yesterday by IBM" was the *Journal's* lead.[33] The *Times* was only slightly more dispassionate in tone: "The International Business Machines Corporation introduced yesterday a typewriter that it believes will eliminate a lot of the drudgery of the secretary's job. It may also eliminate a lot of secretaries."[34] This wanton, even cruel language provided the initial framing for the coverage in both of the major dailies. (The *Lexington Leader* stuck to the facts about the MT/ST's technical innovations.)[35] Of course, as we have seen, IBM itself planned to aggressively market the machine (and justify its price tag) exactly on the basis of such cost-benefit calculations.

The notion that the MT/ST and advances in office technology more generally would decimate secretarial labor also belied another reality: that it was precisely secretaries who were going to have to operate the new equipment, learning how to interact with the machine and leverage its capabilities with a very high degree of sophistication. Yet the importance of such skills was rarely acknowledged. "Women's labor, no matter how much technical dexterity, mental expertise, or training it requires, is usually defined as inferior simply because it is women's labor," argues Juliet Webster, surfacing the tautological logic that underlay the conundrum that a secretary's expertise was required to run the very machine that was supposed to be replacing her.[36] IBM itself realized that it would take an extraordinary effort to retool the secretarial workforce for this "new concept" in office technology.[37] Thus, even as the engineering team was readying the MT/ST for launch, the educational and sales division—the "customer engineers"—were orchestrating an extensive training regimen. Training manuals and workbooks were written, curricula and tutorials were mooted, and exercises were drafted. Dedicated training centers within IBM showrooms were established, with an intensive schedule of classes and demonstrations. There new operators could work "free from ordinary office distractions," and even more importantly, in the company of their peers.[38] But it wasn't just the complexity and intricacy of the machine that compelled attention, or even the allure of "power typing;" it was, as IBM's in-house newspaper put it, the fact that the MT/ST would "lend the aura of its prestige to typists across the country as they learn to use its tape logic."[39] This notion of "tape logic" would quickly be condensed into a distinctive catchphrase, "Think Tape," that appeared throughout the company's instructional literature.

The whole point of the MT/ST technology, after all, was that the real record of the text was stored not on the page but electromagnetically, which is to say on a reel.[40] It wasn't enough, then, for a secretary to master the basic procedures for storing and retrieving text, accomplished with the aid of the push-button console and knobs; she also had to internalize the logic of the tape storage system in order to find her way around a document and bring the machine's full capabilities to bear. Because language was atomized at the level of individual keystrokes, the MT/ST treated any text as simply one long, unbroken string. Spaces between words and carriage returns and the like simply constituted more data, in the form of special formatting codes. Backspacing to correct an error on the page meant rewinding the tape a single character, which would then be overwritten with the new keystroke. The operator, therefore, had to assimilate this linear logic, learning to navigate the tape through recourse to various kinds of codes and internal divisions, which might take the form of an individual word or line or page or some other unit of text that had been invisibly indicated.

But the MT/ST was also an irreducibly material entity. It was noisy, for one thing, filling the room with clicks and clacks and bumps and thumps, to say nothing of the constant background whirr of the motor. A skilled operator would learn to navigate the tape not only by sight and by touch, but also by sound, using aural cues to know, for example, when a reference code had been set successfully (it made a distinctive noise). The most difficult task for an operator to conceptualize was so-called revision typing. Not just backspacing to correct an immediate error, but introducing changes to the text after a draft had already been completed. Inserting new text into an existing document required setting the machine in playback mode (a twist of a knob), and then waiting until the automatic printing cycle (170 words per minute, remember) was approaching the point at which the changes needed to be introduced. The operator would then control the print job manually, advancing the output word by word, and stopping the process altogether to type the new text in the correct location. It was in those situations that Thinking Tape was of paramount importance (all emphases in original):

> Typists are accustomed to thinking in terms of lines of typing. But now, as you go into revision typing, <u>stop</u> thinking about what you see on the printed page. <u>Start</u> thinking about what is recorded on the tape.
>
> As you study the tape logic involved in revision typing, it will be easier to understand what is happening if you learn to <u>visualize the tape.</u> Any time you wonder why the machine acted as it did, or you wonder what will happen when you depress a particular button, <u>visualize the characters on the tape!</u>
>
> At times you should diagram a recording yourself . . . to help "picture" the tape and answer your questions. Sounds simple, but it works. Do it whenever you question why the machine responded as it did.
>
> Think of continuous tape, not printed lines.[41]

The logic of "thinking tape" is thus revealed: the machine is the ultimate authority. The operators must adjust their instincts and expectations accordingly. Attention to individual lines of type must be replaced by awareness of the reel of tape. Not only was the operator encouraged to "visualize" the tape itself, but the workbook helpfully included a full page of blank tape diagrams, divided into cells, that she could pencil in with actual characters for a particularly knotty sequence. On the same page as the paragraphs just quoted, a diagrammatic strip of tape with characters assigned to their individual cells is seamlessly spliced to a photorealistic length of tape unspooled from an actual reel in a tape cartridge.

The injunction to "think tape" therefore allowed the MT/ST's instructors and operators to give voice to the conceptual challenges inherent in a new kind of writing, which they otherwise lacked a coherent vocabulary to express. The surface of the tape on which texts were recorded with character codes (inscribed as magnetic flux reversals) was made visible and legible, partly through careful documentation and explanation, and partly through hands-on visualization routines that functioned as a supplementary writing space in which the human operator could attempt to emulate the logic of the machine.

Of course, IBM's customer engineers and their pupils weren't the only ones then learning to "think tape." Tape, along with its potential for processing (in the form of cutting and splicing), was also the preoccupation of William S. Burroughs and confederates like Ian Sommerville and Brion Gysin. The commercial rollout of the MT/ST coincided almost exactly with Burroughs and Sommerville's experiments with a Philips "Carry Corder," or compact cassette recorder. For them, the fascination was in the clandestine portability of the device: it could be brought into any situation and used to surreptitiously record what was going on, and the record could then be played back at some unexpected time and place. Some of the exploits they proposed were merely prankish—one could record street sounds like a fire engine, for example, and later surprise one's companions with the siren. But one could also use the device to record routine office interactions: "Record your bosses and co-workers analyze their associational patterns learn to imitate their voices oh you'll be a popular man around the office but not easy to compete with," Burroughs wrote in *The Ticket That Exploded* (1967).[42] The emphasis on knowledge of "associational patterns" and the like placed this scenario more or less in line with such recommendations as might have been found in the publications of the American Management Association.

Burroughs and Gysin went a step further in their famous cut-up techniques, splicing tape segments together to create collage effects with voices and sounds—just as an earlier generation of Dadaists created poetry by randomly assembling cut-up newspaper text. Both the cut-up technique and the ethos of tape "processing" (a word Burroughs used explicitly) speak to a common zeitgeist, in which language (whether spoken or written) is subject to habitual decomposition and rearrangement.[43] There were thus more than casual correspondences between what was happening in the avant-garde and what was happening in the office. The tacit knowledge that was required in order to slice and splice the tape cut-ups would have been shared by any office secretary trained on one of the several varieties of equipment used to record form letters on rolls of paper tape; she would have known how to manipulate these tapes to create different combinations of text, sometimes literally cutting and pasting—precisely the kind of task the MT/ST was designed to eliminate.[44] And Burroughs himself, of course, had a deep familial connection to typewriters and office machinery as the grandson of the founder of the Burroughs Corporation. (We have seen that Evelyn Berezin's Redactron Corporation was sold to the Burroughs Corporation in 1975, where its products became the Burroughs Redactor line of word processors.)

More even than "typing on glass," tape was the medium that initially defined word processing. There was clearly enough in the air to tell a discerning author that something fundamental was changing in the world of writing. John Barth was thinking about tape in his novel *Giles Goat-Boy* (1966), which is presented under the fictional conceit of a found text originating on computer tapes from the "WESCAC" mainframe (it includes a "post-tape" as well as a postscript); Harlan Ellison's "I Have No Mouth and Must Scream," with its swaths of tape-punched "talkfields" (so troublesome for the text's compositors), was published a year later in 1967. The Italian poet Nanni Ballestrini was using a computer program to manipulate and recombine texts on tape as early as 1961.[45]

Tape was a very different surface for writing than either the typewritten page or the glowing glass screen. It was possessed of its own irreducible geometry and logic. Not pointing and clicking, not carriage returns, but recording and playback. Why think tape? As the Office Products Division manual had reminded the trainee, a perfect tape meant perfection on the page.[46]

After his MT/ST was delivered and installed in the dramatic fashion described, Deighton would continue his custom of writing while standing at an upright Selectric, fed by a long roll of telex paper. It was Ellenor Handley who set about the task of mastering the new acquisition, just like thousands of secretaries before her in more conventional office settings. There were several false starts and mechanical issues to sort out (the machine could be temperamental), but soon enough it was fully integrated into the workflow for *Bomber.* Deighton's drafting was characterized by a careful system for producing typescript on paper of different colors, corresponding to the different narrative points of view exhibited in the book. This allowed him easy access to individual segments and also the ability to take in, literally at a glance, the balance and distribution of opposing perspectives as the stack of manuscript pages grew. "One might almost think the word processor (as it was eventually named) was built to my requirements," he later commented:

> I am a slow worker so that each book takes well over a year—some took several years—and I had always "constructed" my books rather than written them. Until the IBM machine arrived I used scissors and paste (actually Copydex one of those milk glues) to add paras, dump pages and rearrange sections of material. Having been trained as an illustrator I saw no reason to work from start to finish. I reasoned that a painting is not started in the top left hand corner and finished in the bottom right corner: why should a book be put together in a straight line?[47]

This description of what we might call Deighton's own cut-up technique suggests his familiarity with the MT/ST, but he himself freely admits, "It was nearly always my secretary who used that machine."[48] The reality is probably more complicated than either of his statements can reveal on their own. While Deighton and Handley's division of labor comports with the word processing paradigm as it was then being established in office environments, the two nonetheless worked in close physical proximity to one another, often in the same room (the MT/ST was installed in Deighton's ground-floor study). There must have been the unselfconscious leaning over shoulders, gestures, pointing, and quick staccato conversations of the kind that characterize intense long-term collaborations. There must, in other words, have been moments when Deighton, Handley, and the MT/ST fused together in something like a cybernetic loop—the very essence of word processing in its full systemic sense. Thus, even as literary production modeled itself on corporate practice, it modified and scaled those practices to more human levels where traditional roles and distinctions might erode. This kind of phenomenon doubtless helps account for Handley's feeling like "a real part of the process," even as it also explains Deighton's affection for the intricacies of the hardware.

Bomber was greeted with considerable acclaim upon its publication in 1970. Today (especially in the United Kingdom) it is still regarded as an important work of fiction about the Second World War, praised by Anthony Burgess and Kingsley Amis, among others. The BBC adapted it as a real-time radio drama, aired in 1995 and again in 2011; and it was included on the "lost" Man Booker long list for 1970 mooted in 2010.[49] But it has still another claim on literary history: "There is no doubt in my mind that it was 'groundbreaking' and I'm sure we all felt so at the time," Handley comments, referring to the role of the MT/ST technology.[50]

Clearly Deighton too had sensed something of its significance. Employing the British nomenclature for the MT/ST and its model number, he notes in the afterword to the original edition: "This is perhaps the first book to be entirely recorded on magnetic tape for the IBM 72 IV."[51] His statement acknowledges a fact about the book's production, but it also signals something more: For likely the first time in history, a newly composed work of full-length literary fiction existed not just in however many leaves of manuscript or typescript the author had accumulated in the course of writing it, and not just in the various printings and editions that followed, but in another format as well, a format composed entirely of codes (actually minute fluctuations across a band of magnetic tape coated in iron oxide).[52] These codes were perfectly legible, but not to human eyes—a skilled operator like Handley could perhaps approximate the logic of the machine by "thinking tape," but the magnetically recorded text itself would remain a cipher until output on a plain sheet of paper.

Nor did this feat entirely escape recognition at the time. At least one early report from the American Management Association took note of Deighton's use of a word processor, hailing it as an example of the technology's "unexplored potential" "within the very private and creative world of the professional writer."[53] Why, if Leo Tolstoy had also had an "automatic typewriter," ventured the report's authors, perhaps he might have saved enough time and energy in the revisions to *War and Peace* to have written another masterpiece! (Apparently no one told them that Tolstoy was famously photographed with what were labeled his "three assistants"—two female secretaries and one mechanical typewriter—let alone about *Anna Karenina* and perhaps one or two other masterpieces.) But then, in a brazen act of erasure, the authors omit from the chunks of Deighton's afterword quoted in their report his explicit mention there of Handley as operator of the machine. This cannot be mere happenstance. Handley represented an unwelcome intrusion into the "private" and "creative" world that is the presumptive sanctum of traditional literary authorship. Or as Price and Thurschwell wryly note, "The opposite of genius is typist."[54] So the authors of the AMA report must have felt, preferring not to despoil the image of the high-tech, high-powered novelist at the helm of his machine. I have endeavored to furnish a truer rendering here.

Deighton moved to Ireland shortly after finishing *Bomber.* The MT/ST (which required constant care and upkeep) did not follow him. By 1980, however, he had acquired an Olivetti TES 501 word processor secondhand from the American novelist Richard Condon (best known for *The Manchurian Candidate* [1959]) who was then also living in Ireland;[55] it had a one-line LED display and stored text on a pair of 5¼-inch floppy disks. Handley continued to work for him through the 1970s, eventually also typesetting his manuscripts with a product called the IBM Composer.[56]

In an eerie historical coincidence, Ulrich Steinhilper had flown Messerschmitts for the Luftwaffe in the early years of the Second World War, making ace before being shot down over Kent in England, captured, and shipped off to a prison camp in Ontario (from which he escaped, multiple times). He wrote several books about these exploits after his retirement from IBM, which are still popular among military aviation buffs. He also became a frequent guest at commemorative air shows in Great Britain. Less than a decade after finishing *Bomber* with the assistance of the machine that was the first consumer realization of the concept Steinhilper had survived the war to name, Deighton would author a classic nonfiction account of the Battle of Britain: it would be entitled simply *Fighter* (1977). Given the aviation circles in which they both moved, it is not at all beyond the realm of possibility that Steinhilper and Deighton crossed paths at some point. But it is not likely they understood their significance to one another if so.

Bomber was a harbinger of what we would today call a techno-thriller. While the book broke new ground with its complex portrayals of characters on both sides of the Channel, it is also a book about a new type of warfare that depended on long-range weapons and sensors, the human combatants never seeing one another face to face. "Sometimes," opines one character early on, "I think it's just the machines of Germany fighting the machines of Britain."[57] Those words, of course, indisputably belong to Len Deighton. But the hands that recorded and processed them using the MT/ST belong to Ellenor Handley.

NINE

REVEAL CODES

When John Barth needed to find a way to introduce a collection of his short stories in the aftermath of the September 11 attacks, he turned to a word processing metaphor—WYSIWYG, or "Wys" for short—conferring its name on the "brackish tidewater marsh-nymph" who becomes Barth's interlocutor as he works through his relationship to what would now strike him as eleven "mostly Autumnal and impossibly innocent" pieces of fiction.[1] Joan Didion begins *The Year of Magical Thinking* (2006) by recounting the properties of the Microsoft Word file containing the first words she wrote after her husband's death from a heart attack as he was sitting across the kitchen table from her. The file's seemingly inconsequential technical details are proffered by Didion as a kind of touchstone, an objective anchor for her memories even as she asserts her autonomy over a misleading date stamp: "That would have been a case of my opening the file and reflexively pressing save when I closed it."[2] Paul Kafka's *Love [ENTER]* (1997) features chapters named for its characters with a date stamp and a time stamp and the suffix .doc appended to each of them, for example "10/23/92 23:06 Bou. Doc." Anne Carson embeds computer syntax directly into the title of her 2013 book-length prose poem *Red Doc>*. Elly Bulkin's *Enter Password: Recovery / Re-Enter Password* (1990) uses "recovery"—a word associated with restoring digital information—as both the eponymous password to her electronic journal and a starting point for a meditation about memory. *Microserfs* (1995), Douglas Coupland's fictional ethnography of day-to-day life on the lower rungs of the corporate computer world, takes the form of a journal that its narrator, Daniel Underwood, keeps on his PowerBook laptop;

periodically the digital consciousness of the PowerBook interrupts the narrative with typographic displays of word art in outsized fonts that are data dumps of its internal memory, the zeitgeist of the book's contemporary moment, or both. Meanwhile, Jeanette Winterson's novel *The PowerBook* (2000) is about a writer who works online; its sections are named with familiar but resonant commands like SEARCH, VIEW, and EMPTY TRASH.

Poets likewise took notice of their computers: in a 1988 poem that still reappears regularly on the Internet, Gary Snyder explains to us, "Why I Take Good Care of My Macintosh": "because my computer and me are both brief / in this world, both foolish, and we have earthly fates."[3] Charles Bukowski, besides writing of Osbornes, Kaypros, and Intel 8088 chips, also mentioned his computer frequently in his poetry and letters (and soon enough, email) after getting a Macintosh IIsi for Christmas in 1990.[4] "My Computer" (1997) details his editors' consternation upon hearing the news of this acquisition: "I am aware that a / computer can't create / a poem. / but neither can a / typewriter" is Bukowski's rejoinder.[5] Robert Crumb drew the disheveled Bukowski sitting in front of his keyboard and monitor: "Old writer puts on a sweater, sits down, leers into computer screen, and writes about life. How holy can we get?" reads the caption, thereby demonstrating how fully computers had been assimilated into the stock of cultural imagery around writing and authorship.[6] Jacques Derrida's *le petit Mac*, meanwhile, is memorialized by Michael Fried in a poem that contrasts the abrupt news of Derrida's terminal diagnosis with a vision of the philosopher closing the cover of his laptop.[7]

For writers spending so much time with a keyboard and mouse beneath their fingertips and the glowing pixels of a screen angled in front of their eyes, it was inevitable that they would begin exploring and exploiting word processing technology in their own literary language and technique. This deliberately diverse and eclectic set of examples—and there are dozens, if not hundreds more, from the last three decades of publishing—are all instances wherein the mundane conventions of writers' computers and word processors are invited into the aesthetic or affective space of their work, thereby offering up new reservoirs of images, tropes, and formal devices.[8] At first some authors, like Anne Rice, Tom Clancy, and Umberto Eco, were content merely to sneak mentions of their freshly out-of-the-box computers into their prose. In like manner, Don DeLillo, who remained loyal to his typewriter, cast his writer-protagonist Bill Gray in *Mao II* (1991) as a refusenik; a character who is portrayed as an enthusiastic convert presses him on the question: "I would think for a man who clearly reworks and refines as much as you do, a word processor would be a major blessing." But Gray, like DeLillo, is unmoved.[9] Other writers, such as Stephen King and John Updike, found an unlikely muse to indulge in works like "The Word Processor" (1983) and "INVALID.KEYSTROKE" (1983). John Varley produced a longish short story, "Press Enter ■," whose characters had names like Kluge, Hal Lanier, and Osborne; it was published just a year after King's story, and despite those in-jokes is rather more harrowing.[10] Soon, however, the technology became just another part of the routine scaffolding of plots and settings. Russell Hoban's novel *The Medusa Frequency* (1987) places its nocturnal author protagonist in front of an Apple II where, late at night, the words on the screen "come out of a green dancing and excitation of phosphors."[11]

What these examples have in common is that (unlike earlier, fantastic depictions of writing machines by authors such as Calvino, Leiber, and Lem) they are marked by the presence of actual products and brand names: Apple, WordStar, Wang, TRS-80, Macintosh, and so on. Even when endowed with supernatural powers, the technology of personal computing is otherwise portrayed in a newly realist manner—the Apple II really did have a green monochrome screen, Microsoft Word documents really are called ".docs," and the existential crises of characters were given dimension by actual software functions like ENTER and EXECUTE. Readers, many of whom would have owned the same computers themselves, would have recognized these details as authentic.

Tao Lin, who is often singled out for the ruthlessly minimalistic prose style he achieves through many dozens of iterations of his texts through his word processor, is also known for embedding chat sessions, email, and other forms of electronic communication in his fiction: "I want to change my novel to present tense," says a character in the midst of one such exchange. "Is there some Microsoft Word thing to do that," Sam wonders, doubtless ventriloquizing Lin himself who routinely avails himself of such functions.[12] Taking its title from the venerable WordPerfect command used to display the underlying formatting markup of a text, this chapter will look at some of the ways in which fiction and poetry became occasions for authors to explore word processing's unique formal and imaginative elements—the material codes of a new relationship between technology and the literary.

Writers began taking advantage of computers and word processors in substantive ways, seizing upon their potential to serve as significant devices (*literary* devices) in their fiction. One of the most noteworthy examples comes from Henry Roth and the tetralogy of novels written in the mid-1990s and published collectively as *Mercy of a Rude Stream* (2014).[13] The novels are narrated by Ira Stigman, a thinly veiled stand-in for Roth himself. Stigman is an aged author who is struggling to write the story of his younger self growing up in an early twentieth-century Jewish section of Harlem, New York (the story is also entitled *Mercy of a Rude Stream*). Throughout, Stigman includes asides to his computer, named Ecclesias (his "friend and life support system"), which allows him to work through his writer's block and reservations about the book in progress.[14] Ecclesias, meanwhile, quickly takes on a voice of its own, speaking back to the older Ira with suggestions and provocations. We are not meant to think of Eccelsias as a computer endowed with artificial intelligence like its many progenitors in science fiction; Roth litters the narrative with references to the mundane and material nature of its operations, from the constellation of punctuation marks along the top row of the keyboard to the minutiae of how many pages can fit on a floppy disk or how to optimize its RAM. The computer instead functions as something like an alter ego or secret sharer, a literary device that allows Ira his own interlocutor within the space of the fiction, even as Roth himself has an implicit interlocutor in the character of Ira. Eventually we learn that Ecclesias is an IBM PC Jr. purchased in April 1985, the same computer (unsurprisingly) that Roth himself used.[15]

The most important episode involving Ecclesias, however, comes as Ira is struggling to confide, at the beginning of the second volume, *A Diving Rock on the Hudson* (1995), an incestuous adolescent relationship with his younger sister, Minnie. Heretofore Minnie has been entirely absent from the narrative, but Stigman finds himself unable to proceed across this "stile" without disclosing her existence and the nature of the affair, which he views as the defining event of his life. He considers rewriting everything from the beginning, but discards that idea as impractical; he considers an exculpatory preface or forward, but rejects that as too pat. Finally, in dialogue with Ecclesias, he reaches an accommodation of sorts: "Yes, I have a younger sister. Let it fade," Stigman instructs. Ecclesias, however, is not satisfied: "When will you admit her to the realm of a legitimate character, acting, active, asserting herself, as an individual?" Stigman replies: "I don't know, I don't know if I'll ever be able to write about her in all the emotional dimensions she deserves. But I have to do something. I'll have to: sometime opportune, in passing mention . . . a flake of this terrible, unspeakable, inter . . . inter . . . interlude. Ssss. Interplay, flay, slay, clay, lay. . . . So you have your answer, Ecclesias, at least in part."[16]

The word Stigman cannot bring himself to pronounce—that is, to type—is presumably intercourse, which instead becomes first "interlude" and then "interplay," immediately followed by a cascade of Old Testament utterances rhyming on the second syllable. The wordplay perhaps recalls James Joyce, who for Roth (and Stigman) was a literary hero; but it is occasioned here by the novel device of the computer, which Roth endows with not only a name but a persona. (Much like the Magic Eight-Ball in Stephen King's short story "The Word Processor," the computer here functions in a vaguely oracular way, as reinforced by its quasi-biblical name, this being part of a tradition that also includes Pournelle's Ezekial and Umberto Eco's fictional Abulafia from his novel *Foucault's Pendulum.*)[17] Moreover, the screen on which Stigman expresses his ambivalence allows him to "screen" his conflicting emotions from his notional reader, even as they are also, of course, conveyed to Roth's actual reader through the text of the books themselves, in which the Ecclesias segments are—notably—printed in boldface. "Things as they were," writes Ira (knowingly conjuring a blue guitar), "changed upon the computer."[18]

A lot of what's published nowadays is what we would have called overwritten back before computers," Jerry Pournelle comments. "You never hear anything being called overwritten nowadays, do you? Because in a sense everything is."[19] Word processing, so the conventional wisdom goes, encourages authors to overwrite because it is so easy for them to continue revising and embellishing their prose. The availability of thesauri (a keystroke or mouse click away) and for that matter the Web itself—with the potential for uncovering extraneous detail lurking behind every search box—no doubt exacerbates these tendencies. The charge "overwritten" as brought to bear in this sense is pejorative, shorthand for the combination of efficiency and easy excess that is associated with word processing in the popular imagination.

This prejudice is satirized in Jesse Kellerman's novel *Potboiler,* published to mixed reviews in 2012.[20] His novel is a clever sort of Klein bottle that arguably loses some of its interest once the central premise is revealed. It begins with the death of an author, one William de Vallée, whose popular thrillers feature a special agent named Dick Stapp. De Vallée's lifelong friend Arthur Pfefferkorn is on his way to attend the funeral. In the airport Pfefferkorn spots a Pattersonesque display of his friend's books: "Every ten yards or so Pfefferkorn passed another towering cardboard bin, its top crowned by an enlargement of Bill's jacket photo."[21] Pfefferkorn himself is a novelist, or rather a failed one, tortured by the question of literature's place in the world. Setting past jealousies aside, he comes to an understanding of de Vallée's success: "Literature did not decrease injustice or increase fairness or cure any of the ills that had plagued mankind since time immemorial. It was sufficient, rather, to make one person, however bourgeois, feel slightly less unhappy for a short period of time."[22] Visiting the estate after the funeral, Pfefferkorn happens upon one final, almost-finished manuscript in his old friend's study. Conveniently, it is typewritten, this detail permitting Kellerman the contrivance of its being the only surviving copy. On impulse Pfefferkorn absconds with the manuscript, intending to submit it for publication under his own name. Of course, as a writer of a more sophisticated sort, he also can't resist cleaning up the (overwritten) prose: "It was characteristic of special agent Richard 'Dick' Stapp to perform difficult physical feats in *one fluid motion*. Pfefferkorn didn't care for that expression one bit."[23] Searching and replacing at his computer, Pfefferkorn purges the text of such instances, twenty-four in all, along with numerous other examples of overwriting.

Potboiler thus works from the premise of a writer's revenge fantasy, but in fact all of this is quite literally pretext: Pfefferkorn's "novel" goes on to great commercial success and he is whisked away on a book tour; the royalties roll in, but soon enough he is contacted by de Vallée's agent, who has his own copy of his former client's typescript after all. The "agent" turns out to be a government agent, that is to say, a secret agent—the Dick Stapp novels, we learn, were an elaborate ruse for transmitting messages to operatives in foreign countries when other means of communication were deemed too risky (not for nothing are they in airports all over the globe). Each one contained a set of instructions or directives. "Bill wrote in code," Pfefferkorn is told.[24] Except even that is an overstatement; de Vallée did not write his books at all. They were algorithmically generated by a piece of software called Workbench, nothing less than a state-of-the art "fictocrpytic" upgrade to Calvino's OEPHLW. In espionage terms, de Vallée was a mere cut-out; in poststructuralist terms, he was a pure discursive construct—"a perfect fraud," as his handler, now recruiting Pfefferkorn to take his place, puts it. And he isn't the only one: "Most blockbuster American novelists are on our payroll," the agent tells Pfefferkorn. "Anything with embossed foil letters, that's us."[25] From there things become progressively less interesting as Pfefferkorn is deployed to the Eastern European countries of West and East Zlabia, there to help undo the damage he has done by finding and replacing all of those "*one fluid motions.*" But he also comes to the ironic realization that he has finally accomplished exactly what he has always wanted to do—he has published a book that quite literally changed the world, in a direct and measurable way. When the president of West Zlabia is assassinated, one might say the president has been globally found and replaced as a consequence of Pfefferkorn's electronic interventions in de Vallée's text. "The power of literature," the (secret) agent muses.[26]

Yet even as Pfefferkorn and his word processor are busily winnowing traces of Workbench's overwritten prose, turning the text of what was intended to be de Vallée's next best-seller into "cryptographic Swiss cheese," he is overwriting the novel (much as he overwrites his friend's identity as its author) in another sense as well. In computing parlance, overwriting refers to saving a file using the same name as some other—preexisting—file, thereby replacing the contents of the older file (which might be a version of the same document or something entirely different) with the newer one. In practice what actually happens is that the computer's operating system simply shunts the original file aside until its physical space on the disk is eventually recycled at some later date—this is the reason it is usually possible to restore a "deleted" file relatively soon after it has been erased. Eventually, though, the surface area of the storage media will be reclaimed. Even then, however, there may be traces of the original data left underneath, as in a palimpsest—at one time it was even thought that electron microscopes might be used to restore files bit by bit through careful scrutiny of the underlying data traces.[27] Beyond these technical considerations, however, overwriting can serve to suggest the seemingly endlessly fungible nature of digital text: "We constantly overwrote everything," William Gibson says in *The Difference Engine's* 2011 afterword. "Ourselves, each other, the hods of actual Victorian print media Bruce hauled home from the University of Texas. The result is a text that couldn't have been produced without word processing."[28] We have already looked at what Gibson and Sterling called their airbrush technique for the novel, but here the word "overwrite" repays attention as one of the formal properties of word processing—not in the pejorative sense originally suggested by Pournelle, but as a deliberate composition technique that to a degree mimics the data storage routines of the computer's operating system. In writing *The Difference Engine,* Gibson and Sterling freely edited one another's prose as well as the text of their Victorian source materials until the original identity of the text had vanished under innumerable and indeterminate layers of revision. Gibson still works in much this same way, commenting in his *Paris Review* interview that his daily writing routine involves starting at the beginning of whatever he is currently working on—even novels—and then reading his way through until he reaches its furthest point of completion, editing and revising as he goes. "The beginnings of my books are rewritten many times," he comments. "The endings are only a draft or three, and then they're done."[29]

Other authors have adopted similar practices for equally deliberate ends. When Seth Grahame-Smith was given the commission to write *Pride and Prejudice and Zombies* (2009)—an "expanded edition" of the classic Jane Austen novel in which Elizabeth Bennet is made over as a martial arts prodigy and Mr. Darcy is a monster hunter—he started by downloading an electronic copy of Austen's most popular novel from the Internet. From there he simply copied and pasted the entirety of the text into his word processor and began revising, overwriting Austen's text with the nouveaux storyline but still leaving an estimated 85 percent of it intact. As he worked, he used his word processor's Zoom feature to generate holistic high-altitude views of the document, ascertaining at a glance the balance between his own prose (which was in red) and Austen's.[30] This method recalls the tradition of treated artist's books, like Tom Phillips's *A Humument* (1970) and Johnathan Safran Foer's *Tree of Codes* (2010), both of which use earlier works of fiction as the literal physical basis for (respectively) their watercolor page paintings or laser-cut latticework of words. But whereas Phillips and Foer both employed painstaking physical design processes to achieve their desired effects, Grahame-Smith uses a word processor and a thematically appropriate red font to gradually remake Austen's novel into a rococo yarn set in an alternative nineteenth-century England overrun by the shambling "manky dreadfuls." Though novels, including Jane Austen's, have been the subject of parody and adaptations in various media for many decades, here the revisionary process is operating with far greater semantic granularity, the symbolic character of electronic text permitting the editorial cursor to intercede at any point to change a word, a sentence, a punctuation mark, or an entire passage.[31] The best analogy is perhaps electronic music, where a remix layering in beats and instruments can change the character of a composition, turning a pop ballad into a club track or reggae into calypso. Grahame-Smith's book eventually reached number three on the *New York*

Times best-sellers list, spawning sequels and imitations as well as a graphic novel, an iPhone video game, and a forthcoming feature film.

Gibson and Sterling recognized the similarities between their technique and both music (through sampling) and the twentieth-century avant-garde, especially William S. Burroughs and his cut-ups. At a practical level, word processing thus enables writers to participate in such contemporary creative norms as are commonplace among DJs, musicians, and visual artists. But a work such as *Pride and Prejudice and Zombies* also seems to brazenly embrace the same anxieties about word processing and the automated production of literature that were forecasted by authors like Stanislaw Lem with his U-Write-It, Fritz Leiber with his wordmills, and Italo Calvino with his OEPHLW. In these dystopian scenarios, literature is produced according to type or template, the masterpieces of yesteryear mere fodder for the Markov chains that algorithmically generate the consumer-profiled best-sellers of tomorrow. Yet the reality is that enterprises like Grahame-Smith's are notable precisely because of their comparative scarcity—the book and its handful of derivatives stand out because readers are stimulated rather than jaded by the self-conscious novelty of the prospect of reading a literary remix. There are no indications that the publishing industry as a whole is on the verge of being overtaken by a zombie apocalypse, hordes of undead classics shuffling through the word processors of mindless mechanized hacks to be reanimated as the popular fare of today. Instead, such efforts succeed best when there is strong thematic resonance between the original text and its rewrite. And the kind of attention that this kind of overwriting demands—its sense and sensibility, we might say—invariably yields a unique intimacy with the particulars of the original that is very different from the churnings of the wordmill.[32] "The people in Austen's books are kind of like zombies," Grahame-Smith notes. "No matter what's going on around them in the world, they live in this bubble of privilege. The same thing is true of the people in this book, although it's much more absurd."[33]

In Haruki Murakami's *1Q84* (2011), by contrast, overwriting is overtly fictionalized within the confines of the story for purposes of enabling some of the novel's key plot elements. Widely considered Murakami's most ambitious work to date, *1Q84* is centered on two protagonists living in Tokyo in the year 1984—a female martial artist and sports massage therapist named

Aomame, and an aspiring male writer and university mathematics instructor named Tengo. Early on, Tengo receives an extramural commission to rewrite a short novel from an unknown juvenile author so that the end product may be submitted for an important literary prize. Because the original manuscript was produced on a word processor, Tengo (it is explained) must go out and get one for himself to make the new text look like the old. Despite his initial reservations about the assignment ("Coauthorship is not that unusual," his editor assures him),[34] Tengo warms to the task. His charge is to rewrite the raw, unpolished adolescent prose for style, not for substance. He begins by retyping the first few pages of the text into his own newly acquired word processor; entitled *Air Chrysalis,* this story-within-the-story is a kind of jumbled fable in which the corpse of a blind mountain goat serves as a portal between worlds. Tengo soon finds that his efforts have more than doubled the length of the text: "The original was far more often underwritten than overwritten, so rewriting it for coherence and consistency could not help but increase its volume."[35]

As he works, Tengo switches freely back and forth between the screen of his word processor and printing and revising hard copy with a pencil: "The feel of the words he chose would change depending on whether he was writing them on paper in pencil or typing them on the keyboard. It was imperative to do both."[36] Suffice to say that Tengo's efforts are a success, and the book is greeted with all the promised acclaim. The triumph, however, also gives him occasion to meditate on his own life, which included a childhood encounter with Aomame, his first object of sexual desire; he concludes that his life has largely been a failure, and finds himself wishing to rewrite his own past, even though he knows this is futile. "Time had the power to cancel all changes wrought by human artifice, overwriting all new revisions with further revisions, returning the flow to its original course."[37] Instead he throws himself into a new novel, borrowing details from the setting of *Air Chrysalis,* most notably distinguishing its world from ours with two moons in the sky. Here Murakami offers us one of his most vividly rendered passages: "Like Vladimir Horowitz seated before eighty-eight brand new keys, Tengo curved his ten fingers suspended in space. Then, when he was ready, he began typing characters to fill the word processor's screen."[38]

Aomame, meanwhile, throughout the duration of Tengo's literary renaissance, has been seeing to business of her own, beginning, at the novel's outset, with a descent down a mysterious stairway alongside a freeway where she has stepped out of a taxi stalled in traffic. Now, when she stands outside of her apartment on her balcony at night and looks up, she sees two moons in the sky. What's happening here, of course, is that Tengo and his word processor are somehow writing into being the alternative world Aomame came to inhabit after descending the staircase. His fingers, hooked above the keyboard—like half-moons—are not rewriting his own personal past, they are rewriting a shared present. Tengo himself eventually enters the space of this world, the titular 1Q84, and reunites there with Aomame. Murakami never makes explicit the analogy between overwriting a word processing file and overwriting a world, but one perceptive critic, Andrew Ferguson, does: "By that point he has *overwritten* 1984; it is a previous version of the world-file that is now 1Q84."[39] Just as Murakami plays with the relationship between 1984 and 1Q84—as Ferguson notes, "nine" in Japanese is cognate with our letter "Q"—we, as his readers, are encouraged to test the slender orthographic boundary between words and worlds.

Word processing first came to Japan in the early 1980s (the Fujitsu that Tengo buys was introduced in 1983), and the story may also be taken as Murakami—sensitive, as we have seen, to the differences between words on the screen and words on the page—marking an equally wrenching moment wherein his own literary futures diverged.[40] (Murakami was a comparatively early adopter, beginning with a word processor in 1987 for his novel *Dance Dance Dance* [1988]; but Jay Rubin suggests that Kōbō Abe was likely the first major Japanese writer to switch to a word processor, with his novel *Hakobun sakura maru*—in 1984.)[41] The word processor Tengo carries home from the Shinjuku market does not possess overt supernatural powers like Richard Hagstrom's word processor in Stephen King's short story; instead, Murakami plumbs the relationship between building words and worlds in fiction (Tengo had repeatedly lapsed into carpentry metaphors to describe his rewrite of *Air Chrysalis*). New technology is combined with what is indeed a very old storytelling trope, the ability of the storyteller to spin new worlds into existence with his or her tale. Murakami is thus responding to the way in which the particular powers of all writing technologies seem even more pronounced and amplified with a word processor: text blinks on and off, winking in and out of existence with comparative ease. Just as Umberto Eco toyed with the notion of converting *Gone with the Wind* to *War and Peace* with a few keystrokes, and Seth Grahame-Smith used his word processor to turn Jane Austen's England into a zombie wasteland, Murakami is transfixed by the way in which the binary registers of the computer's memory can flicker into alignment and bring whole new worlds into being—perhaps we might call it *world* processing.

Though there were noteworthy exceptions—Mona Simpson, Douglas Adams—the Apple Macintosh was initially rejected by many creative authors in the immediate aftermath of its debut in 1984. Not only did it look very different from other personal computers then on the market, even while sitting idly on a tabletop (the diminutive "mouse" tethered alongside), its bitmapped graphical display—the realization of concepts pioneered at Xerox PARC a decade earlier—would have been jarring to a constituency just coming to terms with their monochromatic identity as green-screeners. Jerry Pournelle set the tone in his *Byte* magazine column: "The Macintosh is a wonderful toy; but it's not very much more."[42] Most damningly, the MacWrite software that shipped with the original 128K Mac—for a full year the only word processing software generally available for the system—had severe limitations: it would crash if a document exceeded about eight pages, which was fine for light office work but obviously a nonstarter for serious long-form writing.[43]

And yet the Mac was anything but another office computer. On the contrary, its design, deeply informed by Steve Jobs's aesthetic convictions, was intended to wrest computers from their associations with the uninspiring office setting for once and for all.[44] Besides MacWrite, it also shipped with a graphics program called MacPaint. Here Macintosh came into its own: Computers, which initially were imagined as instruments for performing numeric calculations and then began storing and manipulating text, now could do the same with pictures. It was no coincidence that the first digitally produced comic book, Peter Gillis and Mike Saenz's *Shatter*, which appeared later that same year, was designed on a 128K Mac and printed with Apple's dot-matrix Imagewriter (that name too is telling).[45] In language reminiscent of first reactions to word processing, Gillis commented, "By writing and drawing electronically we could control and manipulate every aspect of what we had done, easily. Flop a panel? Darken a figure? Slide a figure over to one side? Rescript a word balloon? No photostatting, erasing, or performing surgery with an X-Acto knife would be necessary—we controlled the image with a touch of the mouse."[46]

Because of the Mac's bitmapped screen, letterforms—fonts—could be created and customized as easily as images. This was no mere happenstance.

Jobs had been in the habit of dropping in on calligraphy classes during his abortive undergraduate career at Reed College, and from that point on he was fascinated by the expressive dimension of writing and typography.[47] The original Mac came with a dozen unique proportional fonts, most of them created by Susan Kane, who had also designed its memorable palette of desktop icons. Even as writers were puzzling over what use there could be for a computer that couldn't manage to save more than a short chapter's worth of text in a single file, graphic designers were embracing what the rest of the system had to offer. The publishers of *Émigré,* a leading journal for experimental type design, acquired a Mac within weeks of its debut. Using an early third-party program called FontEditor, they quickly grasped the potential: "Although it has always been possible to design and draw a typeface, the actual typesetting equipment one needs to set type freely is cumbersome and highly specialized. . . . The Macintosh allowed you to store the data that defined the typeface and then access it through the keyboard. Now, for the first time, it was possible for any individual to design and draw a typeface and then actually use it without restrictions."[48] Even without creating their own custom designs, Mac users were suddenly able to transform the visual appearance of texts at will, merely by pulling down a menu of font selections and clicking the mouse. "The textual surface is now a malleable and self-conscious one," exulted Richard Lanham, an English professor at UCLA. "All kinds of production decisions are now authorial ones."[49] This awesome capability, as Lanham notes, would shortly be termed "desktop publishing."

Some prescient authors and editors had already realized the potential use of computers for publishing and distribution as well as composition. The avant-garde literary and arts journal *Between C&D,* which began publication in New York's East Village in 1983, was printed and distributed on fanfold paper from a dot matrix printer and came packaged in a zip-lock bag. (Authors published by *Between C&D* include Kathy Acker, Dennis Cooper, Susan Daitch, Gary Indiana, Patrick McGrath, and Lynne Tillman.) Editors Joel Rose and Catherine Texier recalled: "The combination of our high-tech look—the computer printout, the fanfold, the dot-matrix print type—in conjunction with handmade art by East Village (or Downtown) artists on the front and back covers, and the ziplock plastic bag binding, along with, needless to say, the featured 'new writing' immediately attracted both readers and writers, from New York City and elsewhere."[50] Such an episode comports with the kinds of arguments scholars such as Harold Love and

Peter Stallybrass have long made about the persistence of scribal and manuscript writing into cultures of printing, whereby the cross-transfer between media (in this case print and the digital) results in the proliferation, rather than the attenuation, of prior forms of production.[51]

But as contemporary commentators like Lanham had understood, with the advent of the Macintosh readers and writers had access to the same digital platform as publishers to create images and cast type as well as to process words—and freely integrate all of these components in their finished work. Journalist and nonfiction writer Gay Talese, who bought the first of his many Macs in 1988, quickly became captivated by this potential: "fiddling with the many Macintosh fonts as I composed, in varying type sizes and shapes, my personal correspondence, fax messages, shopping lists, folder labels, instructional notes to deliverymen, and the outlines for scenes and situations that might appear in a future chapter of my book."[52] But whereas many users made light of such experiments, marking them with a certain degree of self-consciousness or self-deprecation—Talese compared its font libraries to the bounty of flavors available at Baskin-Robbins[53]—none took the Macintosh more seriously than the Barbados-born poet and historian (Edward) Kamau Brathwaite. For Brathwaite, the Macintosh quickly became more than just a writing tool, however powerful; it became an integral part of his poetics, personal identity, and worldview.

Kamau Brathwaite's writing and publishing career has been long, rich, and nuanced, and is increasingly studied. His introduction to the Macintosh coincides with an especially difficult period in his life that he refers to as the Time of Salt, encompassing the death of his first wife, Doris Monica Brathwaite, from cancer in 1986, the severe damage done to his house in Jamaica by Hurricane Gilbert in 1988, and his own close call with death—he believes that he in fact died—in a violent home invasion in 1990. In an interview published in 2001, Brathwaite provides this account of how he discovered computers in the midst of these brutal and traumatic events:

> My writing hand becomes a dumb stump in my head. . . . I mean I can't write or utter a sound or metaphor. But Sycorax comes to me in a dream and she dreams me a Macintosh computer with its winking *io* hiding in its margins which, as you know, are not really margins, but electronic accesses to Random Memory and the Cosmos and the *Iwa*.

> And she dreams me these stories . . . and shows me how to find *jo* to write them out on the computer. And the two together introduce me to fonts and the fonts take me across Mexico to Siqueiros and the Aztec murals and all the way back to ancient Nilotic Egypt to hieroglyphics—allowing me to write in light and to make sound visible as if I'm in a video.[54]

"Writing with light" is a phrase he employed over and over again to describe the experience of word processing. His own Macintosh computer, which he named Sycorax (for the sorceress who is the mother of Caliban in Shakespeare's *The Tempest*), functions as his surrogate archive and memory. He would move from working with the default font set on the Macintosh to designing his own, creating what is typically referred to by reviewers and critics as his "signature" or "trademark" Sycorax Video Style, beginning with the self-published *SHAR / Hurricane Poem* (1990) and then *Middle Passages* (1992), *The Zea Mexican Diary* (1993), *Barabajan Poems* (1994), and *Dream Stories* (1994), and continuing through *ConVERSations with Nathaniel Mackey* (1999), *Ancestors* (2001), *Born to Slow Horses* (2005), *DS (2)* (2007), and *Elegguas* (2010). These titles have appeared variously under Brathwaite's own Savacou imprint and from small presses, university presses, Longman, and New Directions, the range of venues reflecting the practical difficulties of publishing Brathwaite's work in a manner faithful to his vision.[55]

The Sycorax Video Style is intimately related to other aspects of Brathwaite's writing, notably the Caliban story and his own frequently adopted identity and voicings as Caliban in the Caribbean English "nation language" of his poetry.[56] With Sycorax, the Mac's font libraries and layout capabilities are used to visually orchestrate and arrange the language of Brathwaite's poems on the page. All of the works just mentioned are distinguished by a more or less common visual appearance, which involves a medley of modern-looking fonts in different sizes, as well as occasional abstract symbols, glyphs, and clip art. (The casual mention in his interview comments of making sound visible "as if I'm in a video" also suggests a possible analogy to the music videos popularized by MTV in the eighties.) Often the letterforms are so enlarged that their pixelated edges—what typographers would call "jaggies"—are apparent, further emphasizing their digital pedigree.

Occasionally some uninformed and hapless reviewer will glance at Brathwaite's books and assume he must be as frivolous as Lanham once imagined newbie Mac owners to be, giddily playing with the technology. Such reactions grossly underestimate his technical and literary achievement, as well as the depth of thought he brings to computers and their place in his poetry. "The computer," Brathwaite says, "has made it much easier for the illiterate, the Caliban, actually to get himself visible. . . . Because the computer does it all for you."[57] In such statements he is clearly responding to the ease of composition that so many other writers have also remarked upon, an intuition he encapsulates in the oft-invoked phrase "writing in light." But whereas writing with light functioned for other authors as merely a paean to the power and seductions of word processing, for Brathwaite it signaled something more potent or profound: an extension of the oral tradition. "The typewriter is an extension of the pen," he says. "The computer is getting as close as you can to the spoken word."[58] In this he is clearly in sympathy with Walter Ong, whose comments on secondary orality—the immediacy of oral interaction and performance we associate with the digital—we considered in Chapter 4.

Different academic critics have sought to interpret the Sycorax Video Style in different ways. For Stewart Brown, Sycorax is integral to Brathwaite's efforts to present his work on terms of his own choosing, not defined by an implicit comparison to Western poetries; because of the bracing nature of the typographical effects, "we do not unconsciously read this text *against* our expectations of the ways poems conventionally work."[59] In an extremely perceptive essay Carrie Noland notes that Brathwaite's earliest comments on record about computers were critical, aligning them with other Western technologies inimical to the project of nation language.[60] She also observes that many of the typographic gestures that he employs resist vocalization: "How does one perform a period that is one space away from the last word in the sentence . . . ? How does one vocalize 'cd' (originally 'could'). . . . How do you pronounce a '>'? Or 'w / open?' "[61] Still another Brathwaite scholar, Ignacio Infante, recasts the Sycorax Video Style as a "virtual voicing" distinguished by two characteristic features of digital media—delocalization and dematerialization—both of which, Infante argues, Brathwaite cultivates in the service of a postcolonial Caribbean poetry.[62]

Brathwaite's dream narrative is one origin story about his discovery of computers.[63] Another has him borrowing a Macintosh while a visiting fellow at Harvard, an episode to which we will return. But despite his portrayal of himself as a novice, Brathwaite was no stranger to computing and word processing even before his first encounter with a Mac. There were computers in the house in Jamaica, including a Kaypro that belonged to Doris Brathwaite, aka Zea Mexican; she used it to keep accounts and the archives of his literary career.[64] And Brathwaite owned a computer of his own called (portentously) an Eagle. This now-forgotten brand designed to compete with the IBM PC ran first CP/M and then DOS as operating systems; it came bundled with a word processor that was called (even more portentously) Spellbinder. This is the system Brathwaite would have been working on when he wrote "X / Self's Xth Letters from the Thirteen Provinces," the poem first published in *X / Self* (1987) wherein he begins using the phrase writin(g) in light:

> is not one a dem pensive tings like ibm nor bang & ovid
> nor anyting glori. ous like dat!
> but is one a de bess tings since cicero o
> kay?[65]

The Eagle was indeed priced inexpensively as compared to an IBM or Wang, and the product—all hard edges and angles, nothing at all like the Macintosh—would have seemed neither glamorous nor glorious. But like many other authors, Brathwaite felt compelled to meditate on the difference between the computer and his previous writing instruments. Because "X / Self's Xth Letters" takes the form of a letter to his mother, he couches the experience in terms she would apprehend: there is no liquid paper to flake the page like house paint when correcting mistakes, and there are no sheets to squeeze through the rollers like dough. He is also acutely aware of the implications of working with such an overtly Westernized piece of technology, associating it with the *Star Wars* robot R2D2, Jackie Robinson, "mercantilists," and above all Prospero, the embodied figure of colonial power:

> but is like what I tryin to sen / seh &
> seh about muse /
>
> in computer &
> learnin prospero linguage &
>
> ting
>
> not fe dem / not fe dem
> de way Caliban
> done
>
> but fe we
> fe a-we

"X / Self's Xth Letters" was not initially printed in the Sycorax Video Style as such—there are no font or layout experiments beyond what could have been achieved on a manual typewriter like "dat ole remington yu have pun top de war." Even so, Brathwaite glimpsed new possibilities.[66] The poem is unabashedly joyous and playful, the words tumbling out in quick, sharp sentences, punning and bouncing off of one another, repeating, splitting, sparking, sometimes running down the length of the page like a screen scrolling by, until he re-centers himself in the moment, the actual act of what he is doing, "chip / in dis poem onta dis tablet / chiss / ellin darkness writin in light."

By 1988 Brathwaite was a visiting fellow at Harvard University. His wife, Doris, had died two years earlier. Soon, as Hurricane Gilbert approached Jamaica, he would watch helplessly from Cambridge, trying in vain to get someone on the phone to safeguard his house and its irreplaceable archives of Caribbean literature.[67] The Eagle computer was also there, left behind "like an emotional anchor or icon."[68] This latter circumstance may explain why, on a hot August day a month before the storm would form, he arranged to borrow a Macintosh from a student services organization at Harvard. (Our knowledge of this episode comes from the preface Brathwaite wrote to "Dream Chad," one of the texts in the 1994 *Dream Stories* volume.) We cannot know exactly what type of Mac Brathwaite had gotten hold of, but it was likely a Macintosh SE (System Expansion), released in 1987 and a marked improvement over earlier models—it included a 20 MB hard drive—while still being easily portable with the built-in carrying handle. Apple had long solved the problem with MacWrite's infamous lack of ability to store pages. Nonetheless, Brathwaite (in the preface) recounts typing the "Dream Chad" story over and over again, only to have the computer "malfunction" each and every time just as he was on the verge of finishing and saving it.[69] If that were not enough, he also receives a mysterious screen message that "SOMEBODY ELSE WAS USING IT."[70] The most likely explanation is that his computer had somehow connected to another Mac through the system's built-in AppleTalk network. Still, Brathwaite is rattled; he suggests the possibility that the Eagle computer might be jealous that he was "into a Ma(c)"—a notable introspection, given that the poem narrates his encounter with an unnamed female figure resembling the woman who would become his second wife, Beverly, the eponymous "Chad."

Despite these upsets, the experience is formative, laying the groundwork for "what has become a very close relationship w Apple / Mac—in fact in the end I purchased my own Sycorax on which most of the video style stories of this collection have been written."[71] The computer that he was using by then was a Macintosh SE/30 (this model had debuted in 1989) with an accompanying StyleWriter inkjet printer.[72] When he revised "Dream Chad" in the Sycorax Video Style in 1993, he subtitled it "a story."

Today Brathwaite continues to write and publish using the Sycorax Video Style, with many of the fonts that emerged from his experiments in the 1990s still visible and identifiable in his most recent work. There is no question that he is one of the earliest and most significant literary authors to take advantage of the dramatic new creative possibilities that computing, specifically the Macintosh, had to offer. Brathwaite had realized, just as did *Émigré* designer Zuzana Licko, that the technical limitations of the early Macintosh were inseparable from the aesthetic qualities of the new bitmapped fonts they were creating. You couldn't just import existing typefaces from print: "Each medium has its peculiar qualities and thus requires unique designs," Licko says.[73] Yet unlike many writers who self-consciously experimented with computers as avant-garde instruments, Brathwaite seems to have had relatively little interest in what technology commentators often call the cult of the new, the obsession with the next big thing that sets the terms for so much of our day-to-day encounter with the digital. To look at Brathwaite's work now is to encounter a visual aesthetic that appears conspicuously "retro," as though fonts continue to be designed on 16-bit systems and printed from 360 dpi printers without the benefit of anti-aliasing. The effect is especially pronounced when reading his work on contemporary subjects, such as "Hawk," which he wrote in the aftermath of September 11 (Brathwaite had been in New York City, teaching at NYU, at the time):

> but some day certain in the future of New
> York. his magic enigmatic majesty now flower-
> ing the room . his body glow-
> ing *the only word we have* for what is now this glow-
> erring around these future towers of his solo masterpiece
> rising himself again in sound towards the silver cross
> of an approaching jet. dissecting in the blue
> the full white mosque and omen of the moon
> just afternoons ago . high over Berkeley Square. Over Washington Square[74]

To read these lines in the heavily pixelated type in which they appear in *Born to Slow Horses* (2005) is to be confronted with the visible imprint of time and history in relation to technological progress. Computers notoriously operate on fast time, governed by the vicious cycles of product rollouts and seemingly immutable laws about chip speeds doubling their processing speed every eighteen months. It is a timeline of terrible accelerations, radical and terrorizing in their own way, as well as a time of sudden crashes—a word that is everywhere in computer talk. One way to read Brathwaite's more recent Sycorax work is thus as an anachronism. But another way to read Brathwaite is as a poet who has learned the language of Prospero only to speak it in the deliberate manner of Caliban—not just in the phrasings and voicings of the islands, but according to alternative timelines and temporalities, a slow time outside of Western techno-history, admitting neither consumer novelty nor planned obsolescence, but answering instead to the work—the words—of the moment.

Although most authors are content to leave issues of layout and design to their publishers, notable literary work at the boundaries between word processing and desktop publishing has continued to flourish.[75] Mark Z. Danielewski drafted the original manuscript for his breakthrough novel *House of Leaves* (2000) in longhand, then revised it with a word processor. But after he signed a book contract with Pantheon to publish it, he flew (at his own expense) to New York City, set up shop in his publisher's offices, and taught himself QuarkXPress in order to do the typesetting—he didn't wish to entrust any of the staff designers with his vision for the typographic effects so central to the book.[76] Dave Eggers is widely reported to do all his writing in Quark, a carryover from his numerous publishing ventures.[77] Steve Tomasula gives a designer co-credit on the cover of each of his novels. The poet Tan Lin (not to be confused with Tao Lin) used Microsoft Word as the design environment in which to assemble *Heath* (2008), an "outsourced" chapbook composed of snippets of found Web content loosely related to Heath Ledger's death six months earlier. David Daniels, a painter and (before his death in 2008) a fixture in the Berkeley outsider art community, spent over a decade (1988–2000) using Microsoft Word to create his self-published tome of pattern poems entitled *The Gates of Paradise*.[78] Some 400 poems form a variety of shapes—animals like seals, elephants, and birds, household objects, people, and more abstract icons and symbols. His work, which has obvious antecedents in a tradition going back at least as far as the early seventeenth-century British poet George Herbert, is notable in the context of the literary history of word processing because of the extent to which it dramatizes the increasingly permeable boundaries between words and images that is characteristic of the technology. "I've gotten to the point where I can write out pictures on the computer just about the way you could write longhand, because I've done it for so many years. Some take me a week, and some take me a day. It's like weaving or carving," Daniels told a journalist.[79]

But other authors and artists have found some very different ways to reveal the formal and aesthetic codes of word processing. In a volume entitled *Speak! Eyes—En zie!* (2010), Elisabeth Tonnard prints the texts of classic works of literature—Shakespeare's Sonnets, Eliot's *Prufrock and Other Observations* and *The Waste Land,* Dickens's *Bleak House,* works by Poe, and more—as rendered by Word's AutoSummary feature. While the resulting texts are not edited by Tonnard, she reports experimenting with the parameters of the AutoSummary feature itself, which allows users to control the semantic granularity and length of each summary.[80] In this instance, the hand of authorship is visible not in the text so much as in the manipulation of the algorithmic variables used to generate it. "Ham. / Ham. / Ham. / . . ." reads the AutoSummary for *Hamlet,* the stage direction repeated twenty times down the length of the page. The results for *Prufrock and Other Observations* are perhaps more compelling:

> The street lamp sputtered,
> The street lamp muttered,
> The street lamp said,
> The street lamp said,
> Memory!

The project was born of Tonnard's interest in whitespace and eliminating words from texts—a previous book called *Let Us Go Then, You and I* (2003) uses whiteout (tip-ex) to paint over the text from the opening section of "Prufrock"—"a book written in white ink," as she describes it.[81] (The genre here is known as the "erasure," which other contemporary poets, including Mary Jo Bang and Mary Ruefle, also work in.)[82]

Tonnard draws her share of inspiration from Kenneth Goldsmith, the American conceptualist who claims he is a "word processor" and not a writer, poet, or artist. The raw form of Goldsmith's works—verbatim transcriptions of newspapers, traffic reports, news broadcasts, his own speech—overtly reflects the effortless contemporary duplication and proliferation of texts without regard for the volume and mass of words.[83] Language, suggests Goldsmith, has become "completely fluid" in its contemporary digital settings: "It's lifted off the page and therefore able to be poured into so many different forms and takes so many different shapes and really be molded and sculpted in a way that wasn't possible before."[84] Poet Brian Kim Stefans, meanwhile, likewise uses Word's AutoSummary to present a version of Goldsmith's *Soliloquy* (2001)—a transcription of every word Goldsmith spoke for a week—at 2 percent of its original length, in effect doubling down on the conceptualist premise while illustrating what happens when a data dump of the human voice is "poured" through the sieve of an enterprise-level software feature.[85]

Both Tonnard and Stefans are in effect turning Microsoft Word inside out, using its digitally native processes to make what is usually automated, derivative information—known as metadata—into a primary text for the reader's attention. (Stefans's text is not only shorter but also considerably more readable than Goldsmith's original; notably, AutoSummary was removed from Word beginning with the 2010 edition of Office.)[86] Matthew Fuller has gone one step further in an art piece entitled (after Whitman) *A Song for Occupations* (1999–2000). Exhibited at the Norwich School of Design, it consisted of several thousand individual pieces of paper, each containing a single graphical element of Word—buttons, icons, menu items—arranged in clusters on a wall. Fuller used a suite of specialized software engineering tools to effectively disassemble Word: "This is a particularly suitable target since the application is massively overloaded with 'features,' each of which cater[s] to specific cultures of use, many of which do not overlap except in the core functions of the application (i.e., text entry)," he writes in an artist's statement. "When all the composite elements of an interface are brought into the field of vision, the simple accrual of decontextualised detail and its asymmetry with what we 'know,' tricks other ways of understanding software and its machined invisibility into emergence."[87] A similar impulse is at work in Tomoko Takahashi's *WordPerhect* [*sic*] (1999), an online piece (rendered in Flash) that presents the user with a roughly drawn cartoon word processing interface, limned in what appears to be black ink on a white background.[88] Typing generates crude, seemingly handwritten, less-than-perfect characters on the screen. Takahashi's word processor is fully functional, but the interface yields an inversion of the typical user-friendly experience. Clicking on the Mail icon produces the following set of instructions, which appear as a scrap of notepaper "taped" to the screen: "print the document, put into an envelope or ssomething similair [*sic*] which can contain the document. Go to post office and weigh it and buy stamps"—and so on, for another hundred words, including further typos and blemishes.

And then there is Joel Swanson's *Spacebar,* a 24-by-32-inch print that was exhibited at the University of Colorado Boulder's Media Archaeology Lab in 2013. *Spacebar* depicts a plain white unadorned Apple spacebar (set against a white background). "In some ways the spacebar is the only accurate key," notes one commentator on the piece. "Rarely do the typefaces of the letters appearing on your keyboard match the typefaces of the letters appearing on your screen, except in the case of the spacebar. You push the space bar and a blank appears between words, an icon of breath, of vacancy."[89]

By extracting this single element of the computer keyboard—fundamental to language as well as to technology, as scholars such as Paul Saenger have shown[90]—the artifice of texts and their technologies is underscored in a visually striking composition possessed of its own strong aesthetic. The gesture becomes emblematic of many of the works we have discussed in this chapter, each of them in their own way striving to reveal the perennial artifice of the literary, now deeply imprinted by a postmodern poetics of word processing—copying and pasting, finding and replacing, deleting and overwriting.[91]

TEN

WHAT REMAINS

It is said that when Rob Reiner's adaptation of Stephen King's novella *Stand By Me* (1986) was in theaters, some audiences howled in visceral anguish when, at the very end of the film, the adult Gordie, now a writer, switches off the computer he is using to type out the story without any visible evidence of having hit Save. Certainly the specter of losing it all—a bug, a bad sector, a power failure, even just pressing the wrong key—loomed large for any user. Jimmy Carter's experience with lost pages from hitting a wrong key on his Lanier was by no means atypical. So anxious was Len Deighton about blackouts and brownouts in Ireland where he was living at the time that he had Olivetti technicians install a nickel-cadmium battery in his then-current word processor—effectively creating what we would today call an uninterrupted power supply.[1] Frank Herbert and Max Barnard were careful to specify plans for a tape backup system that would take down every keystroke (much like the MT/ST) for their never-built dream machine.[2] Jack Vance was frustrated by frequent crashes with the custom BigEd software that enabled him to write as his vision progressively worsened.[3] Piers Anthony's brand-new system was almost fried by a lightning strike within days of its debut.[4] While working on her Burroughs Redactor III at her Florida home, Robyn Carr was anxious that workmen digging nearby would hit a power line.[5] And in 1981 *Time* magazine, channeling anxieties about computer bugs, told its readers that an environmentalist writer, Michael Parfit, had "recently heard a zap, and his Radio Shack TRS-80 stopped dead. It seemed that ants had crawled into the air vents."[6]

Authors had suddenly taken on a whole new set of woes and anxieties, coexisting with all the old ones. “A child’s illness, relatives coming to stay, a pile-up of unavoidable household jobs, can swallow a work-in-progress as surely as a power failure used to destroy a piece of work in the computer,” Alice Munro wrote in an introduction to one of her short-story collections, also (not incidentally) aligning the threat of high-tech catastrophe with the gendered responsibilities of the domestic sphere.[7] Among her own ten rules of writing, Margaret Atwood commends neophytes to always remember to back up their text with a memory stick.[8] On the other hand, new forms of procrastination could be welcomed: according to one interview, Gabriel García Márquez rejoiced when not-infrequent power failures made turning on his computer and facing the task of writing an impossibility.[9]

“S is for Save: do it often,” wrote Rita Aero and Barbara Elman in their widely reprinted “ABC’s of Word Processing.”[10] In an era with no lithium-ion batteries, no AutoSave, and no hard drives on most consumer products, authors, like all users, quickly internalized the urgency of backing up. Some habitually printed out their work at the end of each day. Anne Rice in 1985: “I print out the work of the day when I finish every day no matter what the hour. I have a superfast printer that can cough up a whole novel in no time.”[11] Rice stored the accumulated drafts in waterproof zip-lock bags stacked next to the computer. When Peter Norton’s Disk Doctor was released in 1988, giving ordinary users the means to recover corrupted or deleted files, it quickly became a best-seller.

But these mechanical failings—and the rituals one takes to ward them away, the brandishing of totems like memory sticks and zip-lock baggies, the acquisition of specialized recovery software—are only one manifestation of a more fundamental question about the nature of literary culture and literary history in an age of word processing. Novelist T. Coraghessen Boyle wrote for the *New Yorker* about what it meant for him to box and ship his literary papers to the Harry Ransom Center at the University of Texas:

> Since I began writing stories and novels in the early nineteen-seventies, I have kept every scrawled-over draft, every letter from friends and readers and my fellow writers, every rejection slip and review, as well as all the correspondence with my editors, agents and publishers around the world—and now I no longer have them. No matter that I never looked at any of them, ever, and that they remained thrust deep in their manila folders in the depths of filing cabinets and yellowing boxes imprinted with the logos of appliances

> long since defunct—they were there and they gave me weight. Of course, there is the consideration that all this paper I've been dragging around down through the years could have met a very different fate, masticated by rats and bugs, rotted, engulfed by mold, earthquakes, mudslides, or the flames of the wildfires that annually threaten this very old wood-frame house on the California coast where I am now sitting at the keyboard.[12]

What someone will one day do with all that paper—destined to be cataloged and housed in the Ransom Center's bunker-like building on the Austin campus—we cannot fully anticipate. Paper, after all, is a rich medium, a ready conduit for all manner of information, not all of it intentionally transmitted. Consider that in early 2011, American literature scholars were abuzz with news of the discovery of thousands of previously unknown papers, written in Walt Whitman's own hand, at the National Archives of the United States. Ken Price, a distinguished literature professor at the University of Nebraska, had followed a hunch and gone to the Archives II campus in College Park looking for federal government documents—red tape, essentially—that might have been produced by the Good Gray Poet during Whitman's tenure in Washington, DC, as one of those much-maligned federal bureaucrats during the tumultuous years 1863–1873.[13]

Price's instincts were correct: he has located and identified some 3,000 official documents originating from Whitman, and believes there may still be more to come. Yet the documents have value not just, or perhaps not even primarily, prima facie because of what Whitman actually wrote. Collectively, they also constitute what Friedrich Kittler once called a discourse network, a way of aligning Whitman's activities in other domains through correlation with the carefully dated entries in the official records. "We can now pinpoint to the exact day when he was thinking about certain issues," Price is quoted as saying in the *Chronicle of Higher Education*.[14]

Put another way, the official government documents written in Whitman's hand are important to us not just as data, but as metadata. They are a reminder that the human life-world all around us bears the ineluctable marks and traces of our passing, mundane more often than not, and that these marks and tracings often take the form of written and inscribed documents, some of which survive, some of which do not, some of which are displayed in helium- and water-vapor filled cases, and some of which lie forgotten in archives, or swept under beds, or glued beneath the end papers of books. Science fiction writer Bruce Sterling, not exactly a Luddite, recently turned over fifteen boxes of personal papers and cyberpunk memorabilia to the Ransom Center, but flatly refused to consider giving over any of his electronic records. "I've never believed in the stability of electronic archives, so I really haven't committed to that stuff," he's quoted as saying.[15] We don't really know, then, to what extent discoveries on the order of Ken Price's will remain possible with the large-scale migration to digital documents and records.

Sterling's brand of techno-fatalism is widespread, and it is not difficult to find jeremiads warning of the coming of a digital dark age, with vast swaths of the human record obliterated by obsolescent storage.[16] So when a writer sits down "at the keyboard," as they so often do now—Boyle's unselfconscious phrasing in the midst of a meditation on safeguarding his own personal legacy is telling—what kind of "papers" remain? Hard-copy printouts, certainly, but what about all of the other things Boyle mentions, the correspondence and other detritus of a literary life now often routinely conducted with the aid of email and social media?[17] And what about manuscript revisions made quickly, silently (Track Changes turned off), invisibly right there on the screen, flickers of thought executed by muscle memory in the fingertips, barely even consciously registered if the writer is deep in the flow? What about the diskettes and the memory sticks? What about the computers themselves, surprisingly intimate objects that are a writer's companion for much of his or her day? Are they part of what gets boxed up and sent to the archives? What gives us weight now? What remains after word processing?

I almost reached my manuscript, typescript, printouts, and disks in time," Maxine Hong Kingston tells us at the start of her 2003 memoir, *The Fifth Book of Peace*.[18] The first three Books of Peace are part of Chinese historical legend, appearing and disappearing across the centuries through migratory tides, invasions, and revolutions; the fourth, we learn, was the working title for what was then to be Kingston's next book, a novel that would conclude with a "happy" ending to the Vietnam War.[19] It was destroyed when just such a wildfire as fellow Californian T. C. Boyle feared swept through her neighborhood in the Oakland hills in October 1991, leaving her house not much more than an empty shell. She describes racing homeward upon hearing the news reports coming on her car radio, and being turned away by authorities because of the danger. Afterward she returned to confront the loss of the book in the aftermath. She passes through the soot-blackened stone archway where her front door once stood and enters her former writing nook. "The unroofed sun shone extra brightly on a book-shaped pile of white ash in the middle of the alcove."[20] She had been 156 pages in, 156 "good, rewritten pages."[21] What remained: "I laid my hands on the silvery vanes of feathers, like white eyelashes. Each vane fanned out into infinitely tinier vanes. . . . I placed my palm on this ghost of my book and my hand sank through it."[22]

In Kingston's writing, this moment of deep loss becomes deeply poetic, building as it does a careful stock of imagery around colors like white and silver and objects like lashes, vanes, feathers, fans, and ash. Yet she also moves quickly on to more practical considerations, noting that her book existed in more than one "form": we learn that she had more than one computer, one upstairs and one downstairs (at least one was an Epson QX10), and different pieces of the book had been "in" each of the different machines.[23] But both had burned—melted down, really—beyond any conceivable hope of data recovery. Kingston found she could not subsequently re-create the book, indeed could not "re-enter fiction," as she put it.[24] A former student offered to hypnotize her, explaining how it would work: "You picture the computer screen scrolling up your book. You read it off the terminal into a tape recorder, and we transcribe it. Voilà."[25] The plan scared her: "What if I got stuck in that word-for-word version?"[26] It's an intriguing question, not further glossed: Does it mean, what if she got fixated on the original version of the text, unable to alter a word that (as the hypnosis might demonstrate) was in fact lodged deep within her subconscious? Or does it mean, what if she got stuck inside of something like an endless computer loop, inside those lines of text scrolling up and down the screen, unable to break free of her hypnogogic state?

It would take Kingston over a decade to write and publish *The Fifth Book of Peace,* a memoir and meditation (and mythos) on the relationship between war and peace, creativity and destruction. The book is divided into four sections: Fire, Water, and Earth, together with Paper. Missing is air.[27] Today we routinely store documents in the "cloud," ostensibly mitigating the potential for disasters such as Kingston suffered, where the flames consumed paper and silicon alike so indiscriminately as to make the medium irrelevant. That kind of positivism, however—faith in the sanctity of the cloud—would be precisely the wrong lesson to draw from this book about (among other things) forces too elemental to resist. For Kingston, the only guarantor of preservation, and of remembering, is time—not because it preserves things inviolate, but because it allows the necessary processes to take hold. "After a loss there will always be mourning," she told an interviewer. "We want it that way because we don't want to forget our feelings for that person or that thing. However, the mourning changes; mourning breaks up into different elements."[28]

If the wildfires are kept at bay, often what remains is paper. Computers, as we know, produce a lot of paper, even today. John Hersey's DECtapes are long gone, but multiple hard copies of *My Petition for More Space*—each indicating the exact date on which they were printed—are safely ensconced in the Beinecke Rare Book and Manuscript Library at Yale. In 1983 Updike wrote "INVALID.KEYSTROKE" directly on his word processor, but we know of its existence only from its publication in a book and from the hard-copy drafts in the collection at the Houghton Library at Harvard. Harold Brodkey's literary papers are likewise at the Houghton; when an article about literary manuscripts in the *New York Times* dismissively referred to the manuscript for *The Runaway Soul* as a mere "computer printout," it drew a sharp response from the curator in charge of the collection: "While certainly Mr. Brodkey at various stages composed on his computer and printed out the text, these versions are densely overwritten, again and again, by hand. The result is a highly complex document in which one can observe the author at work firsthand."[29] As with the far more modest typescript for Updike's poem, we find that the computer (and what happens on its screen) coexists with the work of the hand, the mechanical technology of the printer—yet another instance in which the vocation of a person has become the name of a machine—mediating between the two.

Len Deighton finished the first draft of *Bomber* by October 1969. Ellenor Handley hand-delivered the hefty manuscript package from his home in Merrick Square to the offices of his publisher, Jonathan Cape. By the year's end, he had produced a second, close-to-final, draft. Footnotes to the story—a device he regularly employed in his spy novels—had now been incorporated into the main text, and chapter 1 in particular—a tricky piece of business because it had the burden of introducing the RAF flight crew—had been heavily revised. The MT/ST, with its dual tape reels, would have been a great asset in both tasks. What became of either the tapes or the manuscripts thereafter was thought to be unknown.

In March 2013 I wrote a version of Deighton and Handley's story for *Slate* magazine.[30] Shortly thereafter I received email from an individual in Britain who informed me he was in possession of the second manuscript, what Edward Milward-Oliver refers to (in correspondence) as the "master manuscript"—the last full draft of the book that was printed from the MT/ST's tapes before it was typeset and put into proof the following spring.[31] We were subsequently able to arrange a loan to Milward-Oliver, who studied it in preparation for his forthcoming biography of Deighton. Dated December 1969, this draft is bound in heavy and well-worn magenta-colored card covers. There are 558 foolscap pages, deckled on the top and bottom edges where the sheets were separated. There are occasional annotations and minor revisions marked in red ink in Deighton's hand. Along the right-hand edge of many pages, between two and five bite-like impressions are clearly visible from where the sheets had been gripped by the teeth—or the unseen hands—of the MT/ST's tractor feed. The mechanism had thus left its mark.

There is as yet no blue plaque in Merrick Square commemorating the literary events that (literally) unfolded there as the various manuscript drafts of *Bomber* passed through the MT/ST's tractor feed, under its chattering golfball printing head, and into Ellenor Handley's waiting hands.[32] Deighton's former house is thus not tied into this curbside history of literary London. (By the official criteria for blue plaques, the personage so honored must have been dead for at least twenty years.) But fortunately not all forms of remembering are so staid. Later in 2013 Edward Milward-Oliver was able to offer a showing of the manuscript to Handley. They sat with it on a sunny late summer day at his home near Chichester, her hands turning the same machine-marked pages. The collector, who wishes to remain anonymous, retains the weighty typescript at his home in Britain—he keeps it in a wardrobe, or so I am told.[33]

Lucille Clifton's cherished Magnavox VideoWRITER 250 reposes in Emory University's Manuscript, Archives, and Rare Books Library. It is sometimes brought out for display. Before coming to this well-deserved rest, however, the machine was given one final, essential task perform. While the archivists at Emory were able to copy the data files from the original 3½-inch diskettes Clifton had used (one disk had reportedly come to them still inserted into the drive), they found that they were unable to actually open the files and read their contents without the assistance of the system's own proprietary software. This is a common enough problem for digital preservationists; at Emory they solved it by using the VideoWRITER to access the diskettes one last time, printing out hard copies of the scores of documents they contained, including drafts of the poems in many of her most important books.[34]

I am negotiating to buy Stephen King's Wang," deadpans a character named Ngemi, a collector and connoisseur of antiquarian computers in William Gibson's novel *Pattern Recognition* (2003). "The provenance is immaculate."[35] Alas, the line is a throwaway: a few years ago I spoke with King's longtime personal assistant, Marsha DeFilippo, and inquired about the Wang's fate—apparently it was sent to California (she thinks) for repair and never came back.[36] Somewhere along the way this venerable talisman met with misadventure, or maybe, simply, finally, just gave up the ghost. But what does it mean to possess a writer's computer? Nowadays Jonathan Franzen is said to write with a "heavy, obsolete Dell laptop from which he has scoured any trace of hearts and solitaire, down to the level of the operating system."[37] Someday an archivist may have to contend with this rough beast: is it an object intimately associated with his writing practice (like a typewriter) or a consumer appliance, like a food processor?

"Some computers will become museum pieces," Jacques Derrida once predicted. "Some particular draft that was prepared or printed on some particular software, or some particular disk that stores a stage of a work in progress—these are the kinds of things that will be fetishized in the future."[38] In fact a number of significant writers already have material in major literary archives in digital form. The list of notables includes Lucille Clifton (Emory University), the British poet Wendy Cope (at the British Library), William Dickey (Reed College), Stanley Elkin (Washington University), Jonathan Larson (composer of the musical *RENT;* Library of Congress), Timothy Leary (the New York Public Library), Norman Mailer, Gabriel García Márquez, and Terrence McNally (all at the Ransom Center), Toni Morrison (Princeton), Susan Sontag (UCLA), Natasha Trethewey (Emory), John Updike (the Houghton), Alice Walker (Emory), and David Foster Wallace (the Ransom Center again, though only a handful of diskettes), among others. The best-known example to date is Salman Rushdie. Emory University has four of his Macintosh computers in their collection along with the rest of his "papers."

The earliest of these—a Mac Performa—is currently available as a complete virtual emulation on a dedicated workstation in the rare books reading room. Rushdie's first book written on a computer was *The Moor's Last Sigh* (1995). He had gotten the Mac in early 1992, in the midst of the fatwa that followed the publication of *The Satanic Verses* (1988), when under police protection he had to change residences often. He later commented, "I can't understand why I didn't do it before. Just at the level of writing, this is the best piece of writing I've ever done and I'm sure one of the reasons for this is the removal of the mechanical act of typing. I've been able to revise much more."[39] Patrons at Emory can browse the Performa's desktop file system, seeing, for example, which documents he kept together in the same folder; they can examine born-digital manuscripts for *The Moor's Last Sigh* and other works; they can even take a look at the games on the machine (yes, Rushdie is a gamer—as his *Luka and the Fire of Life* [2010] confirms).[40]

A computer is, as Jay David Bolter observed quite some time ago, a writing space—one that models, by way of numerous telling metaphors, a complete working environment, including desktop, file cabinets, even wallpaper.[41] Gaining access to someone else's computer is therefore like finding a master key to their house, with the freedom to open up the cabinets, cupboards, and desk drawers, to peek at family photo albums, to see what's recently been playing on the stereo or TV, even to sift through what's been left behind in the trash. While it is true that analog writing materials sometimes also embed their own archive—as the example of Updike's typewriter ribbon, which revealed his final messages to his typist, reminds us—there are qualitative differences when we speak of digital media. The principle of storing data in the same medium and format as the programs that make use of it is a bedrock principle of computer architecture, formally instantiated in the so-called Von Neumann model that dominated computer systems design throughout the second half of the twentieth century. (As John von Neumann himself was wont to put it, it was all the same "organ.")[42]

As early as 2001 the State Library of Victoria in Melbourne purchased the Macintosh laptop—reportedly missing its "o" key—that the Australian novelist Peter Carey used to write *The True History of the Kelly Gang* (2000); it is currently on display there under glass, alongside samples from his literary papers.[43] Similar to the work at Emory, archivists there contemplate making a "clone" of the machine available so visitors can explore its electronic innards. The Ransom Center, meanwhile, has computers owned by Norman Mailer (in fact used exclusively by his typist and personal assistant, Judith McNally—the keyboard is covered with nicotine stains), hypertext pioneer Michael Joyce, and now Gabriel García Márquez.[44] Márquez had had a computer at least as early as 1986, when he was working on *The General in His Labyrinth;*[45] after he died, in 2014, the Ransom Center acquired his literary papers—they included two Smith Corona typewriters and five Apple computers.[46] Other computers have become objects of collectors' desire: Stieg Larsson's widow has been widely reported as being in possession of a laptop housing a mostly finished draft of a fourth book that is a sequel to his *Millennium Trilogy*.[47] Whatever the manuscript's fate—if it indeed exists—the fact that the claim can be made at all demonstrates the difference between owning somebody's computer (with the organ of its internal storage) and owning their typewriter.

For systems where data storage was exclusively on external media—like the 8-inch disks King would have used with his Wang—having the computer without any of that associated media takes on a different kind of significance. Derrida may choose to call it fetishization, but museum professionals are known refer to a special class of artifacts as numinous objects, distinguished not by their intrinsic worth but by their close association with a person or event.[48] More than mere fetishes or totems, numinous objects act to concretize individuals and their experiences. Above all, numinous objects tell stories: a prize baseball, for example, whose scruffs and stains evoke the roller-coaster of emotions we felt while watching the big game.

There is a Lanier word processor on display at the Carter Presidential Library in Atlanta, Georgia, though whether or not it is the actual one Carter used is unclear. Stanley Elkin's Lexitron (the Bubble Machine) seemingly no longer survives, though the library at Washington University has the PC clone he switched to afterward. Perhaps best of all, Barry Longyear's Wang System 5 wound up as a prop in a school play—it became the centerpiece of a cockpit for a spaceship.[49]

Zeke died in early 1983. Zeke, of course, was Jerry Pournelle's cherished microcomputer. "Like the wonderful one-horse shay, everything went at once," he reported in his *Byte* magazine column (the reference is to an Oliver Wendell Holmes poem).[50] He gave some of the remaining working parts to Larry Niven as spares for his own system; others apparently went to a local computer club; still others were integrated into a new computer, promptly dubbed Zeke II. Not an integrated system of the sort that was by then commonplace, but instead a customized build combining the best components of the myriad different products Pournelle had reviewed over the years, it was a unique assembly. Like its predecessor, Zeke II was built on a CompuPro S-100 bus and Z-80 chip. It had 64K of RAM and a pair of dual 8-inch floppy disk drives, each double-sided double-density diskette capable of storing 1.2 MB.[51] Pournelle also continued to run his custom word processing software, called simply WRITE: "I have seen nothing better . . . for creative writing," Pournelle declared to his readers in *Byte*.[52] For the monitor, he kept his venerable 15-inch Hitachi display. This was in fact crucial to him: "We took the video chips out of Ezekial and put them in the Ithaca board, so that the display on my big Hitachi 15-inch screen is identical to the old Zeke. I continue to use 16 lines of 64 characters to avoid eyestrain. Also, I'm used to it: after all, a standard manuscript has 60-character lines. A page is usually 25 or 26 lines, so I don't see a whole page at once; but I've noticed an unexpected benefit. Having only 16 lines on a screen tends to make me shorten my (usually too long) paragraphs."[53] The keyboard, meanwhile, came from an Archive computer, the same brand once favored by Arthur C. Clarke. "The Archive has great key feel, a good nonelectronic 'click,' and a really nice (Selectric-style) key layout," Pournelle commented.[54]

Zeke II would serve as Pournelle's workhorse for much of the decade. In 1990 he donated it to the Smithsonian National Museum of American History. The Hitachi monitor, CompuPro case, and Archive keyboard were set up next to an Osborne 1 and publicly displayed as part of the "Information Age" exhibition in a prominent locale on the first floor, where they were seen by thousands of visitors every week. Other items included Samuel Morse's telegraph, a piece of the first transatlantic telegraph cable, Alexander Graham Bell's original telephone, parts from the ENIAC computer, and a German ENIGMA encoder. "This exhibition of more than 700 artifacts explores how information technology has changed our lives—as individuals and as a society—over the past 150 years," read the catalog.[55] In 2006 the Information Age exhibition closed, and Zeke II, like so much of the Smithsonian's holdings, was relegated to storage.

In January 2012 I made my way downtown to the museum. A curator was waiting to receive me. Zeke II is resting in an unassuming white metal cabinet on the fifth floor. Two shelves above are components from the famous IBM Deep Blue computer—so presumably Zeke doesn't lack for intelligent conversation, or at least a good game of chess. All there is to see, though, really, is a featureless beige box with a black metal front sporting an embossed CompuPro logo, a single red square button (unlabeled), and a red power switch. Above the square button there is an adhesive plastic label with raised letters that reads "ZEKE II." (All of this was sufficiently unremarkable that there had been some doubt as to whether it would even remain a candidate for the Information Age exhibition—numinous objects, after all, often being in the eye of the beholder.)[56] The Hitachi monitor and the Archive keyboard also are there, along with the disk drive unit, a handful of the 8-inch disks, and a stack of manuals and documentation for the system. There are no plans to return any of it to the public view at this time.

Pournelle knew I had been planning to go: "Say hello to Zeke for me," he emailed the day before. "Tell him I miss him."[57]

Kamau Brathwaite has always sensed the unusually intimate nature of the linkage between his personal computers and his personal archives. Even in the midst of the terrible stresses from his first wife Doris's terminal illness, he acknowledged the precarious status of the records of their literary journal *Savacou,* "locked up" in her Kaypro.[58] He also attributed the tribulations he suffered while composing "Dream Chad" on the borrowed Harvard Macintosh to early portents of what would be a near miss to his archives (and Eagle computer) at his home in Irish Town, Jamaica, from Hurricane Gilbert a month later. The storm, which resulted in significant property damage regardless, sowed the seeds of a lifetime dread: "I began to think—to realize—to ***fear***—that perhaps the dream had still to be fulfilled."[59] Above all, however, it is the fusion of his personal Macintosh, "ole Sycorax," with what Ignacio Infante identifies as "perhaps Brathwaite's key tidalectical source for his Caribbean aesthetic" in the powerful mythopoeic figure of Sycorax herself, that informs the relationship he has cultivated between the computer and cultural memory.[60] Brathwaite has since rewritten and republished many of his pre-digital works in the Sycorax Video Style, thereby allowing the computer to become, as Infante notes, "a new archival medium"—functioning as such through its power as a platform for his literary identity, as well as literally in the form of the files saved to its hard drive.[61]

But the constant intuition of peril for both collective memory and personal data that Brathwaite lived with came to a head again in 2004 in a bizarre series of incidents he now refers to as a second Time of Salt as well as his "cultural lynching."[62] In brief, starting in late 2004 Brathwaite began noticing evidence of break-ins and intrusions at his Washington Square apartment, with key artifacts and manuscripts from his extensive personal collection of Caribbean writers going missing under mysterious circumstances. Locks were changed, security cameras were installed, but—as Brathwaite insists—all to no avail. At one point he and his second wife, Chad (Beverly), returned home to find her computer, which she remembered shutting down, "awake" and warm as though it had recently been in use. When his Musgrave Medal (Jamaica's preeminent cultural award) disappeared in 2010, it made international news. According to lists Brathwaite has provided, thousands of items are missing in all.[63] While the circumstances behind these events remain murky, what is undeniable is that they served as a source of extreme personal trauma to the couple. By 2011 Brathwaite (his health also suffering) had departed NYU and returned to the property known as Cow Pastor in his native Barbados. About cultural lynching, he would write: "w /out protection . . . of yr boundaries and for yr possessions & soul . . . all yr gains & discoveries & reclamations & dreams of freedom / independence will come to naught; and w/out allies, whatever you may have evolved as yr *culture* will be filtered fall tered [*sic*] infiltrated stolen from poison spat upon & destroyed w/out even the vestige of record or *qual* or eQual or quail to mark that you tried that you xisted that you made a contribution."[64]

Understandably, Brathwaite had long feared that his computer or printer would break down, and indeed one trustworthy source reports that the Mac stopped working in 2006 or 2007; and that Brathwaite has subsequently been unable to reconstitute the fonts he initially created with it.[65] If so, then it is not too much of an exaggeration to say that this would be a kind of second living death for the poet, a development in the midst of his second Time of Salt that mirrors the indelible moment in 1990 when—a gun at his head having miraculously failed to fire—he virtually died for the first time.

Very little seems to remain of Bad Sector's history other than the lone article by Ben Fong-Torres in *Profiles* magazine. Amy Tan has a Kaypro mug worn almost illegible from too many trips through the dishwasher, but nothing else. "Tech is so ephemeral," she replied to my query—on Twitter.[66]

Disposable yet indispensable, obsolescent but enduring, the floppy disk is an exemplar of what we have learned to call, after Raymond Williams, "residual media."[67] A floppy's afterimage is present in the very window in which I am writing this, a tiny thumbnail icon in the privileged top left corner of the screen. Clicking it activates the Save command—itself an increasingly residual function, as word processing software nowadays typically enters into AutoSave mode by default. (Many users had found a kind of comfort in the reflexive gesture of saving their work, a manual rhythm that has itself become increasingly vestigial.)[68] Such persistence of vision should not surprise us: the floppy disk has been eluding its own obsolescence ever since that waggling descriptor was applied to its most widely distributed form factor, the patently inflexible 3½-inch diskettes that had begun appearing by the mid-1980s.

Floppies were to the personal computing industry what the paperback book was to mass-market publishing, and 78 (later 45) rpm singles were to pop music. In the years before ubiquitous hard drives, the floppy was often the sole storage medium, not just for data but for software and operating systems as well. And floppies were also social media: you could give one to a friend without thinking too much about it. You could personalize them, with labels and annotations and decorated sleeves. Yet floppies were finally transitional technologies, inserted in between the industrial regimen of tape and today's silent, invisible, vastly capacious magnetic and solid-state hard drives.

Floppies may be residual media, but they are also remnants or remainders and survive today as material artifacts as well as virtual totems—most of us have a shoebox or two of them tucked away somewhere. As the physical media (widely cited as having a thirty-year life span) decays, information returns to entropy by the natural processes of magnetic degradation and chemical decomposition. Thus the computer historian and archives activist Jason Scott writes on his widely read blog: "Someone has to break it to you, and that person is me. It's over. You waited too long. You procrastinated or made excuses or otherwise didn't think about it or care. . . . With some perseverance and faced against all the odds stacked against you, something might get out of these poor black squares, but I would not count on it."[69]

Writers had to come to terms with the particular characteristics of the diskettes, in particular how many pages could be stored on each—sometimes, especially in the early days, only a single chapter's worth. This meant that searching for a passage often meant physically swapping the disks in and out of the drive while trying to remember or locate the correct one. By the 1990s, however, even a single high-density IBM-formatted diskette was capable of storing 1.44 MB, or many thousands of pages of raw text. These metrics quickly made floppies a storage medium of an entirely different order of magnitude than we are used to confronting when reckoning by print standards. Some writers organized their diskettes meticulously; others reached for whichever one happened to be at hand, files and data accreting in seemingly random arrangements.

The 5¼-inch disks Stanley Elkin used with his Lexitron when writing *George Mills* are preserved in the Special Collections Department at Washington University, St. Louis. There are seventeen of them, each containing some forty to fifty pages, each labeled by the page-range they encompass—not at all unlike the numbered exam booklets that were their predecessor. The city government of Tyohu, the Japanese city where Kōbō Abe lived for much of his life, keeps two of the 3½-inch diskettes used for his novel *Kangaroo Notebook* (1991) on permanent public display, alongside his pens, notebooks, and other personal accessories.[70] There are already notable instances where a writer's diskettes and other computer media have been used to help edit and posthumously publish important works: these include Douglas Adams's unfinished novel *The Salmon of Doubt* (2002), Ralph Ellison's second novel, which has been published as both *Juneteenth* (1999) and in a more comprehensive version entitled *Three Days Before the Shooting . . .* (2010), and David Foster Wallace's *The Pale King* (2011). After his unexpected death from a heart attack in 2001, Douglas Adams's beloved Macintosh computers were inventoried by a close friend and the results delivered to his long-time editor, Peter Guzzardi, on a CD-ROM containing over 2,500 individual files: letters, outlines, research notes, and drafts for a wide variety of projects, as well as "dozens" of versions of his unfinished final novel. This last presented the thorniest dilemmas. Everything about the book—starting with its title—had evolved and changed since Adams had first begun work on it almost a decade earlier, in 1993.[71] Guzzardi made the decision to "[stitch] together the strongest material, regardless of when it was written, much as I might have proposed doing were he still alive."[72]

Not surprisingly, perhaps, David Foster Wallace's relationship with computers was more vexed than Douglas Adams's. He told Sven Birkerts in a letter that he disliked writing directly on the screen, preferring to work longhand and then transcribe.[73] By the time he was writing *Infinite Jest* (1996) he was reportedly in the habit of deliberately erasing rejected passages from his hard drive so as not to be tempted to restore them to the manuscript later on.[74] When Wallace's agent visited his home after his death, she found "hard drives, file folders, three-ring binders, spiral-bound notebooks, and floppy disks," containing "printed chapters, sheaves of handwritten pages, notes, and more."[75] These materials were then used by his longtime editor, Michael Pietsch, to produce the text that was published as *The Pale King: An Unfinished Novel.* Pietsch worked with them exclusively in hard copy, but used printouts of directory listings to help date different versions of chapters and other sections and determine their original order.[76] The extent to which digital files recovered from Ellison's floppy disks played into the reconstruction and editing of his unfinished work is even more complex. Ellison, who as we have seen began working with WordStar on his Osborne 1 as early as 1982, left behind almost 500 computer files distributed across some eighty-three diskettes. The task of making sense of them initially fell to Adam Bradley, now a prominent Ellison and hip-hop scholar in his own right, but then an undergraduate under the supervision of Ellison's literary executor, John Callahan, at Lewis and Clark College in Oregon.[77] They would work together for more than a decade until the book came out as a 1,100-page volume from Modern Library, sixteen years after Ellison's death.[78]

And just like Stieg Larsson's laptop, floppy disks can become objects of literary controversy. After Frank Herbert's death in 1986, his son (and also biographer) Brian Herbert, together with Kevin J. Anderson, took over the writing for the *Dune* franchise, a move that raised concerns among a vocal minority of fans who saw these efforts as opportunistic or else simply inadequate to Herbert's legacy. At the very center of the issue was the long-awaited final installment known in the fan community as "Dune 7," eventually published by Brian Herbert and Anderson as two separate novels, *Hunters of Dune* (2006) and *Sandworms of Dune* (2007). The younger Herbert and Anderson maintain that these books are based on a thirty-page outline and extensive notes by Frank Herbert, recovered from two floppy disks found (with his father's other papers) in the attic of the family home in Port Townsend, Washington. In December 2005, in an effort to settle the debate, Anderson posted low-resolution black-and-white photographs or scans of these two diskettes online—both 5¼-inch, one branded 3M and one branded Tandy, both with labels apparently written in Frank Herbert's hand.[79] ("Will this convince any of the conspiracy crazies? Probably not," he asks and answers in rhetorical mode.)[80] In fact, the move only served to heighten the conspiracy theories: fans questioned how data could have been recovered, or if the disks would have been readable at all (neither scenario is improbable). But unlike posting scans of actual manuscript pages (as Stephen King did with *The Cannibals*), the floppy disks came to function simultaneously as proof positive and further provocation, their content presented as self-evident even as it was manifestly opaque, being recorded on flimsy magnetic media inside the diskette's black plastic sleeve; residual media that, for some, remain tokens of ongoing suspicion and doubt.

John Updike's biographer Adam Begley describes the Houghton Library's collection of his manuscripts and other documentary remainders as "a vast paper trail, possibly the last of its kind."[81] It's a lament that has become something of a repeated trope in increasingly elegiac writing about authors and their literary legacies. Three years earlier, a writer in the *Guardian* had opined that J. G. Ballard's newly opened papers at the British Library would perhaps be "the last solely non-digital literary archive of this stature" (Ballard never used a computer).[82] But at least in Updike's case the characterization is not strictly accurate. The Houghton also safeguards some forty 3½-inch diskettes that belonged to Updike, plus a smattering of CD-ROMs and other storage media (but no hard drives or actual computers). These all date from a later period of Updike's word processing career, after the Wangwriter II, when he had moved to an IBM-compatible computer and software called Lotus Ami Pro.

My experience with Updike's papers had taught me that he was frugal, and reused all manner of material, typing or writing on the backs of drafts, or even envelopes and receipts and other people's correspondence. Certainly his digital working habits appear consistent: his practice was apparently to store multiple versions of a file on the disk, overwriting previous ones with new ones and notating the date on the disk's label after crossing out what was written previously. For some novels, like *Villages* (2004), *Terrorist* (2006), and the *Widows of Eastwick* (2008; sequel to the more famous *Witches*), there are multiple diskettes with relevant material; others contain several dozen shorter pieces such as stories or reviews. The Houghton has migrated the files off of the diskettes and intends to make them available to researchers, so some intrepid scholar may one day wish to undertake the task of comparing their digital content to the numerous hard-copy printouts of work in progress that Updike was also in the habit of producing. Whether or not major insights into Updike's creative life are thus revealed, this digital collation is paradigmatic of what textual scholarship is going to look like in the coming years.

Not at the Houghton, however, are any of the 5¼-inch disks that Updike would have used with the Wangwriter II. These had been presumed lost for good. We cannot know, for example, whether "INVALID.KEYSTROKE" or any of the many more significant texts Updike wrote throughout those years (such as the *Witches of Eastwick*) exist in variant states not represented by corresponding hard copy on deposit at the Houghton with his papers. We cannot know if there are texts never consigned to paper or whose paper instantiation was subsequently lost or discarded. Given what we do know of Updike's habits, we might judge such scenarios unlikely, but the reality is that in the absence of the digital media, which were themselves a distinct and sometimes primary site of composition for Updike, these uncertainties must stand. The simple fact of the Wang's existence, in other words, means that in the absence of its accompanying storage media there is a gap in our knowledge of Updike's manuscripts.

Enter Paul Moran. Moran is the maintainer of what he's named "The Other Updike Archive." Here is how it came about: Starting back around 2006, he had begun making it a weekly habit to bicycle past the Updikes' Beverly Farms residence and pick through their curbside trash. Over the years he accumulated photographs, personal papers, honorary diplomas, greeting cards, drafts of stories, invitations, and—yes—Wang diskettes, a baker's dozen of them.[83] They include one labeled "Witches V" (this has been crossed out and replaced with "Reviews") and two tantalizingly labeled "Poems" (is "INVALID.KEYSTROKE" among them?). Another is labeled "Stories III." That particular disk also has a hard copy directory listing tucked into its outer paper sleeve showing entries for some three dozen items, each a file named with a single word at most eight letters in length: "Acceptan," "Bats," "Bindweed," "Brazil," and so forth. The diskettes, along with many hundreds of other items so retrieved, now reside in Moran's storage unit in Austin, Texas; he keeps a blog on which he posts images from the collection along with his commentary.[84]

It is fair to say that key members of the Updike establishment have taken a dim view of Moran's activities. The *Atlantic* story that originally reported on the "other" archive includes disapproving comments from the curator in charge of the collection at the Houghton, biographer Adam Begley, and the president of the Updike Society.[85] (The *Atlantic* is also quick to point out that Moran seems not to have broken any laws.) Updike himself, as Begley notes in his comments therein, was "assiduous in collecting himself. . . . He was obviously interested in presenting as full an account of his work and habits as possible."[86] To that end Updike routinely brought carefully labeled boxes of papers to the library at Harvard for safekeeping. One can therefore surmise that anything that found itself curbside—from the Wang diskettes to the red leather-bound honorary degrees Moran made off with on his first visit—went there (and not to the Houghton) deliberately. Nonetheless, the Wang diskettes raise questions many of the other scavenged items do not. Updike, as interested in documenting his own writing practice as he was, may nonetheless have reasonably thought them worthless, dismissing any practical hope of recovering anything usable from his poor black squares even if the requisite equipment could somehow have been found that far into the first decade of the twenty-first century (Wang had filed for bankruptcy in 1992). In fact, however, floppy diskettes from throughout the 1980s are often still readable by archivists, and specialized hardware exists to reanimate and spin the old drives—at least long enough to obtain a bit-by-bit copy of everything that was once stored on them. Moreover, while one can question the propriety of fetishizing other mementos and accoutrements of Updike's career, the Wang diskettes are no mere nostalgic gewgaw. They are intimately associated with Updike's literary legacy.[87] (Not beside the point, it was the convention of many early computer systems, including the Wang, to refer to disk storage specifically as an "archive.")

There is thus no escaping the conundrums associated with the now universal Save icon: as computers and their storage media have begun to appear in archives, scholars and archivists alike have engaged in debates over the ethics of recovering deleted files, sometimes using forensic data-recovery techniques borrowed from law enforcement.[88] These activities raise essential questions about authorial intentions, personal privacy, intellectual property, and the ethical responsibilities of both individuals and institutions to collective cultural memory. What about using digital forensics to recover those deleted passages David Foster Wallace supposedly purged from his hard drive, for example? If the proverbial lost Shakespeare manuscript ever turned up, would we read it even if a note from the Bard on the first page said, "Burn after my death, WS"? (Of course we would. We know that Kafka had asked that his diaries and manuscripts be destroyed—it didn't work.) All of us of a certain age have dragged a floppy disk icon across our desktop screen to the trash can, to see the latter bulge satisfyingly in response to the bytes it ingests. This was once a routine activity for countless users. But what happens when the diskette in question belongs to a major author or other historical figure? And what happens when someone comes along and fishes it out of the bin?

When the *New York Times* lamented back in 1981 that future generations would be deprived "of words scratched out, penciled in and transposed with wandering arrows" and that they would "have to make do with electronically perfect texts," the editorial board was actually talking about the fate of manuscripts in the digital age.[89] The implication was clear: a traditional manuscript—that is, something truly written by hand—exposes its material processes. Very much anticipating Vilém Flusser's twentieth-century distinction between inscription and notation, William Blackstone, the eighteenth-century legal scholar, had argued forcefully for the importance of committing legal records to rag linen paper—still a comparatively expensive luxury item in an era before pulp—as opposed to wood, leather, and other surfaces, even stone. Paper, said Blackstone, had the virtue of being fragile enough to readily expose any attempt to tamper with or change it—whereas stone, by contrast, was inviolate and could withstand harsh scraping or other efforts to expunge it without leaving a trace.[90] By virtue of the inscrutable processes unfolding within the hardware and software's inaccessible innards, a document written with a word processor—ironically—has more in common with Blackstone's lapidary surface: "Writing with a word processor," claims Daniel Chandler, "obscures its own evolution."[91] In other words, as the consequence of what Chandler elsewhere termed suspended inscription, all of that scratching out and penciling in and transposition that the *Times* fretted over happens inside the black box of the machine, and the end result is a product that appears to be *sui generis.* Philip Roth, who switched to word processing in the early 1990s, adopting a DOS-based version of WordPerfect that he continues to use to this day, observes: "I'm doing so much changing as I go along that the drafts disappear, as it were, into the rewrites."[92] Similarly, Zadie Smith recalls simply "saving over" the early versions of her novels and stories instead of maintaining separate files, thereby obscuring traces of their composition in the manner anticipated by Chandler.[93]

These are some very significant examples, and it will be of real concern to literary studies and literary history if they become the norm. Students and scholars have long been fascinated by the extent to which access to an author's manuscripts opens a window onto the mysteries of the creative process. Recall the anecdote about Keats verbally working through the permutations of a famous line, until his listener informs him that it will live forever. We know of the episode only because it was recounted and later written down. But with a literary manuscript, the false starts, the words that fall flat, the extra sentence scribbled in haste and crammed into a margin, the stanza struck out—these not only serve to humanize the author, grounding a possibly remote historical figure in a reality we can all identify with, sweating out the next line, wondering if we've gotten just the right turn of phrase at last; they also remind us of what could have been or might have been, remaking literature as a multidimensional possibility space rather than the finality of words printed on the page.

"First we had a couple of feelers down at Tom's place, / There was old Tom, boiled to the eyes, blind." These, of course, are the ringing, instantly recognizable opening lines of T. S. Eliot's *The Waste Land* (1922). Except, of course, they are not: everyone knows the poem really begins, "April is the cruelest month." Except, of course, these actually *are* the poem's opening lines, or at least they once were, being the beginning of the "Burial of the Dead" section in a text Eliot was at the time still calling "He Do the Police in Different Voices." *The Waste Land*'s turbulent textual history has been popularized through the availability of a handsome facsimile edition that reproduces Eliot's original typescripts and manuscripts for the poem, both of which bear the marks of three different hands—his own, his wife Valerie's, and that of Ezra Pound, who famously "edited"—in fact, heavily revised—the text.[94] It takes only a glance to see the facsimile edition as a viscerally exciting work, its pages filled with greyscale reproductions of the original pages, the dramatic cross-outs and (sometimes carefully, sometime not) scrawled additions testifying to the intensity of the creative forces at work (or at war) across the text, which descends to us now bearing the marks and scrapes and scars of those contests. But what remains of such edits and revisions with an electronic document? Is there any trace? What would we find if we *could* read the bits off of one of John Updike's Wang diskettes?

What happens when—as a function of the Web and social media—many hands can edit and revise a literary text instead of just one or two or three?[95]

At the time when Chandler wrote about the propensity of electronic documents to obscure their own evolution, the only conceivable remedies for that outcome were manual—that is, to adopt meticulous work habits and commit to saving multiple versions of every file on disk and / or printing them out. But even to the extent that such solutions might have been practical, where does one draw the line? What constitutes a significant enough revision or intervention in the text to justify printing or saving a new version? Umberto Eco fully grasped the implications for future scholarship, sketching a scenario that results in a "phantom version" of a digitally composed text: "I write my text A on the computer. Print it out. Correct it by hand. Now we have text B, the corrections for which I input onto my computer. Then I print out again and think (as would any philologists of the future) that what I have is text C. But actually, it is text D, because while I was putting the corrections into the computer, I would certainly have taken some spontaneous decisions and made some further changes."[96] Still, some imagined other possibilities. Jerry Pournelle once reported on a conversation with Frank Herbert, who was then in the midst of designing his dream machine. Herbert wanted to save everything, "every single one of the myriad changes writers make while noodling around during a working session."[97] He thought—*pace* David Foster Wallace—that rejected prose should be stored, possibly for use elsewhere, and that it might also be of interest to future scholars.

Pournelle admitted he couldn't quite grasp the idea, indeed couldn't think of an analogy beyond making a photocopy of a typewritten manuscript every time its author backspaced to X-out a word.[98] In fact, however, as word processors matured, they began incorporating increasingly sophisticated version tracking and date / time-stamping features that (at least in theory) could afford unprecedented access to an author's composition process, allowing editors and scholars to watch a particular passage take shape keystroke by keystroke. What is today known to us as the "Track Changes" feature was initially called "Redlining" (here redlining refers to the practice of marking up documents with a red pen or pencil, a convention that originated in architectural drawing). Redlining began as a stylesheet in Microsoft Word 3.11 (released in 1986) and was a fully integrated feature in Word 4.0 the following year; it was then known as "Mark Revisions" before the name changed once more to "Track Changes" with the release of Word 97. But the idea of version tracking for electronic documents did not originate with Microsoft Word; it was already present as a feature in Word's chief competitor, WordPerfect, where it was particularly important for what was by then one of WordPerfect's key markets, the legal community. And even before word processing, version control was indispensable for computer programmers. It was commonplace for multiple people to work on the same piece of code, and changes could often be subtle—something as simple as a comma or a curly brace could completely alter the functionality of a program. The Source Code Control System was thus developed at Bell Labs as early as 1972.

Our current technological moment is marked by a tremendous paradox: as fragile as electronic media are and as fleeting to the historical record as they may be, they create enormous and potentially unprecedented opportunities for scholarship. On the one hand, we have seen from the examples of Roth and Zadie Smith how easily the changes wrought upon electronic manuscripts can swirl away to oblivion; yet on the other hand, with the advent of Track Changes and similar features, every electronic document has the potential to become what scholars term a genetic text, capable of embedding the history of its own making. We know, for example, from the digital files available at Emory University that Rushdie first opened his first file for his first draft of *The Moor's Last Sigh* on April 20, 1992, at 11:58 in the morning (the directory path reads in part "DRAFTS,NOTES / EARLY WORK:ROUGHS").[99] The prospect of such granularity in our literary histories can be intoxicating. But it is also important to be mindful of what we do not know. We do not know, at least not without further confirmation, whether the clock on the computer was set correctly; we do not know whether Rushdie began working first on some other machine, or saved even earlier work to a diskette, now lost; we do not know if he first began drafting longhand. The kind of information we receive from digital files will always have to be evaluated in relation to what we know of a writer's composition habits from other sources, just as, increasingly, what we know from those other sources will have to be evaluated in relation to what is on the computer.

The most dramatic demonstration of principles may come from the Australian author Max Barry and his 2011 novel, *Machine Man*. Fittingly, *Machine Man* is a cyborg fable, the story of a high-tech industrial worker who, following an initial accident on the job, undertakes to voluntarily subtract (read: amputate) successive parts of himself, replacing each organ and limb with prosthetic components that are better and stronger than the bones, sinew, and nerves of his biological self. The book began as a serial self-published on Barry's website, one page a day for the better part of a year. Readers commented, and Barry revised. The novel then was published in expanded form by Vintage. But neither the bionic plotline nor the serial format and publication history is what is most interesting about the book. Barry, who is rather technically inclined (he favors the UNIX-based text editor vi for his writing), was also in the habit of using a tool called Subversion, a so-called Concurrent Versioning System or CVS, to manage his daily drafts.[100] This is the same technology used by professional software developers to manage their code, requiring authors to check documents in and out of a digital repository that maintains a complete, branching revision history—and allowing different users to move up and down the document tree at will, restoring any previous state of any version that they may desire, with all changes date- and time-stamped (and cryptographically authenticated) to the nearest second (Greenwich Mean Time).

As an experiment, Barry elected to make all of the novel's accumulated versions publicly available on his website. "I'm not sure what use this is to anybody, other than for exposing my writerly fumblings in an even more humiliating manner than I've already done. But it was POSSIBLE, so I have DONE IT," he tells us.[101] The result is a well-presented online interface that allows the user to access what (at a glance) might pass for one of the typescript facsimiles in *The Waste Land* edition. Each variant "page" is presented on a drop-shadowed rectangle of slightly worn paper texture, which forms the backdrop for a clean, serif font. Simple cross-outs and colored text indicate deletions and additions, and users can click through each successive rendition of each individual page from its initial to its final state. In effect, what we find is a simulacrum of the very cross-outs and wandering arrows whose loss was mourned by the *Times*—what Flusser would recognize as inscription, rendered not as an intrinsic property of digital media but as a carefully engineered effect, the text's transmission through time given expression by the programmable logic of the digital environment in which it was composed.

Is the hybrid body of this text thus the cybernetic corollary to the novel's protagonist? Perhaps. More interesting, however, is that in this model text becomes less like an object or an artifact and more like an *event*. This is no mere casual assertion, but rather a fairly precise technical descriptor. Programmers long ago figured out that storing a record of what happens to some piece of code is more efficient than storing each successive iteration of the code itself. To retrieve a given version from a given moment in time in fact means re-creating the text—on demand—from the manifest of events documenting its evolution. Novelist (and self-taught programmer) Vikram Chandra explains the concept, what coders call event sourcing, saying, "There is no enduring object state, there are only events."[102] He aligns this dictate with a set of Buddhist beliefs that argue that our sense of "self" is really only the ongoing emergence of a series of temporally discrete entities. Other commentators, however, have pointed out that there needed to be *some* stable, consistent force acting to connect these otherwise distinct events. Chandra quotes the tenth-century philosopher Abhinavagupta on this point: "It is in the power of remembering that the self's ultimate freedom consists. I am free because I remember."[103]

December 1989. Hyattsville, Maryland. A young writer named Bill and his wife, Beth, are enduring unendurable tragedy: the death of their six-month-old daughter from complications of VATER syndrome. One day Bill finds himself sitting in front of the NEC PC-8500 laptop he'd purchased not long before, opening a new file, and typing the date at the top. The NEC is a small machine, awkward and ungainly and lacking the sleek lines and curves of today's technology. It runs WordStar burned directly into its CP/M ROM board. To better see the characters on the LCD, he finds it helpful to dim the lights and draw the blinds. The screen displays only 25 lines at a time, a limitation that perhaps helps him focus on just that moment, suspending the burdens of the past and the weight of the future. The screen becomes a kind of confessional, a small, luminescent window floating in the darkened room, the Korean-manufactured keyboard a near-silent conductor for thoughts, memories, emotions; and pain.

The file he creates, of course, is named for their absent daughter. Over the span of a year, in what will become his first published book, he tells her story. At the end of each day's work he self-consciously enacts the chorded keystroke sequence that lets him do the one thing he couldn't do at the time and the only thing he can think about doing now: SAVE ANNA.[104]

There is a cinematic counterpoint to the ending of *Stand by Me:* the final scene of the film adaptation of Michael Chabon's *Wonder Boys* (2000). Here the writer is played by Michael Douglas. We see him in a well-appointed study, in front of a laptop, and we watch as the camera zooms in on the screen to follow the mouse pointer down the length of the familiar File menu until it hovers over the option to SAVE. *Click.* The screen goes dark, and the credits roll. People collected their coats instead of crying out in distress.

Perhaps Hollywood had simply grown more sophisticated in its portrayal of computers on the big screen in the intervening decade and a half; a writer using a word processor in a mainstream movie was now expected to behave believably. But it is also possible to view the contrast between the final scenes in the two films as a reflection of our changing relationship to digital content and digital culture. Whereas in the mid-1980s the ghostly green text was vulnerable to disappearing in a flash, by the year 2000—Y2K notwithstanding—we had grown accustomed to thinking of digital media as possessed of its own idiosyncratic forms of resilience, not an inscrutable or ephemeral black box but just another technology, subject to our manipulation and control. Grady Tripp had found a life worth living, and he entrusted it to his PowerBook without a second thought.

Today, another fifteen years further on, we live in a world where ubiquitous computing begets ubiquitous storage. Our applications now AutoSave, our devices tether themselves to the cloud, and we ever-increasingly log and blog facets of our daily experience. Whereas some fear a digital dark ages, others warn of equally dystopian scenarios: perfect memory and total recall. This is the premise of Dave Eggers's *The Circle* (2013), set in a near future dominated by the titular social media company whose overtly Orwellian credo consists in catchphrases like "Secrets are lies" and "Privacy is theft." Politicians and ordinary citizens alike elect to go "transparent," eagerly adorning their bodies with cameras and sensors to record (and upload and share) everything from their biometric data to conversations with intimates to their retail preferences. "I'm watching you now," says a client to her interlocutor, Mae, at the Circle's California headquarters, where Mae's camera is transmitting an image of the screen in front of her to millions of followers, the client included. "She couldn't get over the fact that she could watch Mae typing the answer to her query in real-time, on her screen, right next to where she was receiving Mae's typed answer. *Hall of mirrors!!* she wrote."[105] This may (Mae?) be where we are headed in Eggers's view of the future; this may be what it looks like when hitherto unseen hands are always on view.

Benjamin Moser, who is Susan Sontag's biographer, has written of the vertiginous queasiness he felt scrolling through some 17,000 of her emails, gleaned from two hard drives preserved at UCLA's Charles E. Young Library. "To read someone's e-mail is to see her thinking and talking in real time," he writes. "If most e-mails are not interesting ('The car will pick you up at 7:30 if that's ok xxx'), others reveal unexpected qualities that are delightful to discover. (Who would have suspected, for example, that Sontag sent e-mails with the subject heading 'Whassup?')"[106] At stake is more than just access to the casual intimacies of unguarded speech. Researchers can also search the corpus of emails, looking for keywords, individual correspondents, date ranges, and subject headings. They can visualize patterns of contacts and degrees of separation. Experimental software even allows the researcher to sort the messages by sentiment, such as "angry" or "happy" or "sad."

Literary history will not be erased by word processing, but it will be different after word processing. All of these same tools and techniques can be applied to an author's primary corpus: if not to 17,000 emails, then to potentially innumerable versions and variants of their work in progress, each captured by a system such as Max Barry employed, each forensically correlated with other events recorded on the same system but notionally external to the writing of the story—some particular website visited, some particular chat with a friend, the weather on some particular day—all of these data algorithmically sifted and plotted to reveal . . . what? Networks, patterns, affiliations, affinities. This, at least, is one vision. Other, more deliberate models, deeply grounded in the long-standing traditions of fields like philology, textual criticism, and analytical bibliography, are also emerging, however. Not big data but small data—slow data. One literary scholar, for example, is using forensic computing techniques to carefully reconstruct the composition process of works by the German poet Thomas Kling from his last three hard drives, much as Lawrence Rainey once unraveled the composition history of *The Waste Land* through forensic document examination.[107] A scholar working with such materials must be conversant in the antiquarian cants of vanished operating systems, file systems, file formats, and data structures, as well as specialized tools like hex viewers, file carvers, and emulators, just as we expect an early modernist doing book history to know something of signatures and collation formulas.

There will always be things we know and things we don't know about literature and literary history, regardless of the platforms and tools at our disposal. This is neither surprising nor disappointing. We already know so much more than we would seem to have any right to expect: not just the day and hour and minute (and second) when Salman Rushdie (perhaps) began working on one of his novels, or the sequence of keystrokes Max Barry followed to correct a typo, but at least some remote reflection of what it was or is like—with real empathy and appreciation for tacit knowledge—for John Hersey or George R. R. Martin to sit in front of their respective screens, or for Ellenor Handley to "think tape" while working at the MT/ST. This is what we mean by materiality, the word we first used in the Introduction: not just the presence or absence of information, but the lived struggle to reclaim and to recover it, to remember, to experience, and to know—to be known. This, in the end, is always what remains. What else really matters?

AFTER WORD PROCESSING

"The unwritten literary sociology of this century," is how media philosopher Friedrich Kittler once described his own work. "All possible types of industrialization to which writers respond have been thoroughly researched—ranging from the steam engine and the loom to the assembly line and urbanization."[1]

For Kittler only one lacuna remained, and that was typewriting, not word processing, even though he was writing those words in the middle of the decade in which the Apple Macintosh and Microsoft Word both arrived.[2] Together, bitmapped WYSIWYG displays, the desktop metaphor, the mouse, and Microsoft's software would define word processing for more than two and a half decades to follow. (Under this GUI regimen, wrote Umberto Eco in his famous theological comparison between the Macintosh and DOS, "everyone has a right to salvation.")[3] By 1994, Word would command 90 percent of the U.S. market, Microsoft having dispatched first WordStar and then WordPerfect in what some remember as the "word wars."[4]

And in truth, Word was a revelation.[5] When Charles Simonyi moved from Xerox PARC to Microsoft in 1981, he hired a young programmer named Richard Brodie and set about re-creating the principles behind the Alto / Bravo interface. Word 1.0 for DOS, released in November 1983 with a free diskette inserted into that month's issue of *PC World* (the first time the industry would see that particular gimmick), could display boldface, underlining, and italics directly on the screen, something no other word processor was then capable of doing. Like Bravo and Gypsy before it, Word placed a premium on formatting, encouraging the user to use stylesheets to create different looks for different kinds of documents.[6] Peter Rinearson, a technical consultant who had gotten an advance look at Word 1.0 and would go on to write the early books about how to use it, believed that it would have a profound impact on society, because—like the typewriter and the pen before it—it would change people's relationship to language. Word seemed like far more than just another word processor; Rinearson called it a work of the imagination.[7] But it was the Macintosh, with its bitmapped display, not Microsoft DOS, that offered the environment most suited to Word and its aspirations. By the time Word 3.0 for Macintosh was released in 1987, Simonyi would speak of that particular version as an asymptote, that is, something very close to—though never quite fulfilling—his original vision.[8] (Word for Windows would lag behind, not appearing in its first incarnation until 1989.)

Of course, Word has not been universally admired. In one now-classic critical analysis, Matthew Fuller, also the producer of the *Song for Occupations* installation discussed in Chapter 9, narrated his experience trying to write after deliberately activating every toolbar, ribbon, and feature in the software, which collectively consumed much of his screen.[9] Edward Mendelson, a prominent W. H. Auden scholar who began reviewing word processing software for the *New York Review of Books* back in the 1980s, nowadays discovers in Word an "arcane Platonism" rooted in its original conception of a document (rather than the text) as the foundation of the user experience, thus requiring a cascade of "demiurges" to mediate between the practical needs of those trying to use the software and the idealism of its internal forms.[10] Meanwhile, when the prolific science fiction writer Charles Stross published a (self-described) rant entitled "Why Microsoft Word Must Die," it instantly went viral.[11] Stross unsparingly lashed Word as "a tyrant of the imagination, a petty, unimaginative, inconsistent dictator that is ill-suited to any creative writer's use."[12]

It is indisputable that creative writing was never imagined as an important marketplace for word processing. This has been the case from the earliest days of the MT/ST and the Redactron Data Secretary on through to WordStar and Word itself.[13] "Word processors aren't designed for writers," Len Deighton told Ray Hammond three decades ago. "Everything about the program on my Olivetti tells me that this guy who created it has never sat down and written a long piece of text."[14] (One exception may be Nota Bene, which was conceived with the needs of scholars and others doing complex long-form writing in mind.) Yet the reality today is that many writers use Word by default, and many of them use it with its default settings (by default).[15] It is as though Word, having condensed the essence of word processing itself into its conspicuously foreshortened title, has become fully naturalized as the No. 2 pencil of the digital age. And though contra "text," "word" has not become a verb, it has perhaps become something even more powerful: a fully branded click-wrap content creation platform eponymous and synonymous with the *Logos.* Since 1988, meanwhile, Word itself has been bundled with a larger software suite Microsoft calls Office, a gesture both confirming and collapsing the decades of complex labor history we have painstakingly traced.

Only within the last few years has this hegemony—not just Word, but to some extent the desktop metaphor itself—begun to loosen. We may never recapture the experience of the first half of the 1980s when dozens of different competing word processing platforms and programs were on the market vying for the consumer's attention, but today there is more variety and experimentation in word processing than at any time for at least two decades.[16] Nowadays Stross's preferred writing environment is an application called Scrivener, the creation of an aspiring writer turned software developer named Keith Blount who was originally trying to write a novel himself.[17] By his own account, Blount found Word difficult to manage, in particular because he wanted to be able to work on his book in small-size bits and pieces and manipulate all of them the way he would note cards or scraps of paper on a desk.[18] Scrivener's interface allows the user to break long manuscripts up into small-size units, and to view each one in conjunction with any number of others; it also allows writers to manage notes and other research materials in the same space as their manuscript, managing and overseeing everything using the metaphor of a corkboard. The software's website includes testimonials from dozens of published novelists, screenwriters, and playwrights, nearly all of them speaking to its transformative effect on their writing practice.[19] Virginia Heffernan meditates on the contrast between the program's name and the monotheism of Word in a review published in the *New York Times Magazine:* "Scriveners, unlike Word-slaves, have florid psychologies, esoteric requirements and arcane desires. They're *artists.* They're *historians.* With *needs.*"[20]

Other alternatives are also emerging, notably a genre of programs we might think of as austerityware. In stark contrast to Word's bloat, these are lean, computationally efficient programs calculated to do one thing and one thing only—allow a user to write his or her prose free of distraction. Most austerityware (the label is my own) allows a user to write in full-screen mode, in effect transcending the multiwindowed regimen of the desktop in favor of a return to the kind of dedicated display that Jerry Pournelle or John Hersey would have recognized. Typical of these is WriteRoom, whose default mode is green letters on a black screen, thus explicitly mimicking old-school phosphor—not paper or pseudopaper, as Heffernan notes, but a void that seems both cosmic and primordial. In WriteRoom, writes Heffernan, "your sentences unfurl in prehistoric murk."[21] But while WriteRoom is an effective simulacrum (right down to its Monaco font), there is a telling difference. In the bygone era it seeks to emulate, the computer had just "moved in," as *Time* magazine put it: it typically occupied a dedicated room or at least a dedicated space within the home (recall Asimov's dithering over where to put it). Sitting down in front of the computer was a deliberate decision. Most importantly, systems couldn't multitask—they ran one program at a time, and switching between them meant stopping whatever you were currently doing, swapping in a new diskette, and rebooting the machine. Today, of course, computers are fully integrated into our daily routines and have become the site for managing multiple aspects of our lives, the windows on the screen offering us a manuscript draft one moment, an email message or a tweet the next, perhaps a financial statement or a family photograph on Facebook thereafter. Many austerityware programs therefore also include a lock-in option actively prohibiting the user from escaping their writing environment for some set length of time; indeed, some austerityware will also block the user from accessing the Internet and thus, at least notionally, eliminating its distractions (this even as publishers place increasing demands on writers to be proactive on social media in promoting their work).[22]

Scrivener and WriteRoom are both boutique, artisanal products, which is to say they are the work of one or two developers who with a craftlike ethos are committed to maintaining and refining them for their dedicated constituencies. Other, stranger, confections have also emerged: for example, Write or Die, an online editor that begins erasing your text (last word first) if you pause too long to reflect or procrastinate and thus fall below your target pace.[23] (The novelist David Nicholls has reportedly used it successfully.)[24] But the dominance of Word has also come under pressure from Web services and mobile devices, with writers now routinely working in the ubiquitous on-screen editors that are the backend to blogs and other social media platforms. (Many of these now support a syntax known as Markdown, which presents the user with only a very limited set of formatting codes, in contrast to the "markup" of enterprise-level software like Word.)[25] Twitter has become a recognized venue for short-form fiction, meaning that writers have acclimated themselves not only to its notorious 140-character limit but also to the constraints of working directly within its bare-bones interface.[26] In Japan, meanwhile, novels written on and delivered by cell phones (often by young women) have been popular among teens (especially) for over a decade.[27] These devices and services all offer alternatives to the traditional word processing experience through their abbreviated interfaces and the often radical constraints of their platforms.

However, no Web service has competed with Word as directly as Google, whose Docs application has become the most popular "cloud-based" writing tool in the world.[28] Unlike Twitter and its ilk, Google Docs is a fully featured word processor, freely available along with other Google services. Users tend to gravitate toward it because it relieves them of having to worry about which files are stored on which individual machine, while also offering the supposed safety and security of knowing their work is automatically backed up—somewhere, somehow. (Google Docs also allow a user to roll back their version history indefinitely, in effect automatically tracking every change.) Google Docs makes collaboration nearly effortless, encouraging users to invite others to share in the work of composing and editing the same document, potentially even at the same time.[29] Or else they can simply be invited in to watch: in the autumn of 2012, thousands of people logged in to spectate as a British fantasy author named Silvia Hartmann, having accepted a challenge, wrote her next novel live on the screen in front of them, the letters taking shape one by one from the keystrokes of her otherwise invisible fingers.[30]

The increasing plurality of word processing platforms—the Office worker versus the Scrivener, marking up versus Markdown, isolate austerity versus the sociability of the clouds—reflects an even more fundamental profusion of devices and technologies. Computers have not really been typewriters and TV sets for quite some time; laptops—frequently referred to as "notebooks" and various branded permutations of the term—are now further augmented by tablets, phones, and other mobile devices. We write anywhere, anytime, in any posture. We swipe, we tap, and we speak out loud even as our actual output becomes a mélange of predictive autocomplete algorithms and micro-motor gestures. The keyboard and the screen have at last converged in visual space, as virtual keyboards slide in and out of view on our touch-sensitive displays. It is even currently fashionable to design special implementations of these in emulation of typewriters, so that the tablet device emits all of the aural cues of typewriting even as the user is treated to the illusion of a carriage return (which must be duly zinged at the end of each line) trundling across the screen. The best known such app is the Hanx Writer, bankrolled by Tom Hanks himself (he is a typewriter collector and connoisseur).[31] Its delete function can be toggled on and off between a "modern" and an authentic mode: in the latter, backspacing to correct an error leaves a black overstruck "x" on the screen in its wake. Users can choose the Hanx Writer in lieu of the IOS's default text editor, and have the option of outputting their creations (overstrikes and all) to a variety of social media platforms.

At the same time, the stylus, together with so-called smart pens, are presenting what is surely the most serious challenge to the keyboard in some time.[32] Smart pens, which sync to a nearby digital storage device where the handwriting is interpreted and reproduced as machine-readable text, effectively replicate the input logic of the MT/ST and Redactron, allowing what happens on a sheet of paper to leave a symbolically stored record subject to ongoing manipulation and revision. One researcher at Microsoft is outfitting pens and tablet styluses with sensors to differentiate between different kinds of grips; the basic insight is that people hold the pen differently depending on what they intend to do with it—write, draw, poke at an icon on a screen. By using input from the combinations of bones, muscles, and nerves in the hand to predict what the user intends to do just before they do it, there is the potential for changing the way tablet devices work.[33] The best way to read minds may turn out to be through the hand after all; graphology, the centuries-old pseudoscientific practice of handwriting analysis, may thus be validated in the applied domain of human–computer interaction.

Just as the stylus is reemerging to augment the tablet experience, the typewriter is also being reengineered. If the Hanx Writer is a celebrity vanity project, the Hemingwrite has greater ambitions. The Hemingwrite is being marketed as a "distraction-free" writing device, essentially a dedicated word processor with a number of carefully chosen limitations and constraints. It debuted in December 2014 in a well-promoted KickStarter campaign, meaning that its developers went public in search of backers to bankroll the manufacturing run. "We engineered the Hemingwrite to do one thing, and do it sublimely well," the video introducing the fundraising campaign relates.[34] At a glance, it looks like a small, portable typewriter. There is a full-size keyboard and a small display screen using the now familiar e-ink technology that is commonplace in dedicated digital reading devices, which has the virtue of offering high contrast and legibility in direct sunlight. The scratch-built writing software admits no customization whatsoever: no preferences, no notifications, no toolbars, and certainly no anthropomorphic paperclips. It supports backspace and deletion, but there is no copy and paste option—"just like . . . before 1979 (and the debut of WordStar)," the KickStarter page explains.[35] While there is no Web access as such, built-in Wi-Fi allows it to continually sync to a cloud-based storage system. The battery is supposed to be good for four weeks of steady use. There is a carrying handle. Within thirty days KickStarter backers had funded the project at roughly 125 percent of what the developers were seeking.

The concept is not strictly new. The TRS-80 Model 100, a small notebook-style computer first introduced by Radio Shack in 1983, remained popular, especially with journalists, for years after the end of its natural market cycle because of its light weight and long battery life. Similarly, the AlphaSmart Neo, which looks like a keyboard with a narrow LCD screen curling across the top, has found a niche among writers looking for a dedicated word processing machine. But more than either of these, the Hemingwrite was self-consciously designed—*engineered,* as the video insists—as both a finely crafted writing instrument, not unlike Stephen King's Waterman pen, say, and a kind of lifestyle accessory. (Presumably the Hemingwrite's name is equally deliberately engineered, calibrated to evoke the association with the hard-drinking, hardboiled figure whose prose was nonetheless a model of clarity and concision.) The chief premise on which the product's design rests is the separation between composition and editing. The Hemingwrite is designed for pounding out drafts, for something close to what composition theorists term freewriting. Editing, revising, and polishing all come later—after the device has been synched by its cloud storage to some other platform. In this it actually hews very close to the power-typing model initially proposed for the MT/ST, where the emphasis was on the typist getting the words out on the page. That the first draft left behind a typewritten mess in the process of getting the correct sequence of keystrokes onto the tape didn't matter—the perfect printed copy would come later.

Its designers tout the notion that the Hemingwrite combines the best of what word processing, typewriting, and longhand journaling each have to offer to the writer. In fact, it might be better said to *quote* each of these, self-consciously selecting features and affordances to reengineer and reimplement in a freshly fabricated environment. A writing instrument that is always instantly available? That's journaling. A rugged, responsive keyboard? That's typewriting. Backspacing to correct typos in a blink, and never having to worry about where your document is? That's word processing. For this reason the Hemingwrite, more than the Hanx Writer and as much or more than any recent product or software package, serves to communicate the core truth about word processing in the present moment: that it too is already a historical category, as vintage and retro in its way as the hipster's vinyl or Moleskine. Not because we don't use word processors anymore, and not because word processing implies a stable or universal category of activity, but because word processing has—once again—been denaturalized as the way we write. Word processing was, for a time, something new, something different from writing; then, for a time, it simply was writing, threatening to eclipse everything that it had come before it; now it is once again something to be self-consciously emulated and engineered with deliberation, a specialized app rather than an all-encompassing application.

Doubtless some believe we are on the threshold of just the kind of automated, ubiquitously algorithmic future imagined by writers from Jonathan Swift to Italo Calvino.[36] Already we can pick out the signs, from spell-check to autocomplete to entire news stories reportedly written by computer.[37] Others expect word processing will soon begin to blend seamlessly with advanced information management tools like DEVONthink, which rely on data mining and fuzzy logic to build associative networks through vast stores of digital information in the manner anticipated by the Memex.[38] But I think the future of word processing will prove significantly more variegated. If we choose to work in Word nowadays, we do so to the deliberate exclusion of myriad other options, like Medium (a boutique blogging platform) and WriteRoom and Apple's Pages and Google Docs. We sit with our laptop to the deliberate exclusion of our tablet and phone. Or even our typewriter, as these too are making an upstart comeback.[39] Word processing has thus once again become deeply entangled with the affordances of individual platforms. George R. R. Martin's decision to assiduously continue running WordStar is not eccentric but symptomatic. Word processing is not a technology of the office anymore—that particular association seems as dated as the heavy metal Selectrics adorning the desks in *Mad Men*. (As if to prove this point or else to completely redefine its terms, the current Microsoft branding is

Office Everywhere.) Word processing—when we use the phrase at all—is an activity we evoke self-consciously to mean something like powering through a writing task, marshaling the resources of the computing power we keep so casually now at our fingertips. If I'm word processing something, then I'm as equally likely to be "telephoning" you, which is to say choosing to stress the technology's individuality ever so slightly in order to use its vaguely retro aura as a sly form of rhetorical emphasis. Word processing is something we once did, in the same way we once used dial-up modems and studiously hyphenated the triple alliterative of the World-Wide-Web. We still use the Web, of course; dial-up modems not very much.[40] As for word processing, nowadays we mostly just write, here, there, and everywhere, across ever-increasing multitudes of platforms, services, and surfaces.

In this book I have endeavored to collect and relate something like one of Kittler's unwritten literary sociologies—if not in conscious imitation of his distinctive method, then at least out of the shared conviction that our writing technologies do shape our thinking. (And yet throughout, we have struggled to avoid the pitfalls of determinism: Was Nietzsche's well-documented embrace of the aphorism really a consequence of the Malling-Hansen Writing Ball, or of his deteriorating eyesight? Or both, or neither, or—as seems most likely—both in conjunction with myriad other factors?) If, as I suggest above, the "conclusion" to word processing's story has already been written—in the sense that we can now understand word processing as a distinct technological or historical phenomenon, bracketed in historical time—then any final conclusions I can offer here, so soon after the event, run the risk of seeming premature. Nonetheless some underlying features of my narrative are worth revisiting.

Pace Gore Vidal, word processing did not erase literature, not in any sense I can fathom. Neither, of course, did it perfect literature. But like the typewriter before it, word processing changed the face of literary culture and our imagination of literary authorship. The backbone of my argument has been a consideration of authorial labor in the production of writing, in conjunction with the material particulars of various technologies of writing. For word processing's crucial formative period—from the early 1970s on through to the early 1980s, when it made the transition from office equipment to killer application for the personal computer—this backbone was upheld by a social tissue of embodied—gendered—relations. These had everything to do with work and labor and whose work the labor-intensive process of writing was supposed to be. Gender did not cease to be a consideration in either our imagination or our implementation of writing technologies thereafter, of course, but I have found no particularly compelling evidence to suggest that word processing's uptake in the literary sphere broke down along gendered lines once the tipping point had been reached, even if (as the stories I have collected here suggest) the preponderance of the earliest literary adopters appear to have been men.

Did computers really save writers work in the end? I believe it is better to ask what new kinds of work they helped to make. Although the language of labor is inescapable in any number of testimonials about the efficacy of word processing (in particular where revision is concerned), it is clear that for many authors other forms of work arose in turn, whether it was plumbing the arcane mysteries of the machine, keeping up with computers and software, or just doing forms of work (copyediting, even typesetting) previously relegated to others. Similarly, computers did not replace older writing technologies but instead coexisted with them—many or most writers avail themselves of hybrid working habits that result in their texts migrating back and forth across different media in the course of their production. Archivists and scholars, for their part, will also have new work to do. They will have to contend not just with the legacy of computers and digital storage, but also the textual relations inhering amid the extraordinarily dense and diverse constellations of writing practices that can materially coexist in the tiny universe of an author's study or on top of a writer's desk (and desktop). Hard drives and floppy disks will no more erase literature or literary history than word processing itself will, but literary history and literary criticism also will change as new forms of bibliographic analysis and biographical information—perhaps too much information—become available.

But we will never know all that we want to know, regardless of the abundance of our information and how hard we work to track the changes. We may never, for example, know with absolute certitude who was the first author to sit down in front of a digital computer's keyboard and compose a published work of fiction or poetry directly on the screen. Quite possibly it was Jerry Pournelle or maybe it was David Gerrold or even Michael Crichton or Richard Condon; or someone else entirely whom I have overlooked. It probably happened in the year 1977 or 1978 at the latest, and it was almost certainly a popular (as opposed to a highbrow) author; they were probably writing in English because that was (and is) the language in whose syntax the dominant systems and software are programmed. John Hersey, though working earlier than any of the preceding, did not, as we have seen, initially compose *My Petition for More Space* at the PDP-10's terminal, though he certainly edited and revised there extensively—and he did something these others did not: he used a computer to design and typeset the finished book.[41]

But if we jettison our impulse toward what Nick Montfort has usefully termed screen essentialism in our thinking about the norms of digital culture, then my favored candidates for the first published novel written with a word processor remain Len Deighton and Ellenor Handley.[42] It is true that the MT/ST was not called a word processor during its initial roll-out, but it was once IBM had adopted the term as part of its general sales strategy by around 1970; both in concept and in name then, there is no question that the MT/ST was the first word processor. As such, it defamiliarizes our understanding of that term through its reliance on what was in effect a paper screen; what is important about word processing turns out to be not the glowing letters behind the glass but a workable mechanism for suspending the act (that is, the moment) of inscription. Just as importantly, the intensely collaborative process between Deighton and Handley—though there is no doubt that Deighton was the exclusive author of the text in every literary sense—captures the important sociological dimensions of word processing's formation, both in its rigidly apportioned gender roles and the office workflows that shaped and reinforced them. By the same token, however, we also see how the material circumstances of literary production altered the conceit of the unseen hands: the very fact that Deighton and Handley both worked out of the same room makes their professional relationship more nuanced than that of the traditional boss and secretary, whom word processing explicitly sought to separate from one another. And whatever *Bomber's* legacy as a contribution to twentieth-century literature, it is a clear harbinger of the techno-thriller, one of the leading genres today; if some book has to be first, this one—with its machines fighting machines—is perhaps more topically suited to the claim than most.

Move beyond the fixation on firsts, though, and the range and register of our stories expands dramatically. Every impulse that I had to generalize about word processing—that it made books longer, that it made sentences shorter, that it made sentences longer, that it made authors more prolific—was seemingly countered by some equally compelling exemplar suggesting otherwise. There is no question that the popular advent of word processing around the year 1981 was an event of the highest significance in the history of writing. The technology and the changes it portended would have been visible everywhere to a professional author then, whether at writers' conferences or in the pages of newsstand magazines or on late-night talk shows or just watching what the neighbor's kid was doing in front of that strange new glass box. In terms of sensory input alone the arrival of a computer on a writer's desk would have been drama of the highest order—not just the glowing screen (and the eyestrain that ensued), but the whirring and clicking and clacking of the hardware, the hissing sizzle of the modem, the screech and scrape of the printer, the new objects to touch (keyboards not least, with their new "feel" to adjust to), even the smells of plastic and polymers when the system was humming and throwing off heat. Michael Joyce, an Irish American novelist best known for his pioneering experimental hypertext fiction *Afternoon: A Story* (1987), captures that experience in a letter he wrote to a friend just as he was making his own first acquaintance with a computer:

> I am amazed by how compelling the computer is, how freely I am able to write at it (confined to journals as yet, until I get the fine points of Applewriter down), and how easily time goes by at its terminal. But I am aware of its inwardness, a sense that time spent there is burrowing and silent, somehow geometric. Possibly this really is a function of not having the printer as yet, but I doubt so. I think it has more to do with the utterly elemental quality of the information the Apple handles, the really quite pleasing beauty of the rows of chips when the lid is off, the pleasant geographic satisfaction of memory simultaneously on the disk and in memory itself, the ribbed, grey ribbon that links them. I mean, *this* is seductive technology.[43]

Seductive technology: even an off-the-shelf Apple II was capable of inspiring the full measure of imaginative possibility Joyce would eventually bring to the design of Storyspace and the composition of *Afternoon*. But while the personal computer's sudden arrival doubtless seemed seductive to Joyce and his contemporaries, its impact was still more diffuse than we might first think. For every writer who switched over, another one did not; for every writer who composed on the screen, another one continued to work longhand, or worked from printed hard copies, or both; for every writer who became obsessed with the mysteries of the technology, another was content to subsist within the doldrums of their system's every default; for every writer who wrote more, another *rewrote* more; for every writer who finished a book with their word processor, another did not; and so on.

It may be that the rhetorical equivalencies behind some of these contentions will in time be disproven through more rigorous quantitative and formal analysis, what the digital humanities calls distant reading or macroanalysis; regardless, I prefer to err on the side of individual circumstance and plurality rather than hard determinism. Word processing was (and is) an intimate technology, however paradoxically intangible the software that enables it. Writers live with and within their word processors, and thus with and within the system's logics and constraints—these themselves become part of the daily lived experience of writers' working hours, as predictable and proximate as the squeak of a chair or that certain shaft of sunlight that makes its way across the room. We might also remember that, computationally speaking, "processing" is an intermediary state: it is something that happens in between the system's input and output, and thus always in large measure out of reach. So I want to resist the impulse to fill these final paragraphs with proclamations about what word processing does and does not do for or to the grand human enterprise of writing literature.

However, we should not diminish the magnitude of what has changed. A writer—say it is myself, say it is right now—presses a key and completes a circuit and a spark of voltage flashes through a silicon chip to open a logic gate that illuminates a pattern of pixels on a liquid crystal display screen (could the terminology be more wonderful?); the writer presses another key and the process repeats, or maybe that first letter disappears as the electrical charge briefly held within the system's solid-state memory rearranges itself into another configuration as the typo is corrected, or the writer settles on a word just a little more *juste.* This is an extraordinary act, and it is a trivial act. It is also a material act—and in that material act lives history, a history that is part of what literature now is. Others will contribute their own stories and counterstories, or so I very much hope.[44] For myself, I have only one more key to press, and then, just like that, the teeming processes that make up my words within this i5 4200U CPU will come to a punctuated stop. And now, just like that . . . they do.

AUTHOR'S NOTE

Now that *Track Changes* is published, there will surely be persons who wish to come forward and offer additional information about the literary history of word processing—stories of their own that I have overlooked or not included, or tips and contacts for further research. In order to help ensure that a literary history of word processing can continue to evolve, new information from readers will be evaluated and made available as appropriate at http://trackchangesbook.info. I encourage you to contact me there if you have a story to share. Thank you.

NOTES

Preface

1. "The First Writing-Machines" first appeared in 1905. It is a tricky little text consisting of intertwined autobiography, exaggeration, fabrication, and what are likely simply mistakes. Notably, Twain misidentifies the typescript in question as that of *The Adventures of Tom Sawyer,* though scholarship is definitive on the point that it was in fact *Life on the Mississippi.* Only a few pages of the typescript still survive today. See Horst H. Kruse, *Mark Twain and "Life on the Mississippi"* (Amherst: University of Massachusetts Press, 1981), for details of that book's composition and publication. Darren Wershler-Henry, in *The Iron Whim: A Fragmented History of Typewriting* (Ithaca, NY: Cornell University Press, 2005), examines the "Writing-Machines" sketch in some detail (225–230); for another insightful reading of Twain's typewriting, see Lisa Gitelman, "Mississippi MSS: Twain, Typing, and Moving Panorama of Literary Production," in *Residual Media,* ed. Charles R. Acland (Minneapolis: University of Minnesota Press, 2007), 329–343.
2. Oscar Wilde, *The Picture of Dorian Gray: An Annotated, Uncensored Edition,* ed. Nicholas Frankel (Cambridge, MA: Harvard University Press, 2011), 51.
3. The claim about James's dependence on the sound of the typewriter originates with his amanuensis, Theodora Bosanquet, has been repeated by Marshall McLuhan, and is confirmed by his biographer Leon Edel. See Wershler-Henry, *The Iron Whim,* 100–101.
4. Michael R. Williams, "A Preview of Things to Come: Some Remarks on the First Generation of Computers," in *The First Computers: History and*

Architecture, ed. Raúl Rojas and Ulf Hashagen (Cambridge, MA: MIT Press, 2002), 3.

5. Thomas Haigh, "Remembering the Office of the Future: The Origins of Word Processing and Office Automation," *IEEE Annals of the History of Computing* 28, no. 4 (October–December 2006): 12. As early as 1949, a Jesuit priest, Roberto Busa, S.J., began using IBM punched cards to compile a concordance to the 13 million words of St. Thomas Aquinas, working with computers not only to "process" literary data but also to print the results for what would become his *Index Thomisticus*. The definitive account of Busa's early literary computing work is to be found in Steven E. Jones, *Roberto Busa, S.J., and the Emergence of Humanities Computing: The Priest and the Punched Cards* (London: Routledge, 2016).
6. Recent years have seen many popular books on the history of handwriting, typewriting, and writing writ large. For just a sample, see Philip Hensher, *The Missing Ink: The Lost Art of Handwriting* (New York: Farrar, Straus and Giroux, 2012); Wershler-Henry, *The Iron Whim* (2005); and Matthew Battles, *Palimpsest: A History of the Written Word* (New York: W. W. Norton, 2015).
7. Ezra Greenspan and Jonathan Rose, "An Introduction to Book History," *Book History* 1, no. 1 (1998): ix.
8. For one very recent starting point, which also offers a useful synthesis of many previous studies, see Naomi S. Baron, *Words Onscreen: The Fate of Reading in a Digital World* (Oxford: Oxford University Press, 2015).
9. One project of exceptional promise in this regard (researched and written contemporaneously with *Track Changes*) is Tom Mullaney's forthcoming *The Chinese Typewriter: A Global History;* see also Nanette Gottlieb, *Word-Processing Technology in Japan: Kanji and the Keyboard* (London: Routledge, 2000).
10. Please see the Author's Note regarding trackchangesbook.info, a website I have set up in anticipation of such instances.
11. As recounted in Mona Simpson, "A Sister's Eulogy for Steve Jobs," *New York Times*, October 30, 2011, http://www.nytimes.com/2011/10/30/opinion/mona-simpsons-eulogy-for-steve-jobs.html. She writes: "I didn't know much about computers. I still worked on a manual Olivetti typewriter. I told Steve I'd recently considered my first purchase of a computer: something called the Cromemco. Steve told me it was a good thing I'd waited. He said he was making something that was going to be insanely beautiful." Also email to the author, December 3, 2014.
12. As reported in Curt Suplee, "Tapping at the Chamber Door," *PC Magazine*, May 29, 1984, 249. These same AAP statistics are cited by the *Chicago Guide to Preparing Electronic Manuscripts for Authors and Publishers* (Chicago: University of Chicago Press, 1987), 1.

13. But this was by no means true of elsewhere in the world, even other Anglophone nations. By contrast, for example, Joe Moran suggests that in the United Kingdom it was the September 1985 debut of Alan Sugar's Amstrad PCW 8256 that marked "the tipping point when many writers, published and aspiring, made the trek to Dixons, where it was exclusively sold, and joined the computer age." The Amstrad, which Sugar had designed after seeing word processors in Tokyo, came with a twelve-inch screen, a dot matrix printer, and its own word processing program, called LocoScript. It cost £399. See Moran, "Typewriter, You're Fired! How Writers Learned to Love the Computer," *The Guardian*, August 28, 2015, http://www.theguardian.com/books/2015/aug/28/how-amstrad-word-processor-encouraged-writers-use-computers.
14. See Pynchon, "Is It OK to Be a Luddite?," *New York Times Book Review*, October 28, 1984, 1, 40–41; and Vidal, "In Love with the Adverb," *New York Review of Books*, March 29, 1984.
15. According to Michael Crichton, in 1978 there were approximately 5,000 "desktop" computers in the United States, and by 1982 there were 5 million. See Crichton, *Electronic Life: How to Think about Computers* (New York: Knopf, 1983), 3.
16. As it happens, 1984 was also the year illustrator David Levine began to sometimes draw writers with computers in his famous caricatures for the *New York Review of Books*. The first appears to have been Alison Lurie, for the October 11, 1984, issue where her novel *Foreign Affairs* was reviewed. Levine did not routinely give his authors computers until the early 2000s, however, when he outfitted them with minuscule laptops dwarfed by their outsized heads. The depiction of the computer as slim and lightweight—almost nonexistent as opposed to the blocky lines of his typewriters—becomes his visual measure of the difference between the two technologies.
17. David Bain, email to the author, August 16, 2015. The poet William Dickey, writing in his introduction to a symposium on "The Writer and the Computer" appearing in *New England Review / Bread Loaf Quarterly* 10, no. 1 (Autumn 1987), tenders a similar observation, quoting Donald Hall as saying "Now, when [writers] get together what they talk about is their word-processors" (44).

Introduction

1. For the official *Conan* clip, which has garnered over one million views, see "George R. R. Martin Still Uses a DOS Word Processor," YouTube video, 1:40, posted by Team Coco on May 13, 2014, https://www.youtube.com/watch?v=X5REM-3nWHg.

2. Strictly speaking it had been known already. In an entry on his "LiveJournal" page, dated February 17, 2011, Martin notes, "I still do all my writing on an old DOS machine running WordStar 4.0, the Duesenberg of word processing software (very old, but unsurpassed)." *Not a Blog* (blog), http://grrm.livejournal.com/197075.html.
3. Dennis Baron, *A Better Pencil: Readers, Writers, and the Digital Revolution* (Oxford: Oxford University Press, 2009), 106–109.
4. Thomas J. Bergin, "The Origins of Word Processing Software for Personal Computers: 1976–1985," *IEEE Annals of the History of Computing* 28, no. 4 (October–December 2006): 43.
5. Robert J. Sawyer, "WordStar: A Writer's Word Processor," *Science Fiction Writer* (1990, 1996), http://sfwriter.com/wordstar.htm.
6. For these and other details of WordStar's colorful history, see Seymour Rubinstein, "Recollections: The Rise and Fall of WordStar," *IEEE Annals of the History of Computing* 28, no. 4 (October–December 2006), 64–72; see also, Bergin, "Origins of Word Processing Software."
7. Sawyer notes that this was so integral to WordStar's design that the program even included a utility for letting users swap around the behavior of their Caps Lock and Control keys if the default layout of their keyboard placed the latter more conveniently than the former. See http://sfwriter.com/wordstar.htm.
8. Ibid.
9. Ibid.
10. Adam Bradley, *Ralph Ellison in Progress* (New Haven, CT: Yale University Press, 2010), 40.
11. Michael Heim, *Electric Language: A Philosophical Study of Word Processing* (New Haven, CT: Yale University Press, 1987), 136.
12. Ibid., 194.
13. Mark Poster, *The Mode of Information: Poststructuralism and Social Contexts* (Chicago: University of Chicago Press, 1990), 111.
14. Hannah Sullivan, *The Work of Revision* (Cambridge, MA: Harvard University Press, 2013), 256–257.
15. "Give Claire Messud the Right Pen and Paper," *Boston Globe,* May 31, 2014, http://www.bostonglobe.com/arts/books/2014/05/31/new-england-writers-work-claire-messud/ao0kuSDJg4Gixvm31AyqgI/story.html.
16. Bradley, *Ralph Ellison in Progress,* 33. Bradley is willing to characterize word processing's impact on Ellison as "complicated" and "not always positive," but he is dismissive of the notion that there is any single causal element behind the trajectory of Ellison's later career.
17. Thriller author Alec Nevala-Lee, commenting on Martin's use of WordStar, puts it this way: "Much of writing, as I've said many times before, boils

down to habit, and writers are rightly nervous about upsetting the intricate balance of routines and rituals that they've developed over the years. Even the most productive writer knows that he's one bad morning away from the hell of writer's block, and it makes sense to persist in whatever works, when we're surrounded by a universe of doubtful alternatives." ("A Song of DOS and WordStar," *Alec Nevala-Lee* [blog], April 16, 2013, http://nevalalee.wordpress.com/2013/04/16/a-song-of-dos-and-wordstar/.)

18. Scientists sometimes call the manner in which our creativity and productivity depend on networks of seemingly inert objects *distributed cognition.* Spell-check, which was implemented in WordStar 4.0 with a program called SpellStar, is an easy example: but we find all manner of everyday cognitive functions distributed across hybrid human-artificial systems—think about driving a car (especially now with the aid of GPS).
19. "Tom McCarthy: My Desktop," *Writer's Desktops* (blog), *The Guardian,* November 24, 2011, http://www.theguardian.com/books/2011/nov/24/tom-mccarthy-desktop.
20. "The WIRED Diaries," *Wired 7.01* (archived), January 1999, http://archive.wired.com/wired/archive/7.01/diaries_pr.html.
21. I am of course alluding to technology historian Langdon Winner's classic essay, "Do Artifacts Have Politics?," originally published in *Daedalus* 109, no. 1 (Winter 1980): 121–136.
22. In this I follow Edward L. Said, who glosses Vico's principle as "We can really only know what we make" and "To know is to know how a thing is made, to see it from the point of view of its human maker." Said, *Humanism and Democratic Criticism* (New York: Columbia University Press, 2004), 11.
23. Jerome McGann, *The Scholar's Art: Literary Studies in a Managed World* (Chicago: University of Chicago Press, 2006), ix.
24. Evija Trofimova, *Paul Auster's Writing Machine: A Thing to Write With* (New York: Bloomsbury, 2014), xi–xii.
25. See Paul Auster and Sam Messer, *The Story of My Typewriter* (New York: D. A. P., 2002).
26. "The business of fitting the inchoate and intractable plasma of sensation or experience into the brittle containers—word and grammatical forms—with which we convey it to the reader," is how Deborah Eisenberg puts it. Eisenberg, "Resistance," in *The Eleventh Draft: Craft and the Writing Life from the Iowa Writers' Workshop,* ed. Frank Conroy (New York: Harper Collins, 1999), 114.
27. John Durham Peters, "Writing," in *The International Encyclopedia of Media Studies: Media Effects / Media Psychology,* ed. Angharad N. Valdiva and Erica Scharrer (London: Blackwell, 2013), 9.

28. W. Brian Arthur, *The Nature of Technology: What It Is and How It Evolves* (New York: Free Press, 2009), 12.
29. Vikram Chandra, *Geek Sublime: The Beauty of Code, the Code of Beauty* (Minneapolis: Graywolf Press, 2014), 14.
30. Sawyer uses the example of marking a block of text for some later action to illustrate what he perceives as WordStar's advantage over WordPerfect in this regard: "WordPerfect requires that I decide whether I want to cut or copy a block, then immediately mark the beginning of the block, then immediately mark the end of the block, then immediately position the cursor at where I want the block to go, then immediately move the block, and then find my way back to the place where I was originally working. From the moment I decide I might, perhaps, want to do something with a block of text to the moment I actually finish that operation, WordPerfect is in control, dictating what I must do. WordStar, with its long-hand-page metaphor, says, hey, do whatever you want whenever you want to. This is a good spot to mark the beginning of a block? Fine. What would you like to do next? Deal with the block? Continue writing? Use the thesaurus?" (http://sfwriter.com/wordstar.htm.)
31. In addition to their many papers, see especially Daniel Chandler, *The Act of Writing: A Media Theory Approach* (Aberystwyth: University of Wales, 1995); and Christina Haas, *Writing Technology: Studies on The Materiality of Literacy* (Mahwah, NJ: Erlbaum, 1996).
32. Haas, *Writing Technology,* xiv.
33. See Adi Robertson, "King of the Click: The Story of the Greatest Keyboard Ever Made," *The Verge,* October 7, 2014, http://www.theverge.com/2014/10/7/6882427/king-of-keys.
34. Nietzsche's statement has passed into Anglo-American media studies largely through its invocation by Friedrich Kittler, who sources it to a letter written in February 1882. Kittler, *Gramophone, Film, Typewriter,* trans. Geoffrey Winthrop-Young (Stanford, CA: Stanford University Press, 1999), 200.
35. William Dickey, "A Ham Sandwich and Some Hay," *New England Review / Bread Loaf Quarterly* 10, no. 1 (Autumn 1987): 45. Dickey's essay introduces a symposium on "The Writer and the Computer" that also includes contributions from Wendell Berry (who, as we will see in Chapter 7, does not buy a computer), Sven Birkerts, Robert Pinsky (co-author of the interactive fiction game *Mindwheel*), Rob Swigart (author of the adventure game *Portal*), and P. Michael Campbell. Notably, this issue of *NER/BLQ* is also its tenth anniversary issue, and, as its editors remark at the outset, reflects the journal's ongoing commitment to engaging "Subjects That Matter" (v). For more on Dickey's HyperCard

poetry, which remains unpublished, see his essay "Poem Descending a Staircase," in *Hypermedia and Literary Studies,* eds. Paul Delany and George P. Landow (Cambridge: MIT Press, 1994), 143–152.

36. Matt Schudell, "Lucille Clifton, Md. Poet Laureate and National Book Award Winner," *Washington Post,* February 21, 2010, http://www.washingtonpost.com/wp-dyn/content/article/2010/02/20/AR2010022003419.html.
37. Phoebe Hoban, "Beyond the Typewriter," *New York,* March 10, 1986, 14.
38. Based on the contents of the floppy disks included in her papers at Emory University. "Lucille Clifton Papers > Born Digital Materials," Emory Finding Aids, accessed July 17, 2015, findingaids.library.emory.edu/documents/clifton1054/series11/.
39. Susan B. A. Somers-Willet, "'A Music to Language': A Conversation with Lucille Clifton," *American Voice* 49 (Summer 1999): 81.
40. John McPhee, "Structure," *New Yorker,* January 14, 2013, 53.
41. At least one fan apparently did exactly that, using a copy of WordStar to re-create the opening paragraphs of *A Game of Thrones.* See "George R. R. Martin Writes Game of Thrones on an Old School DOS PC," *Celebrities & Entertainment* (blog), *Tweak Town,* May 14, 2014, http://www.tweaktown.com/news/37719/george-r-r-martin-writes-game-of-thrones-on-an-old-school-dos-pc/index.html.
42. This kind of fascination is not, of course, limited to writers. We crave proximity to the creative process in other domains as well, latching on to all manner of minute particulars of craft and instrumentation. Take rock music: There is an anecdote that circulates about the virtuoso electric guitarist Eric Johnson actually being able to hear the difference when a technician replaced the nine-volt batteries in his effects pedals with a different brand. That story has become a fan favorite because it encapsulates Johnson's reputation as a perfectionist who hears what others cannot. His guitar technique may be ineffable, but something as mundane as a particular brand of battery? That we can grasp hold of. (I am grateful to Alan Galey for suggesting this story to me.)
43. Tom McCarthy, "Writing Bytes," *New York Times Book Review,* November 3, 2013, 12.

1. Word Processing as a Literary Subject

1. The episode is recounted in Charles Babbage, *Passages from the Life of a Philosopher* (London: Longman, 1864), 192–194.
2. Ricardo Birmele, "WordPerfect," *Byte,* December 1984, 277.

3. Art Spikol, "Processed Words," *Writer's Digest*, August 1983, 8.
4. Ray Hammond, *The Writer and the Word Processor* (London: Coronet, 1984), 9. Hammond is a science writer and futurologist. In 1983 he had published a similar book, *The Musician and the Micro*.
5. Ibid.
6. John Durham Peters, "Writing," in *The International Encyclopedia of Media Studies: Media Effects / Media Psychology*, ed. Angharad N. Valdiva and Erica Scharrer (London: Blackwell, 2013), 18–19.
7. See Peter Stallybrass, "Printing and the Manuscript Revolution," in *Explorations in Communication and History*, ed. Barbie Zelizer (Abingdon, UK: Routledge, 2008), 111–118; Stallybrass, "'Little Jobs': Broadsides and the Printing Revolution," in *Agent of Change: Print Culture Studies after Elizabeth L. Eisenstein*, ed. Sabrina Alcorn Baron, Eric N. Lindquist, and Eleanor F. Shevlin (Amherst: University of Massachusetts Press, 2007), 315–341; and Lisa Gitelman, *Paper Knowledge: Toward a Media History of Documents* (Durham, NC: Duke University Press, 2014): 21–52.
8. John Guillory, "The Memo and Modernity," *Critical Inquiry* 31 (Autumn 2004): 111.
9. Margaret Atwood, "Amnesty International: An Address," *Second Words: Selected Critical Prose, 1960–1982* (Toronto: House of Anansi Press, 1982), 397.
10. Kenneth Goldsmith has repeated this line on many occasions. For one instance, see "I Look to Theory Only when I Realize That Somebody Has Dedicated Their Entire Life to a Question I Have Only Fleetingly Considered (a Work in Progress)," at *Kenneth Goldsmith* (author page), University of Buffalo, *Electronic Poetry Center*, accessed July 23, 2015, http://wings.buffalo.edu/epc/authors/goldsmith/theory.html.
11. Hammond, *Writer and the Word Processor*, 9–10.
12. See Lawrence Rainey, *Revisiting The Waste Land* (New Haven, CT: Yale University Press, 2005), for both the definitive forensic account of the text's typewritten composition and a deft reading of the typewriter's figure.
13. Hannah Sullivan, *The Work of Revision* (Cambridge, MA: Harvard University Press, 2013), 39–41.
14. For two starting points into this history, see Barrie Tulliett, *Typewriter Art: A Modern Anthology* (London: Laurence King, 2014); and Jerome J. McGann, *Black Riders: The Visible Language of Modernism* (Princeton, NJ: Princeton University Press, 1993).
15. Darren Wershler-Henry, *The Iron Whim: A Fragmented History of Typewriting* (Ithaca, NY: Cornell University Press, 2007), 2.

16. Adam Bradley, *Ralph Ellison in Progress* (Hartford, CT: Yale University Press, 2010), 54.
17. Ibid.
18. Lawrence Grobel, *Conversations with Capote* (Boston: Da Capo Press, 2000), 32.
19. One exception, featured by Matthew Fuller, is Jeff Bridges's performance in the otherwise forgettable action thriller *Blown Away* (1994), wherein Bridges must defuse a bomb set to go off if a certain pace of typing is not maintained by the victim on her computer, whose remaining storage capacity clicks inexorably (and improbably) down to zero. Fuller, "It Looks Like You're Writing a Letter," *Behind the Blip: Essays on the Culture of Software* (Brooklyn: Autonomedia, 2003), 137.
20. David Mamet, "The Art of Theater No. 11," interviewed by John Lahr, *Paris Review,* Spring 1997, http://www.theparisreview.org/interviews/1280/the-art-of-theater-no-11-david-mamet.
21. Anne Fadiman, *Ex Libris: Confessions of a Common Reader* (New York: Farrar, Straus and Giroux, 1998), 93.
22. For just two examples, see Marshall Fisher, "Memoria ex Machina," *Harpers Magazine,* December 2002, http://harpers.org/archive/2002/12/memoria-ex-machina/; and William Pannapacker, "A Type of Nostalgia," *Chronicle of Higher Education,* February 26, 2012, http://m.chronicle.com/article/A-Type-of-Nostalgia/130904/. (In each instance their title gives away the game.)
23. Stacy Conradt, "The Quick 10: 10 Authors and Their Typewriters," *Mental_Floss,* June 15, 2009, http://mentalfloss.com/article/21979/quick-10-10-authors-and-their-typewriters; see also Richard Polt, "Writers and Their Typewriters," *The Classic Typewriter Page,* http://site.xavier.edu/polt/typewriters/typers.html.
24. Florian Cramer, "What Is 'Post-digital'?," *APRJA (A Peer Review Journal About) Post-Digital Research* 3, no. 1 (2014), http://www.aprja.net/?p=1318.
25. See exhibit listing for Steve Soboroff's typewriter collection at The Paley Center for Media on LA Weekly.com, http://www.laweekly.com/event/steve-soboroff-typewriter-collection-5283730.
26. See https://itunes.apple.com/us/app/hanx-writer/id868326899?mt=8.
27. Tom Hanks, "I Am TOM. I Like to TYPE. Hear That?," *New York Times,* August 3, 2013, http://www.nytimes.com/2013/08/04/opinion/sunday/i-am-tom-i-like-to-type-hear-that.html?_r=0.
28. Margalit Fox, "Manson Whitlock, Typewriter Repairman, Dies at 96," *New York Times,* September 8, 2013, http://www.nytimes.com/2013/09/08/nyregion/manson-whitlock-typewriter-repairman-dies-at-96.html?_r=0.

29. See Richard Polt, *The Typewriter Revolution: A Typist's Companion for the 21st Century* (New York: Countryman Press, 2015).
30. For documentation of Tim Youd's work, see "Typewriter Series," *Tim Youd* (author website), http://www.timyoud.com/typewriter-series.html.
31. Frank Conroy, quoted in Peter H. Lewis, "Computer Words: Less Perfect?," *New York Times,* November 1, 1992, http://www.nytimes.com/1992/11/01/education/computer-words-less-perfect.html.
32. See Josh Giesbrecht, "How the Ballpoint Pen Killed Cursive," *The Atlantic,* August 28, 2015, http://www.theatlantic.com/technology/archive/2015/08/ballpoint-pens-object-lesson-history-handwriting/402205/.
33. See Jay David Bolter and Richard Grusin, *Remediation: Understanding New Media* (Cambridge, MA: MIT Press, 1999).
34. Margaret Atwood, "Can I Have a Word?," *Monocle,* October 2013, 86–87.
35. The name was also an explicit reference to a predecessor program, Colossal Typewriter, written for the PDP-1 at nearby BBN Technologies in Cambridge, Massachusetts. A little over a decade later a young programmer named Will Crowther would go to work at BBN and program its successor machine, a PDP-10, to run something he called Colossal Cave, thus inaugurating the genre of the text-based adventure game.
36. See Bruce Bliven Jr., *The Wonderful Writing Machine* (New York: Random House, 1954).
37. Jeffrey Elliott, "Around the Worlds in 80 Ways (Starring Larry Niven)," *Writer's Digest,* January 1980, 47.
38. Anthony recounts the episode in one of his characteristically extended author's notes, this one to his 1986 novel *Wielding a Red Sword,* which (he tells us) was the fourth novel he'd completed on the new system in the last nine months (a DEC Rainbow 100+with software called Professional Text Processor). Other aspects of Anthony's description are also of interest, starting with the opening caveat to the reader: "If you happen to be one who is sick of hearing about how yet another innocent soul has been computerized, and his joys and horrors thereof, skip this account" (282). (It suggests the extent to which what I have termed the conversion narrative was by then already a set piece among literati.) Anthony's conversion narrative is more detailed and more interesting than most—replete with details not only of the (rather involved) Dvorak hack, but also a literally hair-raising episode involving a near-miss from a lightning strike mere days after he had gotten the new system installed, plus an extended meditation on the differences between the CP/M and DOS operating systems. As a document, it testifies to the extent to which writers could be extremely discerning critics and users of the new technology, fully capable

of knowing and demanding what they wanted. "Computer companies," he concludes, "have very little notion of the needs of novelists." See Piers Anthony, *Wielding a Red Sword* (New York: Ballantine Books, 1986), 282–297.

39. Jorie Graham, "The Art of Poetry No. 85," interviewed by Thomas Gardner, *Paris Review*, Spring 2003, http://www.theparisreview.org/interviews/263/the-art-of-poetry-no-85-jorie-graham.
40. "José Saramago, "The Art of Fiction No. 155," interviewed by Donzelina Barroso, *Paris Review*, Winter 1998, http://www.theparisreview.org/interviews/1032/the-art-of-fiction-no-155-jose-saramago.
41. Amos Oz, "The Art of Fiction No. 148," interviewed by Shusha Guppy, *Paris Review*, Fall 1996, http://www.theparisreview.org/interviews/1366/the-art-of-fiction-no-148-amos-oz.
42. Joan Didion, "The Salon Interview: Joan Didion," interviewed by Dave Eggers, *Salon*, October 28, 1996, http://www.salon.com/1996/10/28/interview_11/.
43. Maria Nadotti, "An Interview with Don DeLillo," trans. Peggy Boyers, ed. Don DeLillo, *Salmagundi* 100 (Fall 1993): 86–97.
44. See Patricia Cohen, "No Country for Old Typewriters: A Well-Used One Heads to Auction," *New York Times*, November 30, 2009, http://www.nytimes.com/2009/12/01/books/01typewriter.html?_r=1&.
45. Paul Auster and Sam Messer, *The Story of My Typewriter* (New York: D.A.P., 2002), 10.
46. Paul Auster, "The Art of Fiction No. 178," interviewed by Michael Wood, *Paris Review*, Fall 2003, http://www.theparisreview.org/interviews/121/the-art-of-fiction-no-178-paul-auster.
47. Ibid.
48. Auster and Messer, *Story of My Typewriter*, 21–22. In the course of my research I spoke with a journalist who had recently interviewed Auster. He informed me the author indeed still does all of his writing in notebooks and then types the day's product on the Olympia. An assistant handles his computer work. Andrew Hax, email message to the author, February 7, 2012.
49. See "Joyce Carol Oates," *Daily Routines* (blog), July 30, 2007, http://dailyroutines.typepad.com/daily_routines/2007/07/joyce-carol-oat.html.
50. Joyce Carol Oates, "The Art of Fiction No. 72," interviewed by Robert Phillips, *Paris Review*, Winter 1978, http://www.theparisreview.org/interviews/3441/the-art-of-fiction-no-72-joyce-carol-oates.
51. Ibid.
52. For a discussion of Oates's "maximalism," see Mark McGurl, *The Program Era* (Cambridge, MA: Harvard University Press, 2011), 297–320.

53. Kazuo Ishiguro, "The Art of Fiction No. 196," interviewed by Susannah Hunnerwell, *Paris Review,* Spring 2008, http://www.theparisreview.org/interviews/5829/the-art-of-fiction-no-196-kazuo-ishiguro.
54. Annie Proulx, "The Art of Fiction No. 199," interviewed by Christopher Cox, *Paris Review,* Spring 2009, http://www.theparisreview.org/interviews/5901/the-art-of-fiction-no-199-annie-proulx.
55. See David Stephen Calonne, "Creative Writers and Revision," in *Revision: History, Theory, and Practices,* ed. Alice Horning and Anne Becker (West Lafayette, IN: Parlor Press, 2006), 163. Also see Suzanne Bost, "Messy Archives and Materials that Matter: Making Knowledge with the Gloria Evangelina Anzaldúa Papers," *PMLA* 130 no. 3 (May 2015): 615–630.
56. Philip Harris, "Morrison Lectures, Reads to Delighted Full House," *Phoenix,* April 10, 2014, http://swarthmorephoenix.com/2014/04/10/toni-morrison-talks-invisible-ink/; see also Toni Morrison, "The Art of Fiction No. 134," interviewed by Elissa Schappell, *Paris Review,* Fall 1993, http://www.theparisreview.org/interviews/1888/the-art-of-fiction-no-134-toni-morrison. Morrison's papers at Princeton (not yet open to researchers, as of this writing) also reportedly contain around a hundred 3½-inch diskettes as well as a handful of 5¼-inch floppies dating from the late 1980s through the 1990s and containing chapter drafts in WordPerfect for *Jazz* (1992), *Playing in the Dark* (1992), and *Paradise* (1997). Handwriting on the disks' labels is that of both Morrison and her assistant (emails to the author from Allison L. Hughes, May–June 2015).
57. Michael Ondaatje, interview with the author, April 10, 2012.
58. See "Magic, Mystery, and Mayhem: An Interview with J. K. Rowling," amazon.co.uk., accessed July 26, 2015, http://www.amazon.com/gp/feature.html?docId=6230; and Linda Richards, "J. K. Rowling," *January Magazine,* accessed August 19, 2015, http://januarymagazine.com/profiles/jkrowling.html.
59. Such was Gloria Anzaldúa's practice, even after she began using a word processor. See Bost, "Messy Archives and Materials that Matter," 622.
60. George Steiner, "The Art of Criticism No. 2," interviewed by Ronald A. Sharp, *Paris Review,* Winter 1995, http://www.theparisreview.org/interviews/1506/the-art-of-criticism-no-2-george-steiner.
61. As documented in Mickiel Maandag's "How iA Writer Is Reintroducing the Word Processor Category," *The Brand Bite* (blog), December 26, 2011, http://www.thebrandbite.com/2011/12/26/how-ia-writer-is-reintroducing-the-word-processor-category/.
62. See Isaacson, *The Innovators: How a Group of Hackers, Geniuses, and Geeks Created the Digital Revolution* (New York: Simon and Schuster,

2014) and Gleick, *The Information: A History, a Theory, a Flood* (New York: Pantheon, 2011). Scholarly histories fare somewhat better: Martin Campbell-Kelly covers word processing (albeit primarily from the market's point of view) throughout his *From Airline Reservations to Sonic the Hedgehog: A History of the Software Industry* (Cambridge, MA: MIT Press, 2003). An especially important resource is a special issue of *IEEE Annals of the History of Computing* 28, no. 4 (October–December 2006), which contains articles from Thomas J. Bergin and Thomas Haigh, as well as a personal history from WordStar cofounder Seymour Rubinstein. In their introduction, Paul E. Ceruzzi and Burton Glad assert that "the true impact of the personal computer was that it allowed software, previously a labor-intensive and often customized activity sold to a limited number of customers, to become commoditized and mass-produced," and they identify the two most significant software genres as spreadsheets and word processing, noting that VisiCalc and WordStar both debuted in 1979 (4). See also Ceruzzi's *A History of Modern Computing* (Cambridge, MA: MIT Press, 2000). As of this writing, the only book-length scholarly history of word processing is in German: Till A. Heilmann, *Textverarbeitung: Eine Mediengeschichte des Computers als Schreibmaschine* (Bielefeld, DE: Transcript Verlag, 2012). The book is based on Heilmann's PhD thesis and spans the history of word processing at large, but with no particular emphasis on the literary.

63. Ted Nelson, *Geeks Bearing Gifts: How the Computer World Got This Way: Version 1.1* (Sausalito, CA: Mindful Press, 2009), 128.
64. Jay David Bolter, *Writing Space: The Computer, Hypertext, and the History of Writing* (Hillsdale, NJ: Erlbaum, 1991), 66.
65. The most complete history of hypertext (both as concept and as implementation) is Belinda Barnet, *Memory Machines: The Evolution of Hypertext* (London: Anthem Press, 2013).
66. Michael Joyce has said this many times, but here I quote from his *Of Two Minds: Hypertext Pedagogy and Poetics* (Ann Arbor: University of Michigan Press, 1995), 32.
67. Works produced with HyperCard and Storyspace have been the subject of numerous scholarly studies over the past twenty-five years or so. For accounts of Storyspace's technical development, see Matthew Kirschenbaum, *Mechanisms: New Media and the Forensic Imagination* (Cambridge, MA: MIT Press, 2008); and Barnet, *Memory Machines.*
68. The definitive study of the form is Nick Montfort, *Twisty Little Passages: An Approach to Interactive Fiction* (Cambridge, MA: MIT Press, 2003).
69. See Lori Emerson, *Reading Writing Interfaces: From the Digital to the Bookbound* (Minneapolis: University of Minnesota Press, 2014).

70. The complete run of the *Newsletter* is available online via the ComPile .com database: http://comppile.org/archives/RWPN/RWPnewsletter.htm. The layouts and production values became more sophisticated, offering a kind of archaeology of the form.
71. Bradford Morgan, interview with the author, October 16, 2014.
72. University of Chicago Press, *Chicago Guide to Preparing Electronic Manuscripts for Authors and Publishers* (Chicago: University of Chicago Press, 1983), 1.
73. See the *MLA Newsletter* (Summer 1985), which contains articles on Nota Bene throughout.
74. See Mark McGurl, *The Program Era: Postwar Fiction and the Rise of Creative Writing* (Cambridge, MA: Harvard University Press, 2009).
75. See Doug Reside, "'Last Modified January 1996': The Digital History of *RENT*," *Theatre Survey* 52, no. 2 (2011): 35–40; and Matthew Kirschenbaum and Doug Reside, "Tracking the Changes: Textual Scholarship and the Challenge of the Born Digital," in *The Cambridge Companion to Textual Scholarship*, ed. Neil Fraistat and Julia Flanders (New York: Cambridge University Press, 2013), 257–273.
76. For a case study, see Matthew Kirschenbaum, "Operating Systems of the Mind: Bibliography after Word Processing (The Example of Updike)," *Papers of the Bibliographical Society of America* 108, no. 4 (December 2014): 381–412.
77. For example, Miriam O'Kane Mara's "Nuala O'Faolin: New Departures in Textual and Genetic Criticism," *Irish Studies Review* 21, no. 3 (2013): 342–352, makes a persuasive but ultimately speculative case for the importance of the digital manuscripts on O'Faolin's hard drive (O'Kane Mara does not have access to the files herself).
78. See Sven Birkerts, *The Gutenberg Elegies: The Fate of Reading in an Electronic Age* (New York: Faber and Faber, 1994), 229.
79. Sven Birkerts, "Reading in a Digital Age: Notes on Why the Novel and the Internet Are Opposites, and Why the Latter Both Undermines the Former and Makes It More Necessary," in *The Edge of the Precipice: Why Read Literature in the Digital Age?*, ed. Paul Socken (Montreal: McGill-Queen's University Press, 2013), Kindle loc. 668.
80. On May 21, 2013, Oates tweeted that "word-processing allows for obsessive Mobius-strip revision, as the typewriter never did. Rewriting of rewriting. Return to p. 1. Repeat."
81. John Barth, "The Art of Fiction No. 86," interviewed by George Plimpton, *Paris Review,* Spring 1985, http://www.theparisreview.org/interviews/2910/the-art-of-fiction-no-86-john-barth.
82. Friedrich Kittler, *Gramophone, Film, Typewriter,* trans. Geoffrey Winthrop-Young (Stanford, CA: Stanford University Press, 1999), 216.

83. Gabrielle Dean (Sheridan Libraries, Johns Hopkins University), personal email to the author, September 2, 2014.
84. Hugh Kenner, *Heath/Zenith Z-100 User's Guide* (Bowie, MD: Brady Communications Co., 1984), vi. For Kenner's interest in computer-generated poetry, see Charles O. Hartman, *Virtual Muse: Experiments in Computer Poetry* (Hannover, NH: University Press of New England, 1996). (The book is obviously named in relation to Kenner's own title, *The Mechanic Muse.*)
85. For one discussion of such views, see Bruce J. Hunt, "Lines of Force, Swirls of Ether," in *From Energy to Information: Representation in Science and Technology, Art, and Literature,* ed. Bruce Clarke and Laura Dalrymple Henderson (Stanford, CA: Stanford University Press, 2002), 99–113. Hunt's essay focuses on the theories of Irish physicists George Francis FitzGerald and Oliver J. Lodge, both of whom built on James Clerk Maxwell's work on electromagnetic fields.
86. Daniel Chandler, *The Act of Writing: A Media Theory Approach* (Aberystwyth: University of Wales, 1995), 133.
87. Even if an author delivered an electronic file to a publisher, it was often still the norm to rekey it. John B. Thompson enumerates a number of reasons this was so: many authors didn't follow the press's formatting instructions; "many authors who had grown up on typewriters didn't differentiate between the letter O and zero 0, between hyphens and en-dashes, and so on"; and many authors continued revising on hard copy even after delivery of the digital file—he quotes a rule of thumb that if there were more than eight to ten corrections per page, rekeying was more efficient than making the corrections piecemeal. See Thompson, *Books in the Digital Age: The Transformation of Academic and Higher Education Publishing in Britain and the United States* (Cambridge, MA: Polity Press, 2005), 407.
88. Frederick Pohl, another prolific science fiction author who used WordStar well past its prime (through 2008), relied on the expertise of longtime fan Dick Smith to keep the software running for him. Pohl dedicated his last novel, *All the Lives He Led* (2011), to Smith.
89. Peters, "Writing," 4.
90. Though liberally reproduced around the Web, this text is currently unavailable from its most widely addressed location on Stephenson's own personal website: http://www.nealstephenson.com/content/author_colophon.htm. I reference it instead from http://wiki.zibet.net/wiki.pl/Neal_Stephenson. Stephenson has also confirmed the particulars of his process in several interviews reprinted in *Some Remarks: Essays and Other Writing* (New York: William Morrow, 2012), noting in particular that the LISP program "was nasty and tedious, but in the end reasonably

satisfying" (31). In a separate interview therein he indicates that *Cryptonomicon* (1999) was the last book he wrote directly on the computer and that he switched over to longhand because he found it often helped him when he was blocked. Plus, he says, "I like the fact that it never crashes, you can't lose your work" (262).

91. For example, Jonathan Franzen, "Liking Is for Cowards: Go for What Hurts," *New York Times,* May 28, 2011, http://www.nytimes.com/2011/05/29/opinion/29franzen.html?pagewanted=all&_r=0.
92. Jonathan Franzen, "Scavenging," in *Tolstoy's Dictaphone: Technology and the Muse,* ed. Sven Birkerts (Saint Paul, MN: Graywolf Press, 1996), 12–13.
93. Nicholson Baker, "The Art of Fiction No. 212," interviewed by Sam Anderson, *Paris Review,* Fall 2011, http://www.theparisreview.org/interviews/6097/the-art-of-fiction-no-212-nicholson-baker.

2. Perfect

1. "Screening Process Is Paper Chase," *New York Times,* March 20, 1981, http://www.nytimes.com/1981/03/20/opinion/topics-screening-processis-paper-chase-paper-chase.html.
2. Ironically, the name WordPerfect was initially unpopular with the company's employees. Cofounder W. E. Pete Peterson stuck with it, though—not only for its resonance with "letter-perfect" (an established printing idiom) but because it really did sound, well, . . . perfect. "It was so positive sounding it made any criticism seem untrue," he writes. See Peterson, *AlmostPerfect: How a Bunch of Regular Guys Built WordPerfect Corporation* (Rocklin, CA: Prima, 1998), Kindle loc. 571.
3. Jonathan Sterne has pointed out to me that the trope "perfect" is likely generalizable to other areas of media and computation; he notes its centrality to electronic music systems and software such as Synclavier and Fairlight as well as Performer and ProTools. "You can find musicians talking about perfectionism in the 1980s in the same exact way that writers are" (email to the author, March 23, 2013).
4. Walter A. Kleinschrod, *Word Processing: An AMA Management Briefing* (New York: Amacom, 1974), 4.
5. Thomas J. Anderson and William R. Trotter, *Word Processing* (New York: Amacom, 1974), 10.
6. Thomas Haigh, "Remembering the Office of the Future: The Origins of Word Processing and Office Automation," *IEEE Annals of the History of Computing* 28, no. 4 (October–December 2006): 10.

7. Ivan Flores, *Word Processing Handbook* (New York: Van Nostrand Reinhold, 1983), 5.
8. See Cornelius E. DeLoca and Samuel Jay Kalow, *The Romance Division: A Different Side of IBM* (Wyckoff, NJ: D & K Book Co., 1991), 88; Kleinschrod, *Word Processing;* Anderson and Trotter, *Word Processing.*
9. Kleinschrod, *Word Processing,* 4.
10. *Training Manual 1, IBM Magnetic Tape Selectric Typewriter Model II/ Basic Model IV* (IBM Office Products Division), i.
11. Anderson and Trotter, *Word Processing,* 19.
12. Russell Baker, "Sunday Observer: Computer Fallout," *New York Times,* October 11, 1987, http://www.nytimes.com/1987/10/11/magazine/sunday-observer-computer-fallout.html. The piece is clever in that it proceeds through progressive revisions of that opening line (not unlike the parlor game Telephone) until we arrive at the closing sentence: "Since it is easier to revise and edit with a computer than with a typewriter or pencil, this amazing machine makes it very hard to stop editing and revising long enough to write a readable sentence, much less an entire newspaper column."
13. Peter J. Bailey, "'A Hat where There Never Was a Hat': Stanley Elkin's Fifteenth Interview," *Review of Contemporary Fiction* 15, no. 2 (1995): 15–26.
14. John Updike, "Where Money and Energy Gather: A Writer's View of a Computer Laboratory," in *Research Directions in Computer Science: An MIT Perspective,* ed. Albert R. Meyer et al. (Cambridge, MA: MIT Press, 1991), np.
15. Jeffrey Elliot, "Around the Worlds in 80 Ways (Starring Larry Niven)," *Writer's Digest,* January 1980, 47.
16. This story is related in Sir Sidney Colvin, *John Keats: His Life and Poetry, Friends, Critics and Afterfame* (London: Macmillan, 1917). Colvin states the provenance as "the late B. W. Richardson, professing to quote verbatim . . . from Mr. Stephens' own statement to him in conversation" (176).
17. Peter A. McWilliams, *The Word Processing Book: A Short Course in Computer Literacy* (New York: Ballantine Books, 1983), 23.
18. As we will see in Chapter 7, the gendered pronouns of this sentence are intentional.
19. John Barth, "Virtuality," *Johns Hopkins Magazine* (1994), http://pages.jh.edu/~jhumag/994web/culture1.html.
20. Ibid.
21. Ronald John Donovan, "Writing Made Easier with Personal Computers," *Writer's Digest,* September 1982, 33–42. The letter to the editor appears in the October issue, p. 6.

22. Ray Hammond, *The Writer and the Word Processor* (London: Coronet, 1984), 13.
23. Karen Ray, "The Secret Life of My Computer," *Writer's Digest,* October 1984, 7.
24. Gish Jen, "My Muse Was an Apple Computer," *New York Times,* October 9, 2011, 9–10.
25. Anne Fadiman, *Ex Libris: Confessions of a Common Reader* (New York: Farrar, Straus and Giroux, 1998), 101.
26. Jen, "My Muse."
27. Fritz Leiber, *The Silver Eggheads* (New York: Ballantine Books, 1961). Leiber is shrewd enough to understand that while such anxieties about the automation of fiction writing might make for a good story, they are also part of a much longer tradition of romanticizing authors as creative geniuses. "Writers have been stealing the credit from wordmill-programmers for a century," one of his characters exclaims. "And before that they stole it from editors!" (77).
28. Michael Frayn, *The Tin Men* (Glasgow: William Collins Sons, 1965).
29. Most notably Raymond Queneau's 1961 *Cent Mille Milliards de Poèmes,* which permitted the titular number of sonnets to be constructed from ten bound sheets scored with fourteen lines apiece. For the Lem story, see Stanislaw Lem, *A Perfect Vacuum* (Evanston, IL: Northwestern University Press, 1999), 96–101.
30. See Calvino, *If on a Winter's Night a Traveler,* trans. William Weaver (San Diego: Harcourt Brace Jovanovich, 1982). The novel was originally published in Italian in 1979. In 1983 Calvino selected this episode to read from while lecturing in New York City, the center of the American publishing world. See Herbert Mitgang, "Books," *New York Times,* May 1, 1983, http://www.nytimes.com/1983/05/01/books/reading-and-writing-fabulous-traveler.html.
31. Arthur C. Clarke, "The Steam-Powered Word Processor," *Analog,* September 1986, 175–179.
32. Ibid., 179.
33. See J. Michael Straczynski, "Brave New Words," *Profiles,* May–June 1984, 40.
34. John Varley, "The Unprocessed Word," in *Blue Champagne* (New York: Ace, 1986), 212.
35. Ibid., 224.
36. Jonathan Swift, *Gulliver's Travels into Several Remote Nations of the World* (1726; Project Gutenberg, 2009), http://www.gutenberg.org/ebooks/829?msg=welcome_stranger. Emphasis added.
37. By 1995 a science fiction author named David Langford had completed the permutation with "The Net of Babel," the inevitable rewrite for the

digital age. See Langford, "The Net of Babel," in *The Best of Interzone*, ed. David Pringle (London: HarperCollins, 1997).

38. Robert Lekachman, "Virtuous Men and Perfect Weapons," *New York Times*, July 27, 1986, http://www.nytimes.com/1986/07/27/books/virtuous-men-and-perfect-weapons.html.
39. WordStar was initially designed to run on the CP/M operating system, but it could be used on Apple machines with a specialized piece of hardware called a SoftCard—so there is a valid explanation for Dr. Ryan's setup. That said, it seems equally likely that Clancy was knowingly or unknowingly exercising his creative license by simply commingling what would have been two highly recognizable products.
40. Clancy discussed computers regularly in his personal correspondence with his collaborator Larry Bond—for example, letters dated January 6, 1983, and December 4, 1983.
41. Lynn Neary, "Tom Clancy Dies, Left 'Indelible Mark' on Thriller Genre," *NPR*, October 2, 2013, http://www.npr.org/2013/10/02/228485169/tom-clancy-dies-left-indelible-mark-on-thriller-genre.
42. For examples of the "Word Processor" moniker, see http://www.timeslive.co.za/lifestyle/2010/02/07/the-word-processor and http://moneyweek.com/profile-james-patterson-alex-cross-47137/.
43. For an account of what it is like to write with Patterson, see Charles McGrath, "An Author's Collaborator Goes It Alone," *New York Times*, May 4, 2009, http://www.nytimes.com/2009/05/05/books/05dejo.html?_r=0.
44. "James Patterson's Best-Selling Thrillers," YouTube video, posted by *Sunrise on 7*, December 28, 2012, https://www.youtube.com/watch?v=GhtcCST0dU0.
45. See Jonathan Mahler, "James Patterson Inc.," *New York Times Magazine*, January 20, 2010, http://www.nytimes.com/2010/01/24/magazine/24patterson-t.html.
46. *Time* quoted in "Blockbuster Author Who Churns Out Potboilers—And Rakes in 40M a Year," *MoneyWeek*, January 28, 2010, http://moneyweek.com/profile-james-patterson-alex-cross-47137/.
47. The statistics and other facts and details in this paragraph are drawn from Mahler, "James Patterson Inc." See also *Forbes*'s reporting on Patterson's sales and earnings: "The World's Highest-Paid Celebrities," *Forbes*, June 29, 2015, http://www.forbes.com/profile/james-patterson/.
48. Mahler, "James Patterson Inc."
49. For an overview of these changes, see "Results for: Publishing-industry" at Answers.com, accessed August 19, 2015, http://www.answers.com/topic/publishing-industry. See also Beth Luey, "The Organization of the Book

Publishing Industry," and Laura J. Miller, "Selling the Product," both in *The Enduring Book: Print Culture in Postwar America; A History of the Book in America,* vol. 5, ed. David Paul Nord, Joan Shelley Rubin, and Michael Schudson (Chapel Hill: University of North Carolina Press, 2009), 29–54, 91–106.

50. Patterson himself champions independent bookstores. See his author website for details about his ongoing activities in support of them, http://www.jamespatterson.com/booksellers/.
51. Jason Epstein, *Book Business: Publishing Past, Present, and Future* (New York: Norton, 2002), 33.
52. Gore Vidal, "In Love with the Adverb," *New York Review of Books,* March 29, 1984, http://www.nybooks.com/articles/archives/1984/mar/29/in-love-with-the-adverb/.
53. Ibid.
54. Ibid.
55. Mahler, "James Patterson Inc." See also Edward Morris, "James Patterson: A Jester on Crusade," *BookPage,* March 2003, http://bookpage.com/interviews/8183-james-patterson#.VPxlwS6K0-2.
56. Edwin McDowell, "'No Problem' Machine Poses a Presidential Problem," *New York Times,* March 24, 1981.
57. Automation was indeed a source of anxiety for much of the literary establishment. But by the second half of the 1950s its creative possibilities were embraced by practitioners who were programming mainframe computers to generate text using aleatory operations. Experiments with the form have been ongoing ever since. Computer-generated poetry would have gained the public's general attention in 1962 in an article in *Time* magazine entitled "The Pocketa, Pocketa School" (it featured the output of a so-called "Auto-Beatnik" computer). For the most complete account of this early history, see Christopher T. Funkhouser, "First-Generation Poetry Generators: Establishing Foundations in Form," in *Mainframe Experimentalism: Early Computing and the Foundations of the Digital Arts,* ed. Hannah B. Higgins and Douglas Kahn (Berkeley: University of California Press, 2012), 243–265.
58. Brian Kunde, "A Brief History of Word Processing (through 1986)," December 1986, http://web.stanford.edu/~bkunde/fb-press/articles/wdprhist.html.
59. Hammond, *Writer and the Word Processor,* 16.
60. Carolyn Guyer, "Along the Estuary," in *Tolstoy's Dictaphone: Technology and the Muse,* ed. Sven Birkerts (St. Paul, MN: Graywolf Press, 1996), 158.
61. Many have noted that today all media converge in digital form, owing to the supposed universalism of binary representation (ones and zeros), but it

is also worth remembering how differently from one another words and images behave—even in digital form. Consider, for example, the difference between a PDF file that comes to you as a page scan (that is, an actual image) and one that contains machine-readable text that is subject to search and other forms of string-based manipulation. Despite their underlying binary composition, these two formats behave very differently for the end-user.

62. "My goal at the time was 'to do to things what light does to them,' a verse by the French poet Guillevic," Codrescu continued, speaking on NPR ("The Kaypro II: An Early Computer with a Writer's Heart," sound file, *NPR,* September 16, 2014, http://www.npr.org/blogs/alltechconsidered/2014/09/16/349027131/the-kaypro-ii-an-early-computer-with-a-writers-heart).
63. John Updike, "A Soft Spring Night in Shillington," *New Yorker,* December 24, 1984, 37.
64. See Sean Cubitt, *The Practice of Light: A Genealogy of Visual Technologies from Prints to Pixels* (Cambridge, MA: MIT Press, 2014), 103.
65. Russell Banks, "The Art of Fiction No. 152," interviewed by Robert Faggen, *Paris Review,* Summer 1998, http://www.theparisreview.org/interviews/1104/the-art-of-fiction-no-152-russell-banks.
66. Richard Abowitz, "Still Digging: A William Gass Interview," *Gadfly,* December 1998, http://www.gadflyonline.com/archive/December98/archive-gass.html.
67. J. D. Reed and Jeanne North, "Plugged-In Prose," *Time,* August 10, 1981.
68. Tom Johnson, "Four for the Future," *Writer's Digest,* April 1981, 25.
69. "Michael Crichton on the Coming Computer Revolution," YouTube video, 8:28, televised on the Merv Griffin Show in 1983, posted by "Merv Griffin Show," December 12, 2012, https://www.youtube.com/watch?v=TtYvQvkUp0A.
70. See Daniel Chandler, "The Phenomenology of Writing by Hand," *Intelligent Tutoring Media* 3 (May–August 1992): 65–74.
71. Flores, *Word Processing Handbook,* 7.
72. Ibid.
73. Andrew Johnston has detailed the pervasiveness of the search for real-time control over a variety of twentieth-century media forms, such as the rendering of computer graphics in film and the digital arts. See Johnston, "Models of Code and the Digital Architecture of Time," *Discourse: Journal for Theoretical Studies in Media and Culture* (forthcoming, 2015).
74. Tom Johnson, "Four for the Future," 25.

75. In fact this freedom of action has remained elusive, as software's continued reliance on modes (see the Introduction) demonstrates. Yet at a more fundamental level, even the very process of getting letterforms to display on the screen required enormous forethought and years of dedicated effort. A former Lexitron engineer writes the following, which gives some sense of the complexities involved; it is worth quoting at length because it is representative of the engineering challenges around CRT technology in the early 1970s:

> The scheme for getting characters on the screen in this environment was therefore somewhat messy. We wanted a full-page life-size display. That meant a big screen with lots of characters. We had a CRT custom made with electrostatic deflection plates, and also a place for a deflection yoke. The idea was that the yoke would scan the beam in a coarse raster, one scan line for each line of text. The plates would diddle the beam around to form each character at the proper location. All you had to do was generate X, Y, and Z (intensity) signals.
>
> The first attempt used a sort of analog ROM, invented and built by a professor at Caltech, whose name I have since forgotten. The heart of the unit was a delay line. It was a transmission line where the center conductor was flat, and was placed between two flat shield conductors. The 3 plates were roughly 6-inches square. In order to slow down the speed of propagation, the center conductor was actually a coil wound around a flat sheet of ferrite. Imagine placing a flat conductor, perhaps ¼-inch to ½-inch wide between the center and one shield. It will pick up a signal when you send a pulse down the delay line. Now imagine splitting this strip of conductor with a narrow gap, and letting the gap wander back and forth on both sides of center. If you look at the voltage between the two pieces with a differential amplifier, you will get a signal that varies according to the position of the gap at the time the pulse comes by on the center conductor. Three such stripes will yield X, Y, and Z signals, and if the gap is carefully placed, will paint a character on the screen. The shield and stripes are conveniently made from a pc board. Obviously, one such assembly can have a number of such groups of stripes, and they can be on both sides of the center coil, but the number of characters that will fit is less than the ASCII character set. So, the complete generator needs several of these stacks of plates and coils. The whole thing was clamped together but thick aluminum plates with bolts at the corners, and ended up about 2 or inches thick.

Getting the meandering gap right is a slow and tedious process, with the professor delivering a new layer of characters every few weeks. So, our first demo had to be done using about 1/2 of the alphabet. Fortunately, it was the lower half, so we had a few vowels to work with, but it took hours to come up with a reasonably normal looking sentence.

The characters generated this way looked pretty good—until you got a bunch on the screen. Then the little inaccuracies became really noticeable. It looked sort of like a really badly aligned type-bar typewriter, or perhaps the printed scrawl of a 4-year-old. Some of the characters were a little high, some a little low, some a little twisted, etc. (Ed Kelm, email to the author, December 31, 2011)

76. Friedrich Kittler, "There Is No Software," in *Electronic Culture: Technology and Visual Representation,* ed. Timothy Druckrey (Denville, NJ: Aperture, 1996), 331–337.
77. Ibid., 332.
78. Jacques Derrida, "Word Processor," in *Paper Machine,* trans. Rachel Bowlby (Stanford, CA: Stanford University Press, 1995), 22.
79. Louis Simpson, "Poetry and Word Processing: One or the Other, but Not Both," *New York Times,* January 3, 1988, http://www.nytimes.com/1988/01/03/books/poetry-and-word-processing-one-or-the-other-but-not-both.html; Derrida, "The Word Processor," 24.
80. Buckley wrote, "To accept Simpson's thesis is to suppose that writers (and poets) always feel that the language of the moment is lapidary." See William F. Buckley Jr., "Out, Out Luddites: Word Processors Are for Poets, Too," *Philly.com,* January 21, 1988, http://articles.philly.com/1988-01-21/news/26282435_1_word-processor-poets-louis-simpson. Simpson also drew a thoughtful response from Judson Jerome in *Writer's Digest,* who includes a sidebar pointing out that new thesaurus programs just coming onto the market could help a poet find matches to the *sound* of particular words. See Jerome, "Processing Poetry," *Writer's Digest,* December 1988, 10.
81. See Vilém Flusser, *Does Writing Have a Future?,* trans. Nancy Ann Roth (Minneapolis: University of Minnesota Press, 2011).
82. Ibid., 11–21.
83. Ibid., 13.
84. Anne Rice's Facebook page, posted September 25, 2011, https://www.facebook.com/annericefanpage/posts/10150391869805452. This is one of a number of places where Rice cites the Osborne 1 and WordStar. See also December 26, 2011, on Facebook, for example. However, in an interview

with Don Swaim, recorded October 30, 1985, she mentions also having something called an Apricot, a "wonderful English computer." Apricot was indeed a British company making PC and Macintosh clones during the 1980s. See Swaim, "Audio Interviews with Anne Rice," *Wired for Books,* October 30, 1985, sound file, http://www.wiredforbooks.org/annerice/.

85. Don Swaim, "Audio Interviews with Anne Rice."
86. Stanley Wiater, "Anne Rice," *Writer's Digest,* November 1988, 41.
87. Ibid.
88. Ibid. The statement is perhaps more than just a throwaway. In remarks posted to her website on August 20, 2003, Rice again invokes the language of perfection in relation to her composition process, including why she stopped taking comments from her editor: "I felt that I could not bring to perfection what I saw unless I did it alone." See http://www.annerice.com/sh_MessagesBeach2.htm.

3. Around 1981

1. "Last Word for the Word Processor," *Economist,* July 25, 1981, 62. Exactly how that impressive figure was determined is not disclosed.
2. Len Deighton, "Foreword," in Hammond, *The Writer and the Word Processor* (London: Coronet, 1984), 4.
3. J. D. Reed and Jeanne North, "Plugged-In Prose," *Time,* August 10, 1981.
4. Richard Krajewski, "A Writer's Guide to Word-Processing Software," *Writer's Digest,* September 1983, 40, 42, 52–59.
5. *Whole Earth Software Catalog,* ed. Stewart Brand (New York: Quantum / Doubleday, 1984), 2.
6. For Bukowski and his Macintosh, see Jed Birmingham, "Charles Bukowski, William Burroughs, and the Computer," *RealityStudio,* September 11, 2009, http://realitystudio.org/bibliographic-bunker/charles-bukowski-william-burroughs-and-the-computer/.
7. Charles Bukowski, "16-Bit Intel 8088 Chip," *Aileron* 6, no. 1 (1985).
8. Steven Levy, quoted in *Whole Earth Software Catalog,* 46.
9. Judy Grahn, email to the author, January 2, 2012.
10. See Alexis C. Madrigal, "The Time Exxon Went Into the Semiconductor Business (and Failed)," *Atlantic,* May 17, 2013, http://www.theatlantic.com/technology/archive/2013/05/the-time-exxon-went-into-the-semiconductor-business-and-failed/275993/.
11. Grahn, email to the author, January 2, 2012.
12. Frederic Golden, quoted in John A. Meyers, "A Letter from the Publisher," *Time,* January 3, 1983, 3.

13. In "I Am a Signpost," Isaac Asimov reports that he was initially approached by the editors of *Byte,* which *Popular Computing* was associated with under its original name, *onComputing.* See Asimov, "I Am a Signpost," in *The Roving Mind* (1983; repr., Amherst, NY: Prometheus Books 1997), Kindle loc. 331–333.
14. Scripsit was soon to be superseded by a successor program, SuperScripsit. Though Asimov wasn't bothered by its limitations—he took comfort in them—some of his colleagues felt differently. Stephen Kimmel expresses his disappointment at discovering that the program was incapable of underlining, for example: "I discovered the deficiencies of Scripsit about halfway through chapter 15 of my third novel *Lord of High Places,*" he begins, writing in a special issue of the magazine *Creative Computing* that was devoted to printers and word processing (SuperScripsit remedied this problem). See Kimmel, "Improving on Scripsit," *Creative Computing* 7, no. 7 (July 1981): 28. Other popular choices for the TRS-80 at the time included Electric Pencil (the descendent of perhaps the earliest word processor ever written for a personal computer) and Lazy Writer.
15. Ray Bradbury, "Another Computer Definition," in *Digital Deli,* ed. Steve Ditlea (New York: Workman, 1984), 201.
16. Asimov, "The Word-Processor and I," in *The Roving Mind,* Kindle loc. 334.
17. Ibid.
18. Ibid., 335.
19. Ibid.
20. Ibid., 336.
21. Ibid.
22. Ibid.
23. Isaac Asimov, "A Question of Speed," in *The Roving Mind,* Kindle loc. 338.
24. Ibid., 337.
25. Ibid., 338.
26. Ibid.
27. Ibid., 339.
28. Isaac Asimov, *I, Asimov: A Memoir* (New York: Bantam, 1994), 473.
29. Ibid., 474.
30. Ibid.
31. Asimov, "A Question of Speed," 339.
32. See Edwin McDowell, "Publishing: Top Sellers among Books of 1984," *New York Times,* January 18, 1985, http://www.nytimes.com/1985/01/18/books/publishing-top-sellers-among-books-of-1984.html.

33. Douglas E. Winter, *Stephen King: The Art of Darkness* (New York: Plume, 1986), 139–142.
34. Ibid., 141–142.
35. Stephen King, "Synopsis: The Talisman," Peter Straub Papers, Fales Library, New York University, August 24, 1983.
36. Peter Straub, interview with the author, May 11, 2012.
37. Ibid.
38. Ibid. Straub said of *Floating Dragon*: "So the manuscript is very strange. The first two thirds have staples and glue and the last third was a very clean looking pile of pages."
39. Ibid.
40. Michael Schumacher, "Ghost Storyteller," *Writer's Digest,* January 1985, 32–34.
41. Cornelius E. DeLoca and Samuel Jay Kalow, *The Romance Division: A Different Side of IBM* (Wyckoff, NJ: D & K Book Co., 1991), 197–198. IBM had followed the MT/ST with the MC/ST in 1969, which stored data on magnetic cards instead of magnetic tape. The first IBM word processor with a screen of any sort (known as the Office Systems/6) meanwhile debuted in 1977 and displayed only six lines of text. Intriguingly, according to DeLoca and Kalow, one argument for thus constraining the display was that six lines approximated a typist's view of a page as their sheet of paper curled behind the platen, bent by its own weight (173). This extremely literal remediation of typewriting suggests that full-screen WYSIWYG (of the sort that was even then being pioneered with the Bravo Editor at Xerox PARC—see Chapter 6) was more of a conceptual departure from simply emulating paper than is typically acknowledged.
42. See William Zinsser, *Writing with a Word Processor* (New York: Harper and Row, 1983). Zinsser, who positions himself as a "mechanical boob" (xiii), hits all of the now-familiar notes in his personal journey with the Displaywriter, from his initial skepticism about word processing, to apprehension about trying out different systems in the showroom, to taking it home, setting it up, initial frustration, and then breakthrough and rapturous conversion.
43. Bruce Feirstein, email to the author, January 13, 2012.
44. Marina Endicott, email to the author, February 1, 2012.
45. Michael Schumacher, "Processing *The Talisman,*" *Writer's Digest,* January 1985, 32.
46. Ibid.
47. Peter Straub, interview with the author, May 11, 2012.
48. Ibid.

49. Letter, Stephen King to Peter Straub, Peter Straub Papers, August 2, 1983.
50. For details of Tan's early life and career, see E. D. Huntley, *Amy Tan: A Critical Companion* (Westport, CT: Greenwood Press, 1998), 1–10.
51. Amy Tan (@AmyTan), tweet to the author (@mkirschenbaum), May 16, 2012.
52. Ibid., October 16, 2014.
53. Ibid., May 16, 2012, 9:21 P.M.
54. Sales records indicate that Ellison bought his first Osborne 1 on January 8, 1982; see Adam Bradley, *Ralph Ellison in Progress* (New Haven, CT: Yale University Press, 2010), 217. Chabon also bought his Osborne 1 in 1982 (email to the author, December 26, 2011).
55. Andrei Codrescu, "The Kaypro II: An Early Computer with a Writer's Heart," *NPR*, September 16, 2014, http://www.npr.org/blogs/alltechconsidered/2014/09/16/349027131/the-kaypro-ii-an-early-computer-with-a-writers-heart.
56. Amy Tan (@AmyTan), tweet to the author (@mkirschenbaum), May 16, 2012.
57. Robert Foothorap, interview with the author, October 18, 2014.
58. Ben Fong-Torres, "Good Things from Bad Sector: Reflections of a Small-Is-Beautiful Users' Group," *Profiles* 2, no. 6 (1985): 46–49. Other frequent *Profiles* contributors included fiction writers Robert J. Sawyer, David Gerrold, and Ted Chiang.
59. Fong-Torres, "Good Things from Bad Sector," 46.
60. Ibid., 48.
61. Ibid.
62. *Whole Earth Software Catalog*, 55.
63. Fong-Torres, "Good Things from Bad Sector," 48.
64. Ibid.
65. Amy Tan (@AmyTan), tweet to the author (@mkirschenbaum), October 16, 2014.
66. One of her last projects for IBM was entitled *Telecommunications and You;* sometimes erroneously described as a mere "pamphlet," in fact this illustrated treatise of some 150 pages reflects substantial knowledge of that industry.
67. Robert Foothorap, interview with the author, October 18, 2014.
68. Amy Tan (@AmyTan), tweet to the author, May 16, 2012, 9:21 A.M.
69. Fong-Torres, "Good Things from Bad Sector," 48.
70. Arthur C. Clarke, *2010: Odyssey Two* (New York: Ballantine Books, 1982), 291.
71. Arthur C. Clarke and Peter Hyams, *The Odyssey File* (London: Panther Books, 1984), xiii.

72. Ibid.
73. Ibid., xiv–xv.
74. Ibid., xvi.
75. See David H. Rothman, *The Silicon Jungle* (New York: Ballantine Books, 1985), 249–270.
76. Clarke and Hyams, *The Odyssey File,* xix.
77. David H. Rothman offers additional details of the ordeal required to establish the link, which included a contingency whereby a backup copy of one version of the modem software would be hand-delivered to Clarke by the wife of an American diplomatic attaché in Colombo. See *The Silicon Jungle,* 258–259.
78. Clarke and Hyams, *The Odyssey File,* xx–xxi.
79. Ibid., 10.
80. Ibid., 31.
81. Ibid., 27.
82. Ibid., 37.
83. Ibid., 49.
84. Ibid., xxv.
85. Arthur C. Clarke, quoted in Rothman, *The Silicon Jungle,* 45.
86. Neil McAleer, *Sir Arthur C. Clarke: Odyssey of a Visionary; A Biography* (New York: Rosetta Books, 2013), Kindle loc. 5691.
87. Ibid., "Foreword" by Ray Bradbury.
88. Ibid., Kindle loc. 6571.
89. Codrescu, "The Kaypro II."
90. Robin Perry, letter to the editor, *Writer's Digest,* May 1981, 6. Perry had authored a "Writer's Guide to Word Processors" feature for the magazine the previous month and is here responding to readers.
91. Curt Suplee, "Tapping at the Chamber Door," *PC Magazine,* May 29, 1984, 250.
92. Ibid., 251.
93. Edmund White, *City Boy: My Life in New York during the 1960s and '70s* (New York: Bloomsbury, 2009), 198–199.
94. Harold Brodkey, "The Art of Fiction No. 126," interviewed by James Linville, *Paris Review,* Winter 1991, http://www.theparisreview.org/interviews/2128/the-art-of-fiction-no-126-harold-brodkey.
95. Ibid.
96. James Fallows, interview with the author, April 24, 2012.
97. Michael Chabon, "On *The Mysteries of Pittsburgh,*" *New York Review of Books,* June 9, 2005, http://www.nybooks.com/articles/archives/2005/jun/09/on-the-mysteries-of-pittsburgh/.
98. Ibid.
99. Ibid.

4. North of Boston

1. For all of its notoriety, there is no authoritative source for the exact model of King's Wang, aside from relying on visual identification. Besides the Krementz photograph, however, the System 5 is clearly visible (and identifiable) in several shots from a 1982 documentary filmed by Harry Nevison for the University of Maine at Orono, King's alma mater. (It has been adorned with an "I [Heart] My Wang" bumper sticker; see "Stephen King: 'I Sleep with the Lights On,'" Henry Nevison Productions, embedded Vimeo video, 25:04, posted October 25, 2013, http://henrynevisonproductions.com/stephen-king-i-sleep-with-the-lights-on/.) George Beahm also correctly identifies the word processor as a Wang System 5—see Beahm, *Stephen King from A to Z: An Encyclopedia of His Life and Work* (Kansas City, MO: Andrews McMeel, 1998), 240—but he does not indicate his source for this information. *Writer's Digest,* meanwhile, calls it a "System 3" (Michael Schumacher, "Processing *The Talisman,*" January 1985, 32), but the journalist is almost certainly conflating it with the System 5 Model 3, which was the product's full name. (There was no "Wang System 3." There was a System 30, but it was a multiterminal computer.)
2. Kevin Kopelson, *Neatness Counts: Essays on the Writer's Desk* (Minneapolis: University of Minnesota Press, 2004), 13.
3. Jill Krementz, *The Writer's Desk* (New York: Random House, 1996), viii.
4. See Sarah Weinman, "Publishers Lunch," *People* (blog), October 17, 2012, http://lunch.publishersmarketplace.com/2012/10/people-94/.
5. Krementz, *The Writer's Desk,* 75.
6. As recounted in an appearance on Bob LeDrew, "A Peek Inside the Office with Marsha DeFilippo," podcast, *The King Cast,* January 23, 2011, http://thekingcast.ca/site/?p=129.
7. Stephen King, quoted in Beahm, *Stephen King from A to Z,* 240.
8. Stephen King, *Skeleton Crew* (New York: Putnam, 1985), 20–21.
9. The *Playboy* printing of the story was accompanied by a full-page color drawing depicting a hand poised over the Delete key. There are numerous minor variants—what textual critics what call "accidentals"—between the two published versions. We can surmise that King took advantage of the existence of a digital copy of the text to freely revise it for inclusion in the subsequent *Skeleton Crew* collection.
10. By "realist manner" I mean as compared to the speculative or abstract renditions of writing machines we have previously encountered in science fiction from authors such as Leiber and Lem or the satires of Calvino and Frayn. In King's story, by contrast, the reader would be able to recognize

actual brand names, keyboard commands, and functions of a consumer-grade word processor.

11. King, "Word Processor of the Gods," in *Skeleton Crew,* 307. Subsequent page numbers are given parenthetically in the text.
12. Jim Hargrove, *Dr. An Wang: Computer Pioneer* (Chicago: Children's Press, 1993), 101.
13. 2 Corinthians 3:6.
14. Walter J. Ong, *Orality and Literary: The Technologizing of the Word* (London: Routledge, 1982), 75.
15. The poem is "The Study."
16. Michael Heim, *Electric Language: A Philosophical Study of Word Processing* (New Haven, CT: Yale University Press, 1987), 138.
17. See Hawthorne's introductory essay to *The Scarlet Letter,* "The Custom-House" (1850).
18. See D. H. Lawrence, *Studies in Classic American Literature* (New York: Penguin, 1990).
19. Stephen King, *Dreamcatcher: A Novel* (New York: Scribner, 2001), Kindle loc. 10007.
20. Stephen King, *On Writing: A Memoir of the Craft* (New York: Pocket Books, 2000), 101.
21. See Stephen King's author website, http://stephenking.com/the_office.html.
22. The source for this text is the virtual "letter" displayed on the virtual "computer" one accesses by navigating the recesses of King's online virtual "office" at http://stephenking.com/the_office.html.
23. See http://www.stephenking.com/forums/showthread.php/13698-Writing-Program.
24. The Wangwriter II was a stand-alone CP/M-based system introduced in late 1981 and intended as something of a competitor to the IBM PC. Updike almost always refers to it as a "word processor," rarely as a "Wang," and to my knowledge he never specifically identifies it as a Wangwriter II. However, it is recognizable as such in a 1987 photo by Nancy Crampton that was first published in the May 8, 2014, issue of the *New York Review of Books* in a review of Begley's biography, *Updike* (New York: Harper, 2014); additionally, Leslie Morris of the Houghton Library informs me she has seen an invoice of Updike's for service on a "Wangwriter II" with "128K of [RAM] memory," dated March 3, 1989.
25. "Invalid Keystroke" by John Updike. Copyright © 1984 by John Updike, used by permission of The Wylie Agency LLC. Text of the poem as printed in *Light Year '84,* ed. Robert Wallace (Cleveland: Bits Press, 1983), 110.

26. Begley, *Updike,* Kindle loc. 110.
27. J. D. Reed and Jeanne North, "Plugged-In Prose," *Time,* August 10, 1981.
28. One source informed me that Wang initially provided the machine to Updike as part of a promotional effort, but I have not been able to verify this account.
29. Roger Angell, interview with the author, May 16, 2014.
30. John Updike, "Updike and I," quoted in Begley, *Updike,* Kindle loc. 7826.
31. John Updike, "Where Money and Energy Gather: A Writer's View of a Computer Laboratory," in *Research Directions in Computer Science: An MIT Perspective,* ed. Albert R. Meyer, et al. (Cambridge, MA: MIT Press, 1991), np.
32. Ibid.
33. See Paul Saenger, *Space between Words: The Origins of Silent Reading* (Stanford, CA: Stanford University Press, 1997).
34. Nicholson Baker, "The History of Punctuation," in *The Size of Thoughts: Essays and Other Lumber* (New York: Vintage, 1997), 75.
35. The John Updike Papers (2527), Houghton Library, Harvard University, Cambridge, Massachusetts.
36. Ibid.
37. Ibid.
38. John Updike, quoted in Curt Suplee, "Tapping at the Chamber Door," *PC Magazine,* May 29, 1984, 251.
39. John Updike, "The End of Authorship," *New York Times,* June 25, 2006, http://www.nytimes.com/2006/06/25/books/review/25updike.html?pagewanted=all&_r=0.
40. Updike, "A Writer at Large: Naked Came the Stranger to a Mystery Plot in Cyberspace," *New Yorker,* September 29, 1997, 31–32.
41. Updike, preface to *Odd Jobs: Essays and Criticism* (New York: Random House, 1991), Kindle loc. 163.
42. See Courtney Reed, "In the Galleries: Russell Banks Adapts to a Word Processor," *Cultural Compass* (blog), March 22, 2011, http://blogs.utexas.edu/culturalcompass/2011/03/22/in-the-galleries-russell-banks-adapts-to-a-word-processor/.
43. "NEWLIGHT," Terrence McNally Papers, Harry Ransom Humanities Research Center, Disk 22a (June 10, 1988).
44. Umberto Eco, *Foucault's Pendulum* (New York: Picador, 1989), 24. Eco had gotten his first computer in 1983, and *Foucault's Pendulum* was the first novel he wrote with it, completing at least one early draft one or two years later. See Eco and Jean-Philippe de Tonnac, *This Is Not the End of the Book,* trans. Polly McLean (London: Harvill Secker, 2011), 44, 73–74.
45. Eco, *Foucault's Pendulum,* 26.

46. See John Updike, "The Writer in Winter," in *Higher Gossip: Essays and Criticism* (New York: Random House, 2011), 4.
47. John Updike, introduction to Krementz, *The Writer's Desk*, xi.
48. Begley, *Updike,* Kindle loc. 7065.
49. Letter to "Edith," May 21, 1983. The John Updike Papers (1703), Houghton Library.
50. Details of the typewriter's provenance are from Steve Soboroff, email to the author, January 13, 2014. Miranda Updike never used the typewriter.
51. Very nearly so, anyway. There are a few additional words, mostly garbled and illegible but seemingly pertaining to postage, then a closing "warm regards." What Updike presumably means is not that he will not be doing any more typing himself, but that he will no longer require the services of a *typist,* the latter being what is replaced by the word processor in his workflow. My deepest thanks to Steve Soboroff for making the ribbon available to me to transcribe.

5. Signposts

1. Isaac Asimov, "I Am a Signpost," in *The Roving Mind* (1983; repr., Amherst, NY: Prometheus Books, 1997), Kindle loc. 333.
2. These took the form of full-page ads in computer magazines showing Asimov posed with a TRS-80 and other hardware: "The smartest way to write!" read the copy for one. "An out-of-this-world deal!" trumpeted another. "I may never use a typewriter again!" blared a third, in unabashed contradiction to Asimov's actual preferences. Tandy maintained an aggressive promotional strategy that leaned heavily on well-known writers, reportedly also placing Tandy products in the hands of Robert Heinlein, Arthur C. Clarke, L. Sprague de Camp, and Flora Rheta Schreiber, among others. Atari, meanwhile, retained Robert Ludlum for the same ends: "In fiction, Robert Ludlum's characters use the most incredible computers imaginable. In fact, Mr. Ludlum's personal computer is an Atari," reads one 1981 ad. 3M used Gordon R. Dickson to shill floppy disks. And so on, in the best tradition of Remington's use of Twain's "First Writing-Machine" sketch.
3. Asimov, "I Am a Signpost," 333.
4. David Gerrold, "Star Trek Author Discovers Word Processors," *InfoWorld,* March 30, 1981, 10.
5. Asimov, "I Am a Signpost," 332.
6. The story originally appeared in *Analog,* June 1964, 78–80. It has more than a touch of Marxist dialectic: "Willy," says the robot named I, BEM of his human servant, "says our time is coming. He says when any species

develops the lower-order means to which it can assign its physical labor, it will get kicked out of its nest by that lower-order creation" (80).

7. See *The Essential Ellison*, ed. Terry Dowling (Beverly Hills, CA: Morpheus International, 1991), 166.
8. See "Answer 'Affirmative' or 'Negative,'" *Analog: Science Fiction, Science Fact* (April 1972): 152–167.
9. Harlan Ellison, quoted in J. Michael Straczynski, "Brave New Words," *Profiles*, May / June 1984, 38. Emphasis in original. He was no more generous to electric typewriters.
10. A point made more generally by Brian Stableford, "Computers," in John Clute and Peter Nicholls, *The Encyclopedia of Science Fiction* (New York: St. Martin's Press, 1995); with a very few exceptions, he notes, science fiction writers "failed utterly to foresee the eventual development of the microprocessor" (243).
11. Ibid.
12. David Hartwell, interview with the author, December 7, 2014.
13. A former public relations representative for Tandy reports having supervised the delivery of over 500 machines to various authors: "I either saw the machines in these writer's homes, or hand-delivered a machine to them, or spoke with them in person about a technical problem relating to the computer." She notes favoring science fiction and fantasy authors because she enjoyed the genre, but she also singles out Flora Rheta Schreiber because she had taken a class from her (Deb McAlister-Holland, email to the author, December 28, 2011).
14. Harold Bloom, "The Art of Criticism No. 1," interviewed by Antonio Weiss, *Paris Review*, Spring 1991, http://www.theparisreview.org/interviews/2225/the-art-of-criticism-no-1-harold-bloom. Bloom adds: "I can only write with a ballpoint pen, with a Rolling Writer, they're called, a black Rolling Writer on a lined yellow legal pad on a certain kind of clipboard. And then someone else types it."
15. WordPerfect also shared this color scheme. Jerry Pournelle, interview with the author, December 27, 2011. See also "Blue Background with White Text," *Fog Creek Software*, http://discuss.fogcreek.com/joelonsoftware/default.asp?cmd=show&ixPost=103262.
16. That incident is chronicled in Brad DeLong, "How Jerry Pournelle Got Kicked Off the ARPANET," *Grasping Reality*, July 5, 2013, http://delong.typepad.com/sdj/2013/07/how-jerry-pournelle-got-kicked-off-the-arpanet.html.
17. After introducing him to publisher Jim Baen, Pournelle wrote the foreword to Gingrich's *Window of Opportunity: A Blueprint for the Future* (New York: Tor, 1984) in which space exploration and computing

are both central themes. "The jump from an electric typewriter to the interactive computer systems now on the market is fully as great as the jump from the quill pen to the typewriter," Gingrich opined (81).

18. "A professor in Maryland has an article in the *New York Times* about word processors and novelist[s]. He doesn't seem to have done any homework at all." So begins Pournelle's characteristically pugnacious response to an article covering my early research. See Pournelle, "Early Days of Word Processing," *Chaos Manor* (blog), December 26, 2011, http://www.jerrypournelle.com/chaosmanor/early-days-of-word-processing/.
19. For a representative instance of the statement, see Pournelle, "I Think I Started Something," *Asimov*, August 3, 1981, 55.
20. See "How to Get My Job," *JerryPournelle.com* (blog), https://www.jerrypournelle.com/slowchange/myjob.html.
21. See Cristopher Hennessey-DeRose, "Joss Whedon—ScFi.com talks to SF Author Jerry Pournelle," *Whedon.info*, September 7, 2003, http://www.whedon.info/Joss-Whedon-SciFi-com-talks-to-SF.html.
22. Jerry Pournelle, "Introduction," *Starswarm: A Jupiter Novel* (New York: Tor, 1998), Kindle loc. 133.
23. Authorities point to the 1958 breakup of the American News Service, a distributor for many early science fiction magazines, as the tipping point. With the resulting demise of dozens of titles, science fiction "ceased to be identified primarily as a magazine form. Shorter forms, from short-short stories to novellas, gave way to novels and even multi-volume series." See Brian Attebery, "The Magazine Era: 1920–1960," in *The Cambridge Companion to Science Fiction*, ed. Edward James and Farah Mendlesohn (Cambridge: Cambridge University Press, 2003), 46. See also Barry N. Malzberg, "The Fifties," Library of America, accessed August 19, 2015, http://www.loa.org/sciencefiction/why_malzberg.jsp.
24. Pournelle, "Introduction," Kindle loc. 143. "Nobody likes to type, and most writers hated it," he echoed in his interview with the author, December 27, 2011.
25. Jerry Pournelle, *User's Guide to Small Computers* (Wake Forest, NC: Baen, 1984), 92. (The book reprints the original *onComputing* article.)
26. The individual upon whom Pournelle frequently bestowed this epithet was a retired government official and intelligence officer named Dan MacLean. See John Riley, "Lord of Chaos Manor: Hoping for a Message from a Long-Lost Friend," *Los Angeles Times*, October 27, 1985, http://articles.latimes.com/1985-10-27/magazine/tm-13029_1_chaos-manor.
27. Pournelle, *User's Guide*, 8.
28. Jerry Pournelle, interview with the author, December 27, 2011. The other components of the system were as follows: a Cromemco S-100 board

carrying a Z-80 chip with the CP/M operating system that we have seen before; 16K of memory; a fifteen-inch black-and-white Hitachi display screen; twin 8-inch floppy disk drives; a Memorex keyboard capable of emulating the familiar IBM Selectric layout (Pournelle was very particular about keyboards, and insisted that writers find one they liked); and a Diablo 1620 printer, then the gold standard for letter-quality impact printing.

29. Jerry Pournelle, interview with the author, December 27, 2011.
30. Pournelle, *User's Guide,* 12.
31. Ibid., 52. This is obviously a bold claim, and it is in stark contrast to the experience of an author such as Paul Auster, for whom the mechanical work of retyping, physically demanding though it is, is an integral part of the creative process. Speaking about his first novel, *The Mezzanine,* which he wrote on a Kaypro II, Nicholson Baker expresses a similar sentiment: "I retyped the whole book. I was always a believer, even with word processing, that there's something useful about having to retrace your steps from the beginning" ("The Art of Fiction No. 212," interviewed by Sam Anderson, *Paris Review,* Fall 2011, http://www.theparisreview.org/interviews/6097/the-art-of-fiction-no-212-nicholson-baker).
32. Jerry Pournelle, ed., *The Endless Frontier* (New York: Ace, 1979), 26.
33. In conversation with me, Pournelle verbally dated his acquisition of the Cromemco system to 1976. He also indicated that *The Mote in God's Eye* (1975) was the last book he wrote on paper: "Everything after that was written on a computer." But if "Spirals" (the story he twice singles out for mention in conjunction with learning to write on the computer) was composed not *too* long before its 1979 publication date (also the year Pournelle wrote the article for *onComputing*), then that timeline seems off—we would have to accept that either the system lay dormant for as much as two years before Pournelle did any fiction writing on it, or else that for whatever reason he omits mention of any earlier projects when discussing the computer in relation to his writing. My best conjecture is that he acquired the computer in probably late 1977 or perhaps even 1978 (but not 1976) and regardless of what was the Ur-text written with it, "Spirals" and the *onComputing* article were both certainly among the very first—shortly followed by the novel-length collaboration with Niven on *Oath of Fealty* (1981), which he remembers distinctly. In his computer guidebook *Adventures in Microland* (New York: Baen, 1985), Pournelle dates the acquisition of his system to 1977 (2); he says the same in various other publications of that era. The technological record also suggests a date later than 1976: Electric

Pencil was first released in December of that year, and the Cromemco Z-2 (Zeke's specific system model) was not available before 1977. Pournelle's literary papers and other personal records were not available to me, and so I can't be certain what other documentation may survive.

34. For a video of the interview, see "Tom Snyder Interviews Durk Pearson and Jerry Pournelle," YouTube video, 10:07, posted by r06u3AP, August 1, 2007, https://www.youtube.com/watch?v=BJ7lHBnlKQM.
35. Michael Shrayer, *The Electric Pencil Word Processor: Operator's Manual,* 1977, 6.
36. Ibid., 10.
37. Pournelle, *User's Guide,* 64.
38. Ibid., 70.
39. Ibid., 60.
40. Ibid., 4.
41. Ibid., 4, 12.
42. Ibid., 4.
43. Frank Herbert with Max Barnard, *Without Me You're Nothing: The Essential Guide to Home Computers* (New York: Simon and Schuster, 1980), 200–201.
44. One recent author who would recognize Pournelle's and Herbert's ambitions in this regard is Andy Weir, whose best-selling novel *The Martian* (2012) has been hailed as a landmark of contemporary hard science fiction. It includes details of a spacecraft's flight to Mars. Weir concluded he needed not just static calculations but actual working models. A trained software engineer, he proceeded to write his own custom programs. "I set it up so I could mess around with the angle of thrust at any point along the journey, as well as turn the engine on or off. . . . The app can also play back the motions of all bodies as an animation, rather than show the whole course at once." Quoted in Angela Watercutter, "The App That Helped Write a Breakout Sci-Fi Novel," *Wired,* February 14, 2011, http://www.wired.com/2014/02/the-martian-software/.
45. Larry Niven, email to the author, December 31, 2011. Niven identifies his transition to the computer as coming in the middle of his writing his novel *The Ringworld Engineers,* published in 1979—lending further corroboration to the likelihood that Pournelle acquired his own system in late 1977 or 1978.
46. Pournelle, *User's Guide,* 74.
47. Ibid., 82.
48. Ibid., 80.

49. Ibid., 67.
50. Nonetheless, Baen was apparently irritated with the PC's "European"-style keyboard, which was designed to facilitate the typing of umlauts and such at the expense of an extra key between Shift and Z and making the Return key smaller and harder to reach. According to Pournelle, Baen hired a programmer to create a utility called Magic Keyboard to correct these shortcomings. The program proved popular and was distributed through Simon and Schuster. See Pournelle, *Adventures in Microland,* 104–105.
51. Pournelle, *User's Guide,* 15.
52. David Hartwell, Herbert's editor at Berkley Books at the time, recalls seeing the computer in Herbert's home and believes it may have been a "Radio Shack" system. Hartwell also has a vivid memory of receiving manuscripts from Herbert printed on a dot-matrix printer (interview with the author, December 7, 2014). Moreover, several diskettes belonging to Herbert survive (see Chapter 10) and bear the Tandy label. Interestingly, however, Herbert's biographer and son, Brian Herbert (who is generally attentive to such details), makes no mention of a Tandy or any other brand-name computer in the household when his father was writing *God-Emperor of Dune* (the book was originally titled *Sandworm of Dune*).
53. Brian Herbert, *Dreamer of Dune: The Biography of Frank Herbert* (New York: Tor, 2003), 284.
54. Herbert, *Dreamer of Dune,* 326; the dollar figure is suggested by David Gerrold in "Star Trek Author," 9.
55. Herbert with Barnard, *Without Me You're Nothing,* 43.
56. Ibid., 14.
57. Ibid., 16.
58. Ibid., 17.
59. Herbert, *Dreamer of Dune,* 127.
60. Ibid., 297.
61. Herbert with Barnard, *Without Me You're Nothing,* 202.
62. Ibid., 200.
63. Ibid., 182.
64. "Graphic Computer System and Keyboard," US 4546435 A, October 8, 1985.
65. David Hartwell, interview with the author, December 7, 2014.
66. Jerry Pournelle, "WORMs and Friends: Words for the Ages and Pictures for the Millions," *InfoWorld,* March 7, 1988, 41. Pournelle related the same explanation in his interview with me.
67. Herbert with Barnard, *Without Me You're Nothing,* 204.

68. "Frank's technological visions far outstripped the practical ability of people to actually implement them," comments his editor David Hartwell. Interview with the author, December 7, 2014.
69. Algis Budrys, "Advice to a New Writer on the Choice of a First Word-Processor," *Science Fiction Review* 53 (Winter 1984): 34–38.
70. Herbert with Barnard, *Without Me You're Nothing*, 197.
71. Barry Longyear, "Wanging It," *Asimov*, August 3, 1981, 44–52.
72. Barry Longyear, email to the author, December 15, 2014. In this message Longyear describes his system as a "Wang Writer," but that was a separate product and was not released until 1981. The "Wanging It" article identifies it as the Wang System 5 Model 1, and this is doubtless correct.
73. Ibid.
74. Barry B. Longyear, *The Write Stuff* (New Sharon, ME: Enchanteds, 2011), 422.
75. Ibid.
76. Ibid., 427–428.
77. Longyear, "Wanging It," 50.
78. Forward's earliest mention of working with a computer, found in the manuscript annotations included with his papers at the University of California Riverside, is June 1981. An annotation on that date regarding his editing of a complete manuscript "printout" suggests that he had been using the computer for some time already. See "Guide to the Robert L. Forward Papers," Online Archive of California, updated 2012, http://www.oac.cdlib.org/findaid/ark:/13030/c8f76bh3/entire_text/.
79. His papers at the University of California Riverside include this annotation accompanying one of the *Rocheworld* [*Flight of the Dragonfly*] manuscripts: "This was the printout of the largest electronic version of *Rocheworld* that I could find. It was on USC tape 5WARD as file ROCHEWOR.81JUN24. Jacqueline Stafsudd finally was able to read the ancient tape and transfer it to an IBM compatible disk." Ibid.
80. Similarly, author and editor Kathryn Cramer observes, "The writers who had the exposure to the technology early on would then have been much earlier adopters of word processing." Interview with the author, April 18, 2012.
81. Eileen Gunn, interview with the author, May 10, 2012.
82. Eileen Gunn, email to the author, May 6, 2012.
83. Eileen Gunn, interview with the author.
84. See Beth Winter, "About the Author," *The Discworld Compendium*, August 19, 2015, http://www.extenuation.net/disc/terryp.html.
85. See Robin Roberts, *Anne McCaffrey: A Life with Dragons* (Jackson: University Press of Mississippi, 2007), Kindle loc. 2248.

86. Douglas Adams, "Beyond the Brochure, or Build It and We Will Come," *Hitchhiker's Guide to the Galaxy* website, last updated January 28, 2002, http://h2g2.com/entry/A281701.
87. M. J. Simpson, *Hitchhiker: A Biography of Douglas Adams* (Boston: Justin Charles and Co., 2003), 184. Simpson's careful attention to Adams's personal computing history (on which this paragraph is heavily dependent) is exemplary, and all too rare among literary biographers, many of whom are still content to simply use the generic terms "computers" and "word processors," if they mention them at all.
88. Ibid., 185.
89. The most comprehensive study to date of interactive fiction is Nick Montfort, *Twisty Little Passages: An Approach to Interactive Fiction* (Cambridge, MA: MIT Press, 2003).
90. Simpson, *Hitchhiker,* 186. As Simpson notes, this claim is disputed by Stephen Fry, an actor and author.
91. Ibid., 185.
92. Indeed, tablets themselves already had a real-world corollary in Alan Kay's ideas for a Dynabook. See Kay, *A Personal Computer for Children of All Ages* (Palo Alto, CA: Xerox PARC, 1972), http://www.mprove.de/diplom/gui/kay72.html, pdf.
93. See David B. Williams, Biographical Sketch, pt. 2., Vance Museum, http://www.vancemuseum.com/vance_bio_2.htm.
94. As detailed in John Vance, "*Lurulu* Completed," *Cosmopolis,* February 2003, 1, http://www.integralarchive.org/cosmo/Cosmopolis-35.pdf. John Vance is his son.
95. David Gerrold, "Science-Fiction Authors Reappraise the Role of Computers," *InfoWorld,* July 5, 1982, 15.
96. Samuel R. Delany, *1984* (Rutherford, NJ: Voyant, 2000), 229, 265.
97. Ibid., 304. The letter is written to Robert S. Bravard and dated November 28, 1984.
98. See "Octavia Butler's 'Kindred'—anyone know of any quote- legitimate-unquote reviews?" *Dreamwidth* (forum), posted by tzikeh, May 2, 2010, http://tzikeh.dreamwidth.org/550650.html?style=light&thread=6989306.
99. Carl Freedman, *Conversations with Ursula K. Le Guin* (Jackson: University Press of Mississippi, 2008), 169.
100. See Octavia E. Butler, "Parable of the Typewriter," interviewed by Tom Knapp, Rambles, February 1996, http://www.rambles.net/butler_typewriter.html. See also "'Congratulations! You've Just Won $295,000!': An Interview with Octavia E. Butler," interviewed by Joan Fry, *Joan Fry* (author's website), accessed August 19, 2015, http://www.joanfry.com/congratulations-youve-just-won-295000/. See also Octavia E. Butler

Papers, 2013, Online Archive of California, Huntington Library, San Marino, http://pdf.oac.cdlib.org/pdf/huntington/mss/butler.pdf. I am grateful to Gerry Canavan for furnishing additional information in this regard.

101. The first actual appearance of the term "cyberspace" was in Gibson's short story "Burning Chrome," published that same year in *Omni* magazine.
102. For an essential reading of the significance of this seeming trivia, see Scott Bukatman, "Gibson's Typewriter," *South Atlantic Quarterly* 92, no. 4 (Fall 1993): 627–645.
103. William Gibson, *Neuromancer* (New York: Ace, 1984), 51.
104. William Gibson, "The Art of Fiction No. 211," interviewed by David Wallace-Wells, *Paris Review,* Summer 2011, http://www.theparisreview.org/interviews/6089/the-art-of-fiction-no-211-william-gibson.
105. William Gibson, email to the author, March 30, 2012.
106. Larry McCaffrey, "Interview with William Gibson," in *Storming the Reality Studio: A Casebook of Cyberpunk and Postmodern Science Fiction* (Durham, NC: Duke University Press, 1991), 270.
107. Bruce Sterling, email to the author, January 1, 2012; see also the *Paris Review* interview, where Gibson recalls the exchange with Sterling: "This changes everything!" I said, "What?" He said, "My Dad gave me his Apple II. You have to get one of these things!" I said, "Why?" He said, "Automation—it automates the process of writing!"
108. Gibson, email to the author.
109. Unlike Arthur C. Clarke and Peter Hyams or Stephen King and Peter Straub, Gibson and Sterling recall but a single abortive attempt to transfer text by modem; see William Gibson and Bruce Sterling, *The Difference Engine* (New York: Ballantine Books, 1990, 2011), 488. Nonetheless, one character in the book, Michael Godwin, is named for the Austin journalist and soon-to-be Electronic Frontier Foundation lawyer who helped them try to figure out the arrangement.
110. Daniel Fischlin et al., "'The Charisma Leak': A Conversation with William Gibson and Bruce Sterling," in *Conversations with William Gibson,* ed. Patrick A. Smith (Jackson: University of Mississippi Press, 2014), Kindle loc. 2033.
111. Ibid., 2054.
112. Gibson and Sterling, *The Difference Engine,* 489.
113. David Hartwell, interview with the author, December 7, 2014.
114. Kathryn Cramer, interview with the author, April 18, 2012.
115. Barry Longyear, email to the author, December 17, 2014.
116. David Brin, email to the author, July 3, 2015.

117. Sheila Finch, email to the author, July 2, 2015.
118. Malcom Ross-MacDonald, email to the author, June 28, 2012. The salesman's advice notwithstanding, MacDonald was working with computers by the early 1980s, starting with the wonderfully named Exidy Sorcerer; he used them not only for word processing but to prepare camera-ready copy for his publishers.
119. Stuart Woods, email to the author, December 26, 2011. Subsequent details about Woods are taken from this same email.
120. See Gary K. Wolfe, "Science Fiction and Its Editors," in James and Mendlesohn, *Cambridge Companion to Science Fiction*, 108.
121. See Vonda N. McIntyre, afterword to *Dreamsnake, Book View Café*, August 19, 2015, http://www.bookviewcafe.com/index.php /Dreamsnake14.
122. Some historical perspective is also useful here. Today, for example, we think of a novella primarily as a stunted or less ambitious novel, but historically novellas came first—they evolved into the "novel" only in the eighteenth century, when authors such as Richardson and Fielding remarked with some trepidation on the prodigious length of their books. The Victorian period, meanwhile, was an era of massive three-volume or "triple-decker" novels, whose length was often acknowledged to be a function of a different kind of literary technology: not a steampunk word processor but serialization, the practice of publishing novels in inexpensive monthly installments. (Thus the myth that Charles Dickens was "paid by the word.") In this era there was truly "no frigate like a book," as Emily Dickinson was to say—a comment about the imagination, to be sure, but also about books' material heft. Virginia Woolf, meanwhile, observed the "fecundity" of her Victorian predecessors and called for "shorter, more concentrated" books, "framed so that they do not need long hours of steady, uninterrupted work." Quoted in Steve Ellis, *Virginia Woolf and the Victorians* (Cambridge: Cambridge University Press, 2007), 40–41.
123. "Frustrated by writer's block, frustrated by blood pressure medication that she felt inhibited her creativity and vitality, and frustrated by the sense that she had no story for *Trickster*, only a 'situation,' Butler started and stopped the novel over and over again from 1989 until her death, never getting far from the beginning," says Gerry Canavan. See Canavan, "'There's Nothing New / Under The Sun, / But There Are New Suns': Recovering Octavia E. Butler's Lost Parables," *LA Review of Books*, June 9, 2014, https://lareviewofbooks.org/essay/theres-nothing-new-sun-new -suns-recovering-octavia-e-butlers-lost-parables. Drafts of the unfinished novel exist as both typescripts and printouts.

6. Typing on Glass

1. Wilfred A. Beeching, *Century of the Typewriter* (New York: St. Martin's Press, 1974), 267.
2. Walter Isaacson, *Steve Jobs* (New York: Simon and Schuster, 2011), 60.
3. Fred Moore, Homebrew Computer Club invitation addressed to Steve Dompier (February 17, 1975), uploaded on Wikipedia by Gotanero, November 12, 2013, http://en.wikipedia.org/wiki/Homebrew_Computer_Club#mediaviewer/File:Invitation_to_First_Homebrew_Computer_Club_meeting.jpg.
4. Copies of von Neumann's famous "First Draft of a Report on the EDVAC," dated June 30, 1945, are easily found online. For a discussion of the considerations around attributing the stored program concept to von Neumann alone, see Paul E. Ceruzzi, *A History of Modern Computing* (Cambridge, MA: MIT Press, 1998), 20–22.
5. Beeching, *Century of the Typewriter,* 127.
6. See Ceruzzi, *History of Modern Computing,* 152.
7. Peter Straub, interview with the author, May 10, 2012.
8. The fullest account of van Dam and Nelson's collaboration is to be found in Belinda Barnet, *Memory Machines: The Evolution of Hypertext* (London: Anthem Press, 2013). Barnet's account is based on extensive personal interviews with both individuals.
9. For the origins of the term "hypertext," see http://faculty.vassar.edu/mijoyce/Ted_sed.html, accessed August 19, 2015.
10. Andries van Dam, interview with the author, March 15, 2012.
11. See James V. Catano, "Poetry and Computers: Experimenting with the Communal Text," *Computers and the Humanities* 13, no. 4 (October–December 1979): 269–275; see also Catano, "Computer-Based Writing: Navigating the Fluid Text," *College Composition and Communication* 36, no. 3 (October 1985): 309–316.
12. Larry Tesler, interview with the author, October 11, 2013.
13. Ibid.
14. Ibid.
15. The best overview of Xerox PARC's history and associated innovations is Michael A. Hiltzik, *Dealers of Lightning: Xerox PARC and the Dawn of the Computer Age* (New York: Harper, 1999).
16. This episode (and Apple's subsequent development of the technologies) is recounted in detail in Steven Levy, *Insanely Great: The Life and Times of Macintosh, the Computer That Changed Everything* (New York: Penguin, 1994), 77–103. For video of Tesler describing the demo in 2011, see Philip Elmer-DeWitt, *Fortune,* August 24, 2014, embedded YouTube video,

http://fortune.com/2014/08/24/raw-footage-larry-tesler-on-steve-jobs-visit-to-xerox-parc/.

17. Michael Shrayer was an Altair owner for whom the blinking LEDs were simply not satisfactory—he quickly managed to make the computer the centerpiece of his own TV Typewriter unit, and a year later released what is generally regarded as the first word processing program for a microcomputer, Electric Pencil. (At the time, Shrayer claims, he had never even heard the term "word processing"; see Paul Freiberger, "Electric Pencil, First Micro Word Processor," *InfoWorld,* May 10, 1982, 12.) Electric Pencil was rough around the edges, even by the standards of software that would soon follow, and Shrayer found he had little interest in running a commercial software company. But Electric Pencil was quickly embraced by home computer hobbyists (and early adopters like Pournelle and Niven), a constituency not yet served by either academic research or the high-end office word processing systems.
18. Hiltzik, *Dealers of Lightning,* xxiii.
19. Ibid., 169–171.
20. Charles Simonyi, quoted in ibid., 198.
21. Andries van Dam and David E. Rice, "On-Line Text Editing: A Survey," *Computing Surveys* 3, no. 3 (September 1971): 113.
22. The relevant documentation is the Federal Standard 1037C, *Telecommunications: Glossary of Telecommunication Terms.*
23. For an excellent complement to van Dam and Rice's then-contemporary survey, see Thomas Haigh, "Remembering the Office of the Future," *IEEE Computer Society* 28, no. 4 (2006), which includes further discussion of both teletype technologies and the distinctions (such as they are) between line editing, program editing, text editing, and word processing. Though the latter two are typically differentiated by their degree of attentiveness to hard-copy formatting, Haigh concludes that "the key distinction is more cultural than technical: text editors are used by programmers to write programs and systems files; word processors are used by everyone else to do everything else" (14). See also Ceruzzi, *History of Modern Computing,* for discussion of time-sharing, teletypes, and text editing in their historical contexts.
24. See van Dam and Rice, "On-Line Text Editing," 96.
25. Ibid., 97.
26. Douglas R. Hofstadter, "Preface to GEB's Twentieth-Anniversary Edition," in *Gödel, Escher, Bach: An Eternal Golden Braid* (New York: Basic Books, 1999), P-12. Hofstadter was also captivated by the on-screen formatting of his prose made possible by TV-EDIT, a fixation that would lead him to the ordeal of producing his own camera-ready films for Basic

Books using an early computer typesetting system designed to work with TV-EDIT.

27. Charles Simonyi, interview with the author, March 13, 2012.
28. Ibid.; see also Hiltzik, *Dealers of Lightning*, 200.
29. Charles Simonyi, interview with the author.
30. See Hiltzik, *Dealers of Lightning*, 208; also Tesler, interview with the author.
31. Tesler, interview with the author.
32. Ibid.
33. Ibid.
34. Ibid. The same story is also recounted from personal correspondence with Tesler in Ted Nelson, *Geeks Bearing Gifts: How the Computer World Got This Way; Version 1.1* (Sausalito, CA: Mindful Press, 2009), 128.
35. Quotations are from a copy of the note provided to the author by Tesler; emphasis added.
36. Ibid. The first sentence is all-capitalized in the original. Underlining (evidence of Gypsy's capabilities) in original.
37. See Hiltzik, *Dealers of Lightning*, and Levy, *Insanely Great*, for coverage of Kay's renowned career.
38. Bonnie MacBird, quoted in Mike Gencarelli, "Bonnie MacBird Talks about Co-writing 1982's 'TRON,'" *Media Mikes*, September 27, 2011, http://www.mediamikes.com/2011/09/interview-with-trons-bonnie-macbird/.
39. Bonnie MacBird, email to the author December 27, 2011.
40. Ibid.
41. MacBird, email to the author, January 18, 2012.
42. MacBird, email to the author, December 27, 2011.
43. MacBird, email to the author, January 18, 2012.
44. The best technical account of the computer graphics techniques in *Tron* remains Richard Patterson, "The Making of TRON," *American Cinematographer*, August 1982, 792–819.
45. See Gencarelli, "Bonnie MacBird Talks.'"
46. See "Yale Book Arts," the Honorable Company of College Printers, accessed August 19, 2015, http://www.fiveroses.org/YaleBookTradition.htm.
47. See "Pierson Press," the Honorable Company of College Printers, accessed August 19, 2015, http://www.fiveroses.org/Presses/PCPress.htm.
48. See "The College Presses," Yale University Library online, last modified November 16, 2001, http://www.library.yale.edu/aob/printing/collegepresses.html.
49. Some details of Hersey's time at Pierson College (Yale) are drawn from Zara Kessler, *John Hersey's Yale Education* (May 2012), http://www.library.yale.edu/~nkuhl/YCALStudentWork/Hersey-Kessler.pdf.

50. Editors, "To Our Readers," *New Yorker,* August 31, 1946.
51. John Hersey, "The Art of Fiction No. 92," interviewed by Jonathan Dee, *Paris Review,* Summer–Fall 1986, http://www.theparisreview.org/interviews/2756/the-art-of-fiction-no-92-john-hersey.
52. Ibid.
53. Peter Weiner, interview with the author, March 20, 2015.
54. Hersey, "The Art of Fiction."
55. Weiner, interview with the author.
56. See Craig A. Finseth, *The Craft of Text Editing* (Berlin, DE: Springer-Verlag, 1991; Cambridge, MA: MIT Press, 1999). E-book, chap. 4, http://www.mit.edu/~yandros/doc/craft-text-editing/.
57. R. Stockton Gaines, quoted in Willis H. Ware, *RAND and the Information Revolution: A History in Essays and Vignettes* (Santa Monica, CA: RAND Corporation, 2008), 122.
58. Edgar T. Irons and Frans M. Djorup, "A CRT Editing System," *Communications of the ACM* 15, no. 1 (January 1972): 17.
59. Weiner, interview with the author.
60. Unless otherwise noted, details of what likely constituted a typical writing session with the Yale Editor on the PDP-10 in this paragraph and those that follow have been reconstructed from various sources in the John Hersey Papers in the Beinecke Rare Book and Manuscript Library, Yale University. These sources (in boxes 53–54 and 58–59) include documentation for the PDP-10 and Yale Editor, as well as Hersey's own notes, drafts, and jottings. Conversation with Peter Weiner has also been helpful in this respect.
61. Hersey, "The Art of Fiction."
62. Ibid.
63. Hannah Sullivan, *The Work of Revision* (Cambridge, MA: Harvard University Press, 2013), 15–16.
64. See "Typesetting: When It Changed," *Making Light* (blog), June 18, 2004, http://nielsenhayden.com/makinglight/archives/005370.html.
65. "Computer-Aided Composition," in *Encyclopedia of Computer Science and Technology,* ed. Jack Belzer, Albert G. Holzma, and Allen Kent, vol. 5. (Boca Raton, FL: CRC Press, 1976), 334. One influential example: The Coach House Press in Toronto acquired a Mergenthaler in 1974, and soon afterward a CTC Datapoint 2200 computer to control it, thus inaugurating decades of experimentation and innovation in digital typesetting and electronic publishing under its imprint. John W. Maxwell is actively researching the technological history of Coach House—I am grateful to him for this information. See also http://hpcanpub.mcmaster.ca/case-study/coach-house-press-crucible-electronic-publishing-technology.

66. LINTRN documentation, John Hersey Papers, Beinecke Library, box 54, April 24, 1973.
67. Autograph draft, John Hersey Papers, Beinecke Library, box 58, nd.
68. Ibid.
69. The PDP-10 was able to render up at least one other novelty for Hersey: an alphabetical listing of every word in the book and the number of times it was used (John Hersey Papers, box 58). A few years later, perhaps thinking it was still safely in the realm of fiction for literary authorship, Italo Calvino would satirically include a similar word frequency list in *If on a Winter's Night a Traveler* (1979).
70. Rene Kuhn Bryant, "More Worrier than Philosopher," *National Review,* December 6, 1974, 1421–1422.
71. Hersey, "The Art of Fiction."
72. According to Peter Weiner, interview with the author; and Weiner is emphatic on this point.

7. Unseen Hands

1. Berry's essay, "Why I Am Not Going to Buy a Computer," was originally published in the *New England Review / Bread Loaf Quarterly* 10, no. 1 (1987): 112–113; emphasis in original. It was reprinted as "Against PCs," *Harper's Magazine,* September 1988. The essay and the exchange that followed its appearance in *Harper's* are both available online through the *Jesus Radicals* website, http://www.jesusradicals.com/uploads/2/6/3/8/26388433/computer.pdf.
2. Gordon Inkeles, "Technological Fundamentalism," *Harper's Magazine,* December 1988, 6.
3. Berry, "Why I Am Not," 112.
4. See Ted Friedman, *Electric Dreams: Computers in American Culture* (New York: NYU Press, 2005), 103.
5. Robert X. Cringley, quoted in ibid., 104.
6. Steven Levy, "A Spreadsheet Way of Knowledge," *Harper's Magazine,* November 1984; reprinted online at *Backchannel,* October 24, 2014, https://medium.com/backchannel/a-spreadsheet-way-of-knowledge-8de60af7146e.
7. See Juliet Webster, *Shaping Women's Work: Gender, Employment and Information Technology* (New York: Longman, 1996). Importantly, Webster sees the relationship between gender and technology as bidirectional: "Gender relations must be regarded as the context within which information technologies are evolving, assuming the mark of their social

and gendered context, and in turn acting upon this gendered social context" (33). Natalia Cecire, meanwhile, frames typing as a form of the "not-reading" she also historically aligns with women's work; see Cecire, "Ways of Not Reading Gertrude Stein," *ELH* 82 (2015): 281–312.

8. Christopher Latham Sholes, quoted in Bruce Bliven Jr., *The Wonderful Writing Machine* (New York: Random House, 1954), 15. Of course, the genealogy of the typewriter's gender is much more nuanced and diverse. Kittler's *Gramophone, Film, Typewriter*, trans. Geoffrey Winthrop-Young (Stanford, CA: Stanford University Press, 1999), and Lisa Gitelman's *Scripts, Grooves, and Writing Machines: Representing Technology in the Edison Era* (Stanford, CA: Stanford University Press, 1999) are both essential as starting points.
9. See especially Christopher Keep, "The Cultural Work of the Type-Writer Girl," *Victorian Studies* 40, no. 3 (Spring 1997): 401–426; Darren Wershler-Henry, *The Iron Whim: A Fragmented History of Typewriting* (Ithaca, NY: Cornell University Press, 2007), 85–104; and Scott Bukatman, "Gibson's Typewriter," *South Atlantic Quarterly* 92, no. 4 (Fall 1993): 636–640.
10. Quoted in Jeanette Hoffman, "Writers, Texts, and Writing Acts: Gendered User Images in Word Processing Software," in *The Social Shaping of Technology*, 2nd ed., ed. Donald MacKenzie and Judy Wajcman (Buckingham, UK: Open University Press, 1999), 224.
11. J. C. R. Licklider, an influential early computer and information scientist, was notably skeptical that men (specifically men) in positions of power and authority could ever submit themselves to a keyboard. See Licklider's classic essay (with its telling title), "Man-Computer Symbiosis," *IRE Transactions in Human Factors in Electronics*, HFE-1, nos. 4–11 (March 1960). I am grateful to Liz Losh for the reminder about Licklider's views on typing.
12. Barry Longyear, "Wanging It," *Asimov*, August 3, 1981, 51.
13. Hoffman, "Writers, Texts, and Writing Acts," 223.
14. Ibid., 227.
15. Juliet Webster is unequivocal on this point: "There are still very few men whose entire jobs are concerned with processing other people's words." See Webster, "From the Word Processor to the Micro: Gender Issues in the Development of Information Technology in the Office," in *Gendered by Design? Information Technology and Office Systems*, ed. Eileen Green, Jenny Owen, and Den Pain (London: Taylor and Francis, 1993), 120.
16. The poem is published in a chapbook with the same title by Ackerman from Thin Ice Press in 1990. The colophon reads, "A book composed by

hand in 10 point Garamond Bold, printed by the poet, me, on a Vandercook 219, at the New College Print Shop, sometime in June and July, wanting to finish." (That last phrase an echo of the title.) There are no other overt references to word processing among the fifteen other poems in the book. The copy I consulted is in the Fales Library at New York University.

17. Shoshana Zuboff, *In the Age of the Smart Machine: The Future of Work and Power* (New York: Basic Books, 1988), 179–180.
18. Thomas Haigh, "Remembering the Office of the Future," *IEEE Computer Society* 28, no. 4 (2006): 7.
19. Thomas J. Anderson and William R. Trotter, *Word Processing* (New York: Amacom, 1974), 5; emphasis in original.
20. George R. Simpson, quoted in ibid., 5.
21. Ibid., 1.
22. Walter A. Kleinschrod, *Word Processing: An AMA Management Briefing* (New York: Amacom, 1974), 2.
23. This list is derived from Anderson and Trotter, *Word Processing*, 5.
24. Ibid.
25. Ibid.
26. For example, Kleinschrod, *Word Processing*, 4. Jane Barker and Hazel Downing make the same point in "Word Processing and the Transformation of the Patriarchal Relations of Control in the Office," *Capital & Class* 4, no. 1 (Spring 1980): 81–82.
27. Kleinschrod, *Word Processing*, 4.
28. Anderson and Trotter, *Word Processing*, 5.
29. Mary Kathleen Benét, *The Secretarial Ghetto* (New York: McGraw-Hill, 1973), 13; emphasis in original.
30. Ibid., 10.
31. Ibid., 6–7.
32. Kleinschrod, quoted in Haigh, "Remembering the Office," 8.
33. Other scholars writing about the history of women and information technology have come to similar conclusions. See, for example, Sadie Plant, *Zeroes +Ones* (London: Fourth Estate, 1997), esp. "automata"; and Anne Balsamo, *Technologies of the Gendered Body: Reading Cyborg Women* (Durham: Duke University Press, 1996), esp. "Feminism for the Incurably Informed." That chapter (previously published in a landmark special issue of *South Atlantic Quarterly* on cyberspace in 1993) is also the point of departure for N. Katherine Hayles in *My Mother Was a Computer* (Chicago: Chicago University Press, 2005). Finally, see Elizabeth Losh, *Virtualpolitik: An Electronic History of Government Media-Making in a Time of War, Scandal, Disaster, Miscommunication,*

and Mistakes (Cambridge, MA: MIT Press, 2009), esp. the concluding chapter.

34. See Alan Liu, *The Laws of Cool: Knowledge Work and the Culture of Information* (Chicago: University of Chicago Press, 2004), 108–109.
35. Barker and Downing, "Word Processing," 92.
36. Evelyn Berezin, interview with the author, April 5, 2015.
37. See "Evelyn Berezin: WITI Hall of Fame 2011 Induction @ WITI's Women Powering Technology Summit," YouTube video, uploaded October 21, 2011, embedded in WITI Hall of Fame online, featured profile on http://www.witi.com/center/witimuseum/halloffame/303047/Evelyn-Berezin-Management-Consultant-Brookhaven-Science-Associates/.
38. See Evelyn Berezin biography, Computer History Museum online, accessed August 19, 2015, http://www.computerhistory.org/fellowawards/hall/bios/Evelyn,Berezin/.
39. Evelyn Berezin, interview with the author.
40. According to the Berezin biography provided in the Computer History Museum online, http://www.computerhistory.org/fellowawards/hall/bios/Evelyn,Berezin/. But see also Martin Campbell-Kelly on SABRE, which makes clear the extent of its improvement over early systems like Teleregister's Reservisor. Campbell-Kelly, *From Airline Reservations to Sonic the Hedgehog: A History of the Software Industry* (Cambridge, MA: MIT Press, 2003), 41–45.
41. Evelyn Berezin, interview with the author. Berezin also relates this story in a video interview recorded by the Computer History Museum in 2015; see "Evelyn Berezin—Computer History Museum 2015 Fellow Award Recipient," YouTube video, 6:51, posted by Computer History, May 7, 2015, https://www.youtube.com/watch?v=tb-v65mOBEw.
42. Ibid.
43. Ibid.
44. Ibid.
45. This description is from Gwyn Headley. See "Goodbye Old Friend, or Gwyn Donates a Computer to a Museum," *fotolibrarian* (blog), May 12, 2014, http://fotolibrarian.fotolibra.com/?p=1115. Headley subsequently donated this same machine to the Computer History Museum in Mountain View, California.
46. Evelyn Berezin, interview with the author.
47. Ibid.
48. Gloria Steinem, "Sisterhood," *New York*, December 20, 1971, 48.
49. All subsequent quotations are to the text of the ad copy, *New York*, December 20, 1971, 21.

50. For additional discussion of the ad and the history of the preview issue of *Ms.*, see Amy Erdman Farrell, *Yours in Sisterhood: Ms. Magazine and the Promise of Popular Feminism* (Chapel Hill: University of North Carolina Press, 1998), 29–36.
51. Evelyn Berezin, interview with the author.
52. Webster, *Shaping Women's Work*, 115.
53. See also Barker and Downing, "Word Processing": "It is within the invisible culture of the office that we find the development of forms of resistance that are peculiarly 'feminine.' . . . Because the work in the office is boring and alienating, and because 'work,' that is, waged work, is not traditionally seen as being central to women's lives by both employers and women themselves, it is not surprising that women tend to bring their domestic lives into work with them" (82).
54. Marina Endicott, email to the author, February 1, 2012.
55. Daniel Max, "McMillan's Millions," *New York Times*, August 9, 1992, http://www.nytimes.com/1992/08/09/magazine/mcmillan-s-millions.html.
56. A noteworthy story here is that of Barbara Elman, a former secretary who, after being asked by her employer to learn a word processing system, opened up her own word processing consulting business and began editing the bimonthly *Word Processing News* out of Burbank, California, in 1982. (Despite the similar title and time frame, it is not to be confused with Bradford Morgan's more academically oriented publication.) *WPNews*, as it was known, catered to screenwriters, journalists, and other professional writers with product reviews, classifieds, and other features. For additional details, see Patricia Keefe, "Panel Considers DP Rich with Career Options," *Computerworld*, May 23, 1983, 20. One subscriber came to call her the "mother confessor" for writers seeking advice about word processing (quoted in Peggy Watt, "Magazines Pursue Vertical Markets," *InfoWord*, February 27, 1984, 29).
57. Samuel J. Kalow, "Word Processing Opens New Career Avenues," *Modern Office Procedures*, nd.
58. See Haigh, "Remembering the Office," 11–12.
59. For a history of *Processed World* as written by its founders, as well as for scans of much of the content, see Chris Carlsson et al., "Some History of Processed World," *Processed World*, 1989–1991, accessed August 19, 2015, http://www.processedworld.com/History/history.html. See also Haigh, "Remembering the Office," 23–24; and Liu, *The Laws of Cool*, 278–282. "What *Processed World* believed in, above all," Liu comments, "was 'bad attitude,' which was its motto, and also the title of a 1990 retrospective anthology of the magazine's articles, readers' letters, and illustrations" (278).

60. Evelyn Berezin, interview with the author.
61. Cornelius E. DeLoca and Samuel Jay Kalow, *The Romance Division: A Different Side of IBM* (Wyckoff, NJ: D & K Book Co., 1991), 60.
62. Haigh, "Remembering the Office," 8–9.
63. Helen M. McCabe and Estelle L. Popham, *Word Processing: A Systems Approach to the Office* (New York: Harcourt, Brace, Jovanovich, 1977), 22.
64. Ibid., 26.
65. Richard Matheson, "A World of His Own," *The Twilight Zone,* season 1, episode 36, directed by Ralph Nelson, aired July 1, 1960 (New York: Columbia Broadcasting System).
66. Geoffrey Winthrop-Young, *Kittler and the Media* (Cambridge: Polity, 2011), 63.
67. Richard Powers, "How to Speak a Book," *New York Times,* January 7, 2007, http://www.nytimes.com/2007/01/07/books/review/Powers2.t.html?_r=3&.
68. Richard Powers, email to the author, May 29, 2012.
69. Ibid.
70. Ibid.
71. Richard Powers, "The Art of Fiction No. 175," interviewed by Kevin Berger, *Paris Review,* Winter 2002–2003, http://www.theparisreview.org/interviews/298/the-art-of-fiction-no-175-richard-powers.
72. See Wershler-Henry, *The Iron Whim,* esp. pt. 3, "Amanuensis: Typewriting and Dictation," 73–131; see also Gitelman, *Scripts, Grooves, and Writing Machines,* esp. chap. 5, "Automatic Writing," 184–218.
73. James Fallows, "Living with a Computer," *Atlantic,* July 1, 1982, http://www.theatlantic.com/magazine/archive/1982/07/living-with-a-computer/306063/.
74. Longyear, "Wanging It," 51.
75. Jerry Pournelle also placed notable emphasis on the necessity of the tractor feed in his advice to writers buying their first computer: "Your printer will have a friction feed much like a typewriter. That's fine for letters and other stuff that wants single sheets, but not useful if you're trying to print off a novel. Therefore, when you buy your printer, buy a 'tractor drive,' also known as a 'pinfeed.' This is a small gadget of metal that will cost an outrageous amount. . . . Grumble at the price, but pay it." See Pournelle, *User's Guide to Small Computers* (Wake Forest, NC: Baen, 1984), 59.
76. Kittler, *Gramophone, Film, Typewriter,* 198.
77. See ibid., 203; and Gitelman, *Scripts, Grooves, and Writing Machines:* "The machine's upstrike design seemed to refute the possibility of error, however unrealistically, and in removing the act of inscription from the

human eye seemed to underscore its character as a newly technological and automatic event" (205–206).

78. Denise Levertov, quoted in "Word Processing: Boon or Bane for Writers?," *3:17 am* (blog) posted by RaisoirJ, November 10, 2011, http://317am.org/2011/11/word-processing-boon-or-bane-for-writers/.
79. Bukatman, "Gibson's Typewriter," 630–631, 635.
80. See K. David Jackson et al., eds., *Experimental—Visual—Concrete: Avant-Garde Poetry since the 1960s* (Amsterdam: Rodopi, 1996). For an even more expansive reading of twentieth-century poetry as a response to the informationalization of language see Paul Stephens, *The Poetics of Information Overload: From Gertrude Stein to Conceptual Writing* (Minneapolis: University of Minnesota Press, 2015).
81. Don Swaim, "Audio Interviews with Anne Rice," sound file, *Wired for Books,* http://www.wiredforbooks.org/annerice/.
82. Ray Hammond, *The Writer and the Word Processor* (London: Coronet, 1984), 12.
83. See Margaret Atwood et al., "System, Method and Computer Program, for Enabling Entry into Transactions on a Remote Basis," EP 2030363 A1 (May 10, 2007).
84. Margaret Atwood, quoted in Christos Tsirbas, "The LongPen: From World-Famous Novelist to High-Tech Entrepreneur," *Daily Galaxy* (blog), December 3, 2007, http://www.dailygalaxy.com/my_weblog/2007/12/the-longpen—fr.html.
85. Alan Galey notes, "The rhetoric of liberation that has long shaped discourse about electronic texts helps to enable the fantasy that books transmit texts in simple and straightforward ways, as emanations from the author's mind to the reader's. Margaret Atwood . . . confirmed this view to an audience of electronic publishing enthusiasts and practitioners in her keynote address at the 2011 O'Reilly Tools of Change for Publishing conference, where she defined publishing as 'transfer from brain to brain, via some sort of tool.'" See Galey, "The Enkindling Reciter: E-Books in the Bibliographical Imagination," *Book History* 15 (2012): 210–247.
86. Joao Medeiros, "Giving Stephen Hawking a Voice," *Wired,* December 2, 2014, http://www.wired.co.uk/magazine/archive/2015/01/features/giving-hawking-a-voice. See also Hélène Mialet, *Hawking Incorporated: Stephen Hawking and the Anthropology of the Knowing Subject* (Chicago: University of Chicago Press, 2012).
87. Many previous scholars have written perceptively about the Memex. For just one example, see Losh, *Virtualpolitik,* 312–317.
88. Vannevar Bush, "As We May Think," *Atlantic,* July 1, 1945, http://www.theatlantic.com/magazine/archive/1945/07/as-we-may-think/303881/.

89. Frank Herbert with Max Barnard, *Without Me You're Nothing: The Essential Guide to Home Computers* (New York: Simon and Schuster, 1980), 206.
90. Surfdaddy Orca, "By Thought Alone: Mind over Keyboard," *Humanity+*, December 21, 2009, http://hplusmagazine.com/2009/12/21/thought-alone-mind-over-keyboard/.
91. Ibid.
92. Ibid.
93. See David C. Dougherty, *Shouting Down the Silence: A Biography of Stanley Elkin* (Champaign-Urbana: University of Illinois Press, 2010), 6, 192.
94. These observations are based on my inspection of the exam booklets in Elkin's papers in the Special Collections Department at Washington University in St. Louis in October 2014.
95. Dougherty, *Shouting Down the Silence,* 192.
96. In conversation with me, Joan Elkin emphasized the role of Stanley Elkin's chronic heart condition as opposed to the multiple sclerosis in the decision to adopt the word processor; however Elkin's biographer, David Dougherty, characterizes Elkin's handwriting as "undecipherable" by this point (6). We can assume both were contributing circumstances.
97. Dan Shea, email to the author, October 6, 2014.
98. Dougherty, *Shouting Down the Silence,* 192.
99. Elkin's confederate was Al Lebowitz, a lawyer who was also an aspiring writer and had thus bought a Lexitron of his own (doubtless he had first encountered them at his office). He was Elkin's chief technical consultant throughout his life (Joan Elkin, interview with the author, October 7, 2014).
100. Letter from John Macrae III to Elkin, March 9, 1982, Stanley Elkin Papers, Special Collections Department, Washington University in St. Louis.
101. Dougherty, *Shouting Down the Silence,* 193.
102. Stanley Elkin, *The Franchiser* (New York: Farrar, Straus and Giroux, 1976).
103. These details from "Stanley Elkin: First Person Singular," directed by James F. Scott, *Artists in Residence Series* (February 26, 1982). Video. Courtesy of Washington University in St. Louis, Department of Special Collections. Elkin wrote about the soap collection in an essay, "Pieces of Soap," in *Pieces of Soap: Essays* (New York: Simon and Schuster, 1992).
104. Peter J. Bailey, "A Conversation with Stanley Elkin," *Review of Contemporary Fiction* 15, no. 2 (Summer 1995), http://www.dalkeyarchive.com/a-conversation-with-stanley-elkin-by-peter-j-bailey/.

105. Elkin hung on to the Lexitron for over a decade, eventually moving on to a PC clone and a piece of software called *Just Type!*, manufactured by a Lexitron subsidiary after the company was bought out by Raytheon. On one occasion, side effects from medication he was taking for complications from the multiple sclerosis led him to somehow lock himself out of the Lexitron for several days, perhaps by forgetting a password or the start-up procedure. It was a traumatic event that upset him badly. He describes it in a letter to Sherman Paul, dated May 7, 1991 (Stanley Elkin Papers, Special Collections, Washington University in St. Louis).
106. Elkin did not have a printer in the house, and his wife, Joan, recalls taking the diskettes across town to get files printed at a local office. So, in a very real sense she was his hands as well, going where he could not (the MS would eventually leave him confined to a wheelchair), overseeing the crucial step of producing his hard copy, even if she was not actually the one typing it.

8. Think Tape

1. Details in this paragraph are drawn from Edward Milward-Oliver's chapter "Tracing a Masterwork: Len Deighton's *Bomber*" from his in-progress biography of Deighton. I am deeply grateful to him for sharing his research with me.
2. Conrad Knickerbocker, "The Spies Who Come In from Next Door: The Spy Novel Syndrome," *Life*, April 30, 1965, 13.
3. Leah Price and Pamela Thurschwell, eds., *Literary Secretaries/Secretarial Culture* (Hampshire, UK: Ashgate, 2005), 1.
4. Ellenor Handley, email to the author, February 5, 2012.
5. Edward Milward-Oliver, "Tracing a Masterwork," unpublished manuscript, 11; emphasis in original.
6. Len Deighton, email to the author, October 21, 2012.
7. Gordon M. Moodie, "Our Newest Product," *IBM News*, June 29, 1964, 2.
8. Ibid.
9. In his afterword to the original edition, Deighton comments that the hardware enabled him to redraft chapters at will and that he could select passages "at only a moment's notice" by means of "memory-coding" (Len Deighton, afterword, *Bomber* [London: Jonathan Cape, 1970], 473–474).
10. Leon Cooper, interview with the author, April 25, 2014.
11. Ibid.
12. "Chemical Products Devises Special Tape Cartridge to Fill Supply Need," *IBM News*, June 29, 1964, 4.

13. This was in contrast to several existing technologies that used rolls of punched paper to record a pattern of keystrokes for form letters, player-piano fashion.
14. Cornelius E. DeLoca and Samuel Jay Kalow, *The Romance Division: A Different Side of IBM* (Wyckoff, NJ: D & K Book Co., 1991), 86.
15. Ibid., 87–88.
16. Ibid., 88–89.
17. Vannevar Bush, "As We May Think," *Atlantic,* July 1945, http://www.theatlantic.com/magazine/archive/1945/07/as-we-may-think/303881/.
18. Abigail J. Sellen and Richard H. R. Harper, *The Myth of the Paperless Office* (Cambridge, MA: MIT Press, 2002), 17.
19. Thomas J. Anderson and William R. Trotter, *Word Processing* (New York: American Management Association, 1974), 3.
20. Ibid., 4.
21. The film is available on YouTube, "The Paperwork Explosion," Jim Henson Company, March 30, 2010, https://www.youtube.com/watch?v=_IZw2CoYztk. All quotations are transcribed from this copy of the film.
22. Ben Kafka perceptively scrutinizes the film in his *The Demon of Writing: Powers and Failures of Paperwork* (New York: Zone Books, 2012), and places particular emphasis on the closing catechism: "What if we shifted the emphasis just a little bit?" asks Kafka. "From '*machines* should *work, people* should *think*' to 'machines *should* work, people *should* think.' Is it possible that the film might be trying to warn us against its own techno-utopianism?" (146–150).
23. Frank M. Knox, *Managing Paperwork: A Key to Productivity* (New York: Van Nostrand Reinhold, 1980), ix.
24. See Zuboff, *In the Age of the Smart Machine: The Future of Work and Power* (New York: Basic Books, 1988), 301–310.
25. Knox, *Managing Paperwork,* ix.
26. See Thomas Haigh, "Remembering the Office of the Future," *IEEE Computer Society* 28, no.4 (2006): 8.
27. DeLoca and Kalow, *The Romance Division,* 72.
28. Whether or not Steinhilper also introduced the English-language term "word processing" at the same time is difficult to ascertain conclusively; the claim that he did so is tendered in his own memoir, *Don't Talk—Do It: From Flying to Word Processing* (Bromley, UK: Independent Books, 2006).
29. See Haigh, "Remembering the Office," 7–8.
30. See Erickson et al., *How Reason Almost Lost Its Mind: The Strange Career of Cold War Rationality* (Chicago: University of Chicago Press, 2013). The notion of "Cold War rationality," they write, was characterized

by a formal, often algorithmic approach: difficult problems and situations were broken down into individual, discrete steps that lent themselves to mechanized, often computational implementation (3–4). Examples include cybernetics, military operations research, game theory, and the GRIT technique in psychology. The point here, in the context of Steinhilper's career association with the largest computer manufacturer in the world, is the extent to which word processing or *textverarbeitung* can be said to have a foundation in a larger complex of ideas.

31. Haigh, "Remembering the Office," 9.
32. Ibid.
33. "IBM Shows a 'Brainier' Typewriter," *Wall Street Journal,* June 30, 1964.
34. "Typewriter Fed by Magnetic Tape," *New York Times,* June 30, 1964.
35. "IBM Reveals New Tape Typewriting Process," *Lexington Leader,* June 29, 1964.
36. Juliet Webster, "From the Word Processor to the Micro: Gender Issues in the Development of Information Technology in the Office," in *Gendered by Design? Information Technology and Office Systems,* ed. Eileen Green, Jenny Owen, and Den Pain (London: Taylor and Francis, 1993), 119.
37. "Intensive Sales Training Program Precedes MT/ST Announcement," *IBM News,* June 29, 1964, 3.
38. "Ed Reps to Give Demonstrations at MT/ST Centers," *IBM News,* June 29, 1964, 1, 2.
39. Ibid.
40. In this the MT/ST bears some resemblance to Monotype technology, which had been available in the typesetting industry for decades. With a Monotype system, the operator uses a typewriter-style keyboard to punch codes onto a paper ribbon or tape, which then becomes the basis for casting individual sorts for printing. (The most immediate contrast is to linotype printing, where hot metal slugs are cast directly from the keyboard's input.) The Monotype thus operated according to the principles of what we have been calling suspended inscription. Operators would need to be thinking about their tapes while they made and corrected errors at the keyboard.
41. *Training Manual 1, IBM Magnetic Tape Selectric Typewriter Model II/ Basic Model IV* (IBM Office Products Division), 46.
42. William S. Burroughs, *The Ticket That Exploded* (New York: Grove Press, 1967), 208–209. The quotation is from "The Invisible Generation" section, first published in the *Los Angeles Free Press* the year before.
43. Ibid., 207, for example.

44. See DeLoca and Kalow, *The Romance Division,* 87.
45. For more on Ballistrini and the "Tape Mark I" poems, see Christopher T. Funkhouser, *Prehistoric Digital Poetry: An Archeology of Forms, 1959–1995* (Tuscaloosa: University of Alabama Press, 2007), 12, 41–42.
46. *Training Manual 1,* i.
47. Deighton, email to the author, October 21, 2012.
48. Ray Hammond, *The Writer and the Word Processor* (London: Coronet, 1984), 217.
49. See TheManBookerPrize.com, July 29, 2015, http://www.themanbookerprize.com/press-releases/man-booker-prize-announces-2015-longlist.
50. Ellenor Handley, email to the author, February 6, 2012.
51. Deighton, *Bomber,* 473–474.
52. Although ASCII existed, the MT/ST instead used a proprietary seven-bit encoding standard known as "Kentucky code." This also limited its compatibility with IBM's mainframe computer systems, which used yet another proprietary encoding schema, EBCDIC. See DeLoca and Kalow, *The Romance Division,* 90.
53. Anderson and Trotter, *Word Processing,* 19.
54. Price and Thurschwell, *Literary Secretaries/Secretarial Culture,* 2.
55. Hammond, *The Writer and the Word Processor,* 217–218.
56. Handley, email to the author, February 5, 2012.
57. Deighton, *Bomber,* 86.

9. Reveal Codes

1. John Barth, *The Book of Ten Nights and a Night* (New York: Houghton Mifflin, 2004), 4–5. In two of the stories, "Click" (1997) and "The Rest of Your Life" (2000), the characters' personal computers figure as plot devices.
2. Joan Didion, *The Year of Magical Thinking* (New York: Vintage, 2006), 3.
3. See Gary Snyder, "Why I Take Good Care of My Macintosh," *IT Times* 6, no. 4 (January 1998), http://ittimes.ucdavis.edu/v6n4jan98/snyder.html. The poem originally appeared in *Turn-Around Times* in March 1988. At that time Snyder worked on a Macintosh Plus, and he has been a loyal Apple user ever since. See also John Markoff, "Digital Muse for Beat Poet," *New York Times,* January 22, 2010, http://www.nytimes.com/2010/01/22/technology/personaltech/22sfbriefs.html), which reports on Snyder's looking forward to receiving an iPad—and incorrectly states that the "Macintosh" poem is previously unpublished.

4. See Charles Bukowski, *The Captain Is Out to Lunch and the Sailors Have Taken Over the Ship* (New York: HarperCollins, 1998), for numerous examples.
5. Charles Bukowski, "My Computer," *Bone Palace Ballet* (Boston: Black Sparrow Press, 1997), 309. It is followed by a poem entitled "Thanks to the Computer."
6. The drawing, signed 1995, first appeared in Bukowski, *Captain Is Out to Lunch,* 56.
7. Michael Fried, "Macintosh," in *The Late Derrida,* ed. W. J. T. Mitchell and Arnold I. Davidson (Chicago: University of Chicago Press, 2007), 208.
8. A development not strictly limited to word processing software. Jennifer Egan's *A Visit from the Goon Squad* (2010) includes a chapter written as a PowerPoint presentation. There have been several epistolary email novels, such as Matt Beaumont's *E* (2000). Jessica Grose's *Sad Desk Salad* (2012) incorporates chat balloons and other online iconography.
9. Don DeLillo, *Mao II* (New York: Penguin, 1991), 137–138.
10. Really almost a novella, it is reprinted in Varley, *Blue Champagne* (New York: Ace, 1986), 230–290.
11. Russell Hoban, *The Medusa Frequency* (New York: Atlantic Monthly Press, 1987), 8.
12. Tao Lin, *Shoplifting from American Apparel* (Brooklyn: Melville House, 2009), 27. Similarly, in an online video interview he explains how the characters in his novel *Richard Yates* (2010) came to be named after child-actor stars Haley Joel Osment and Dakota Fanning: "About a year into writing it the characters names were Dan and Michelle or something like that. And I just probably was talking on Gmail chat to someone and I was like, 'I should just name them Haley Joel Osment and Dakota Fanning.' And they seemed into that. And I seemed into that. So I did Microsoft Find and Replace and did that." See Emily Gould, "Cooking the Books—Episode 15—Tao Lin," October 13, 2010, https://www.youtube.com/watch?v=d2BJSV8Q1Yw.
13. The four novels composing the tetralogy were initially published separately, beginning with *A Star Shines over Mt. Morris Park* in 1994 (the last two appeared posthumously after Roth's death in 1995). All citations in the paragraph that follows are to the collected edition, *Mercy of a Rude Stream: The Complete Novels* (New York: Liveright, 2014), Kindle ed.
14. Roth, *Mercy,* Kindle loc. 6802.
15. Ibid., Kindle loc. 4847. Adam Bradley has noted that Roth's history with his word processor represents a counterpoint to the career of Ralph Ellison. Roth's first novel, *Call It Sleep* (1934), had been hailed by Irving Howe and other critics as a major contribution to American modernism as

well as Jewish American literature. As with Ellison—notes Bradley—there was widespread anticipation for a second novel, which as the years (and decades) passed seemed increasingly unlikely to appear. Unlike Ellison, however, Roth seems to have finally found his voice with the aid of his word processor, producing many thousands of pages in the final decade of his life, enough for six additional novels (four of which compose the *Mercy* cycle), all of which have since been published. See Bradley, *Ralph Ellison in Progress* (New Haven, CT: Yale University Press, 2010), 34–35. Confirmation of Roth's use of an IBM PC Jr. is from Felicia Steele, email to the author, August 10, 2015.

16. Roth, *Mercy,* Kindle loc. 8272.
17. The naming of typewriters was not completely unknown, but the practice is far more commonplace with computers, not least because manufacturers encourage users to bestow a unique name onto their hard drives and computers for purposes of network identification.
18. Roth, *Mercy,* Kindle loc. 4793. The next line mentions Wallace Stevens.
19. Jerry Pournelle, interview with the author, December 27, 2011.
20. Jesse Kellerman, *Potboiler* (New York: Putnam's, 2012). All references are to this edition.
21. Ibid., Kindle loc. 270.
22. Ibid., 763.
23. Ibid., 954.
24. Ibid., 1590.
25. Ibid., 1606.
26. Ibid., 1644.
27. For more on the particulars of erasing magnetic media, see Matthew Kirschenbaum, *Mechanisms: New Media and the Forensic Imagination* (Cambridge, MA: MIT Press, 2008), 25–71.
28. William Gibson and Bruce Sterling, *The Difference Engine* (New York: Ballantine Books, 2011), 488.
29. William Gibson, "The Art of Fiction No. 211," interviewed by David Wallace-Wells, *Paris Review,* Summer 2011, http://www.theparisreview.org/interviews/6089/the-art-of-fiction-no-211-william-gibson.
30. See Carolyn Kellogg, "*Pride and Prejudice and Zombies,* by Seth Grahame-Smith," *LA Times,* April 4, 2009, http://www.latimes.com/entertainment/la-et-zombies4-2009apr04-story.html.
31. In July 2015 *New York Times* columnist David Brooks wrote what many readers took to be a condescending review addressed to Ta-Nehisi Coates for his recent nonfiction book *Between the World and Me* (2015). Brooks had unselfconsciously arrogated the device of the open letter that Coates had used to frame his own work, and this may have exacerbated online

reaction. Parody texts quickly emerged, with bloggers copying and pasting Brooks's column and then thematically modifying the prose while retaining its key signposting language. One of the best-known of the parodies has Brooks writing a review of a Thai restaurant on Yelp: "Maybe the right white response is just silence for a change. In any case, Pad Basil has filled my belly unforgettably. Three stars," it concludes, bitingly. See http://avidly.lareviewofbooks.org/2015/07/17/david-brooks-writes-a-yelp-review-of-the-new-thai-restaurant/. The point, once again, is that the facility with which the underlying textual operations can be conducted—copying and pasting from an electronic source, then modifying the prose locally throughout by working directly from the original base text—encourages distinct forms of textual expression, in this case in the service of parody and critique. The Thai restaurant review and its sibling texts were instantly recognizable as Brooks's column, even as their particulars amused with their seemingly arbitrary implausibility. These productions are the textual equivalent of the animated GIFs and YouTube memes that routinely circulate on the Internet in the wake of a politician's gaffe or some celebrity outrage.

32. Interestingly, much the same observation is manifest in the practices of educators who have experimented with role-playing and other mechanisms as surrogates for traditional literary interpretation. Perhaps the best-known exemplar is the *Ivanhoe* game, originally conceptualized by Johanna Drucker and Jerome McGann. Playing *Ivanhoe* can take a variety of forms, but in practice it requires participants to overwrite portions of a source text—such as Sir Walter Scott's eponymous novel—in order to effect "interventions, changes, additions, and commentaries in the discourse field of an imaginative work." The point of such interventions, insists McGann, "is not to develop readings or interpretations of *Ivanhoe* but to refashion and reshape its materials in ways that replicate the explanatory operations played out in the original story and in the material works that transmit the story to us." See McGann and Drucker, "The Ivanhoe Game," nd., http://www2.iath.virginia.edu/jjm2f/old/IGamehtm.html. See also McGann, *Radiant Textuality: Literature after the World Wide Web* (New York: Palgrave Macmillan, 2001), 209–231; and Drucker, *Speclab: Digital Aesthetics and Projects in Speculative Computing* (Chicago: University of Chicago Press, 2009), 65–97.
33. Seth Grahame-Smith, quoted in Kellogg, "*Pride, Prejudice and Zombies.*"
34. Haruki Murakami, *1Q84* (New York: Knopf, 2011), 25.
35. Ibid., 68.
36. Ibid., 69.

37. Ibid., 364.
38. Ibid., 365.
39. Andrew Ferguson, "Word Processing and Parallel Worlds in Haruki Murakami's *1Q84*," Science Fiction Research Association (April 2013); emphasis in original. Ferguson's fine paper, from which my discussion of *1Q84* has greatly benefitted, adds many additional details derived from both a close reading of the novel and the technical particulars of word processing with Japanese character sets. I am grateful for his sharing it with me.
40. See Nanette Gottlieb, *Word-Processing Technology in Japan: Kanji and the Keyboard* (London: Routledge, 2000), 63–64.
41. See Jay Rubin, *Haruki Murakami and the Music of Words* (London: Vintage, 2002), 167, 191; Gottlieb, *Word-Processing Technology in Japan,* 149, concurs with regard to Kōbō Abe.
42. Jerry Pournelle, *User's Guide to Small Computers* (Wake Forest, NC: Baen, 1984), 318.
43. Journalist Steve Levy attests to sitting in on a New York City users' group meeting when fiction writers discussed workarounds for this particular problem. The best solution, they agreed, would be to write very short chapters. See Levy, *Insanely Great: The Life and Times of Macintosh, the Computer That Changed Everything* (New York: Penguin, 1994), 189.
44. See Levy, *Insanely Great,* for one of the standard histories of Macintosh.
45. 1984 was also the year William Gibson published *Neuromancer. Shatter,* too, very much spoke to the sensibility of the cyberpunk genre. In an afterword to a reprint edition written in 1988, Gillis says about their own book: "It's science fiction of a different sort, too—telling a story about a world changed by technology in a way that was simply impossible two years previously. The future, after all, is not just another stage setting for stories; like the man said, it's where we're going to spend the rest of our lives. And if we're not careful (and who is?) we'll find the future is cropping up all around us." Quoted in Peter Gillis and Mike Saenz, *Shatter* (San Francisco: AiT/Planet Lar, 2006), np.
46. Ibid.
47. The story is recounted in Walter Isaacson, *Steve Jobs* (New York: Simon and Schuster, 2011), 130–131.
48. Rudy VanderLans and Zuzana Licko, with Mary E. Gray, *Émigré (the Book): Graphic Design into the Digital Realm* (New York: Van Nostrand Reinhold, 1993), 18.
49. Richard A. Lanham, *The Electronic Word: Democracy, Technology, and the Arts* (Chicago: University of Chicago Press, 1993), 4–5.

50. Joel Rose and Catherine Texier, introduction to *Between C&D: New Writing from the Lower East Side Fiction Magazine* (London: Penguin, 1988), ix.
51. See Harold Love, *Scribal Publication in Seventeenth-Century England* (Oxford: Clarendon Press; New York: Oxford University Press, 1993); Peter Stallybrass, "Printing and the Manuscript Revolution," in *Explorations in Communication and History,* ed. Barbie Zelizer (Abingdon, UK: Routledge, 2008), 111–118; and Stallybrass, "'Little Jobs': Broadsides and the Printing Revolution," in *Agent of Change: Print Culture Studies after Elizabeth L. Eisenstein,* ed. Sabrina Alcorn Baron, Eric N. Lindquist, and Eleanor F. Shevlin (Amherst: University of Massachusetts Press, 2007), 315–341.
52. Gay Talese, *A Writer's Life* (New York: Random House, 2006), 45.
53. Ibid., 42.
54. Interview with Kamau Brathwaite, *Talk Yuh Talk: Interviews with Anglophone Caribbean Poets,* ed. Kwame Dawes (Charlottesville: University Press of Virginia, 2001), 37.
55. Publishers who take on the challenge of printing Brathwaite's work typically use scans of camera-ready proofs that he provides, resulting in a photo-offset process. But Brathwaite is also emphatic about the physical size of his books as an element in their presentation, and has often had to compromise to accept standardized dimensions; thus his use of his own Savacou imprint for books such as *SHAR / Hurricane Poem* and *Barabajan Poems.* The tension between the computer as an instrument of composition and the constraints of normalized publishing practices in the industry at large is thematized in Brathwaite's work as part of Caliban's fraught relationship to Prospero.
56. For a thorough discussion of the Prospero and Caliban figures in Brathwaite's poetics, see Elaine Savory, "Returning to Sycorax / Prospero's Response: Kamau Brathwaite's Word Journey," in *The Art of Kamau Brathwaite,* ed. Stewart Brown (Mid Glamorgan: Poetry Wales Press, 1995), 208–230.
57. Stewart Brown, interview with Kamau Brathwaite, *Kyk-over-al* 40 (1989): 84–93. Quoted in Stewart Brown, "'Writin in Light': Orality-thru-Typography, Kamau Brathwaite's Sycorax Video Style," in *The Pressures of the Text: Orality, Texts, and the Telling of Tales,* ed. Stewart Brown, Birmingham University African Studies Series No. 4 (Birmingham: Centre of West African Studies, University of Birmingham, 1995), Kindle loc. 2823.
58. Brown, "'Writin in Light,'" Kindle loc. 2823.
59. Ibid., 2994.
60. Carrie Noland, "Remediation and Diaspora: Kamau Brathwaite's Video Style," in *Diasporic Avant-Gardes: Experimental Poetics and Cultural*

Displacement, ed. Carrie Noland and Barrett Watten (New York: Palgrave, 2009), 78.

61. Ibid., 86–87.
62. Ignacio Infante, *After Translation: The Transfer and Circulation of Modern Poetics across the Atlantic* (New York: Fordham University Press, 2013), 167–168. Infante also explicitly identifies Brathwaite's writing as "electronic poetry" (despite the fact that readers always encounter it printed on paper); similarly, Cynthia James aligns Brathwaite's work with hypertext, a digital form usually associated with white Western authors. See James, "Caliban in Y2K? Hypertext and New Pathways," in *For the Geography of a Soul: Emerging Perspectives on Kamau Brathwaite,* ed. Timothy J. Reiss (Trenton, NJ: Africa World Press, 2001), 351–361.
63. Brathwaite also sometimes identifies his computer as Stark, which doubles as the name of Caliban's sister in his writings. See Rhonda Cobham, "K / Ka / Kam / Kama / Kamau: Brathwaite's Project of Self-Naming," in Reiss, *For the Geography of a Soul,* 306–307.
64. The Kaypro is mentioned in several places in Brathwaite's writing, notably in *The Zea Mexican Diary: 7 Sept 1926—7 Sept 1986* (Madison: University of Wisconsin Press, 1993), 32–33; there he relates its frequent use by Doris until her condition grew too severe for her to sit at the keyboard. On the seemingly dormant website for the Caribbean literary journal *Savacou* that Brathwaite founded in 1970, there is a tantalizing note appended to the description of the final issue (14/15) published in 1979: "This is the first Savacou publication generated by Sycorax, in this case Zea Mexican's Kaypro computer—the first such printing / book design / publishing technical development in the Caribbean." This claim, unfortunately, is ahistorical—the first Kaypro model was not released until 1982. Perhaps there was an even earlier computer of unknown pedigree that is here misidentified as Doris's Kaypro; regrettably, I have been unable to arrive at a confident explanation.
65. Kamau Brathwaite, "X / Self's Xth Letters from the Thirteen Provinces," *X / Self* (Oxford: Oxford University Press, 1987), 80–87. As is Brathwaite's practice, the poem has been remediated (I follow Carrie Noland in using this term) at least two other times: in *Middle Passages* (1994) under the title "Letter Sycora X" and again in *Ancestors* (2001); both of these latter presentations incorporate the Sycorax Video Style. All citations here are from the 1987 text. See Noland, "Remediation and Diaspora," for a close reading of the differences between versions, including Noland's experimental attempts to reproduce Brathwaite's layouts on her own computer. See also Kelly Baker Josephs, "Versions of X / Self: Kamau Brathwaite's Caribbean Discourse," *Anthurium: A Caribbean Studies Journal* 1, no. 1 (December 2003): art. 4.

66. Whether the typographically normalized presentation was a function of the only-gradual maturation of the Sycorax Video Style (then at its inception) or of limitations imposed by Brathwaite's publisher at the time, Oxford University Press, has been a subject of debate. See Elaine Savory, "Returning to Sycorax / Prospero's Response: Kamau Brathwaite's Word Journey," in Brown, *Art of Kamau Brathwaite,* 218–222.
67. As recounted in *SHAR / Hurricane Poem* (Kingston, JM: Savacou Publications, 1990).
68. Kamau Brathwaite, "Dream Chad," in *Dream Stories* (New York: Longman, 1994), 48.
69. Ibid.
70. Ibid., 49.
71. Ibid.
72. The colophon to Brathwaite's *Barabajan Poems* (1994) indicates that he worked on that manuscript's first version with his Eagle through January 1989 and began using "Sycorax"—his Macintosh—for a second version in April of the same year.
73. Vanderlans and Licko, *Émigré,* 23.
74. Kamau Brathwaite, "Hawk," in *Born to Slow Horses* (Middletown, CT: Wesleyan University Press, 2005), 97.
75. For one recent consideration of this phenomenon see Alexander Starre, *Metamedia: American Book Fictions and Literary Print Culture after Digitization* (Iowa City: University of Iowa Press, 2015). Starre discusses Danielewski, Eggers, and Foer, among other authors and book designers.
76. See Lily Brewer, "The Function of Kittler's 'Ceasura' in Mark Z. Danielewski's *House of Leaves,*" August 4, 2013, http://www.lilybrewer.com/the-function-of-kittlers-caesura-in-mark-z-danielewskis-house-of-leaves.
77. See, for example, Michael Bierut, "McSweeney's No. 13 and the Revenge of the Nerds," *The Design Observer Group* [blog], May 29, 2004, http://designobserver.com/feature/mcsweeneys-no-13-and-the-revenge-of-the-nerds/2247.
78. *The Gates of Paradise* can be accessed in its entirety online. See http://www.thegatesofparadise.com/, accessed August 19, 2015.
79. David Daniels, quoted in John Strausbaugh, "David Daniels: The Shapes of Things," *New York Press,* August 2000, http://www.thegatesofparadise.com/John%20Strausbaugh%20Review.htm.
80. Elisabeth Tonnard, email to the author, December 27, 2011.
81. See Elisabeth Tonnard, "Let us go then, you and I," *Elisabeth Tonnard* (blog), March 17, 2008, http://elisabethtonnard.com/works/let-us-go-then-you-and-i/.

82. See Bang's sequence of poems composed from *Mrs. Dalloway,* "Let's Say Yes," in *The Last Two Seconds* (Minneapolis: Graywolf Press, 2015) and Ruefle, *A Little White Shadow* (Seattle: Wave Books, 2006).
83. For more on Goldsmith and conceptual poetry, see Scott Pound, "Kenneth Goldsmith and the Poetics of Information," *PMLA* 130, no. 2 (March 2015): 315–330; Marjorie Perloff, *Unoriginal Genius: Poetry by Other Means in the New Century* (Chicago: University of Chicago Press, 2010); and Goldsmith, *Uncreative Writing* (New York: Columbia University Press, 2011). Goldsmith's claim that he presents his conceptualist texts verbatim has come under scrutiny in the wake of his extremely controversial reading of Michael Brown's autopsy report—edited for "literary" effect, by Goldsmith's own admission—at a poetry event at Brown University in March 2015.
84. Kenneth Goldsmith, quoted in Perloff, *Unoriginal Genius,* 164.
85. See Brian Kim Stefans, *Kluge: A Meditation and Other Works* (New York: Roof Books, 2007), 71–78.
86. See "Changes in Word 2010," Microsoft Office, Technet, last modified May 5, 2012, https://technet.microsoft.com/en-us/library/cc179199.
87. Matthew Fuller, artist's statement as emailed to the author, March 20, 2012.
88. See *Word Perhect* online at http://wordperhect.net/.
89. James P. Ascher, "Bibliographical Awareness in Art: Joel Swanson's *Spacebar,*" Media Archaeology Lab, December 3, 2013, http://mediaarchaeologylab.com/bibliographical-awareness-art-joel-swansons-spacebar/.
90. See Paul Saenger, *Space between Words: The Origins of Silent Reading* (Stanford, CA: Stanford University Press, 1997).
91. In this word processing is surely fully compatible with one of the most widely read recent testaments to the "perennial artifice of the literary," as I have made bold to call it: David Shields's *Reality Hunger: A Manifesto* (New York: Knopf, 2010), a book-length collage- and montage-based exposition of what he terms "the simultaneous bypassing and stalking of artifice-making machinery" (172).

10. What Remains

1. Ray Hammond, *The Writer and the Word Processor* (London: Coronet, 1984), 220.
2. Frank Herbert with Max Barnard, *Without Me You're Nothing: The Essential Guide to Home Computers* (New York: Simon and Schuster, 1980), 202.

3. See John Vance, "*Lurulu* Completed," February 2003, *The Vance Integral Edition Project* (VIE; archived PDF), accessed August 19, 2015, http://www.integralarchive.org/cosmo/Cosmopolis-35.pdf.
4. See Piers Anthony, "Author's Note," in *Wielding a Red Sword* (New York: Ballantine Books, 1986), 289–290.
5. See J. D. Reed and Jeanne North, "Plugged-In Prose," *Time,* August 10, 1981.
6. Ibid.
7. Alice Munro, quoted in Cara Feinberg, "Bringing Life to Life," *Atlantic,* December 1, 2001, http://www.theatlantic.com/entertainment/archive/2001/12/bringing-life-to-life/378234/.
8. "Ten Rules for Writing Fiction," *The Guardian,* February 19, 2010, http://www.theguardian.com/books/2010/feb/20/ten-rules-for-writing-fiction-part-one.
9. Anthony Day and Marjorie Miller, "Gabo Talks," *Los Angeles Times,* September 2, 1990, http://articles.latimes.com/1990-09-02/magazine/tm-2003_1_his-labyrinth-nobel-prize-novel.
10. Rita Aero and Barbara Elman, "The ABC's of Word Processing," in *Digital Deli,* ed. Steve Ditlea (New York: Workman, 1984), 166.
11. See "A Conversation with Anne Rice," TheVampireLestat.net, accessed August 19, 2015, last modified, September 6, 2002, http://www.thevampirelestat.net/page%2032.htm. It appears that Rice eventually got out of this habit. See Janet McConnaughey, "Rice Bids Farewell to Lestat," *Los Angeles Times,* January 2, 2004; she apparently still works in WordStar, but she claims not to print anything out until the end. The change in routine may have cost her. On Facebook on May 8, 2012, she commented: "I was writing one of the scariest most intense scenes I've written of late. I was saving regularly, but I was several pages out there and a dark film came down over my screen with the message: 'You need to Restart your computer.' Nothing else was possible. Did the computer know how intense this scene was? Well, I lost only a few paragraphs. But why did this happen? Never before I have gotten such a message from the mysterious innards of my monstrous Mac computer."
12. T. Coraghessen Boyle, "Boxing Up," *New Yorker,* April 9, 2012, http://www.newyorker.com/books/page-turner/boxing-up.
13. Jennifer Howard, "In Electric Discovery, Scholar Finds Trove of Walt Whitman Documents in National Archives," *Chronicle of Higher Education,* April 12, 2011, http://chronicle.com/article/In-Electric-Discovery-Scholar/127096/.
14. Ibid.
15. Jennifer Howard, "U. of Texas Snags Archive of 'Cyberpunk' Literary Pioneer Bruce Sterling," *Chronicle of Higher Education,* March 8, 2011, http://m.chronicle.com/article/U-of-Texas-Snags-Archive-of/126654/.

16. For example, Bryan Bergeron, *Dark Ages II: When the Digital Data Die* (New York: Pearson, 2001).
17. See Rachel Donadio, "Literary Letters, Lost in Cyberspace," *New York Times Book Review*, September 4, 2005, http://www.nytimes.com/2005/09/04/books/review/04DONADIO.html.
18. Maxine Hong Kingston, *The Fifth Book of Peace* (New York: Vintage, 2003), 3.
19. Ibid., 61.
20. Ibid., 34.
21. Ibid., 61.
22. Ibid., 31.
23. Ibid. The detail about the Epson comes from Neila C. Seshachari, "Reinventing Peace: Conversations with Tripmaster Maxine Hong Kingston," *Weber: The Contemporary West* 12, no. 1 (Winter 1995), http://weberstudies.weber.edu/archive/archive%20B%20Vol.%2011-16.1/Vol.%2012.1/12.1KingstonInterview.htm. Kingston also relates that after the fire, "a woman was sitting in church, and she said a vision came to her in church. The vision was, she remembered she saw me on television and I was using an Epson QX10 computer and she had an old Epson QX10 at home she wasn't using. A vision came to her in church, Give Maxine your computer. So here's this woman at the door. She gave me a computer."
24. Kingston, *The Fifth Book of Peace*, 61.
25. Ibid., 39.
26. Ibid.
27. Media philosopher John Durham Peters has adopted the air as something of his native element. See his *Speaking into the Air: A History of the Idea of Communication* (Chicago: University of Chicago Press, 1999), and especially his *The Marvelous Clouds: Toward a Philosophy of Elemental Media* (Chicago: University of Chicago Press, 2015).
28. Seshachari, "Reinventing Peace."
29. Leslie A. Morris, "Harold Brodkey's Printout," *New York Times*, July 18, 1993, http://www.nytimes.com/1993/07/18/books/l-harold-brodkey-s-printout-069093.html.
30. See Matthew Kirschenbaum, "The Book-Writing Machine," *Slate*, March 1, 2013, http://www.slate.com/articles/arts/books/2013/03/len_deighton_s_bomber_the_first_book_ever_written_on_a_word_processor.single.html.
31. Edward Milward-Oliver, email to the author, March 26, 2013.
32. Ellenor Handley recalls: "One of the early 'faults' was a tendency for the paper to continue rolling after it should have finished with hundreds of 'carriage returns' setting up a sort of . . . Sorcerer's Apprentice effect." Email to the author, February 5, 2012.

33. Details in these last three paragraphs are based on my email correspondence with Milward-Oliver, Ellenor Handley, and the anonymous collector, March–September, 2013.
34. As described in the online finding aid for Clifton's papers at MARBL. The hard copies were in turn scanned in order to create searchable PDF files for researchers to access. See findingaids.library.emory.edu/documents/clifton1054/series11/.
35. William Gibson, *Pattern Recognition* (New York: Putnam, 2002), 216.
36. Marsha DeFilippo, interview with the author, March 20, 2012.
37. Lev Grossman, "Jonathan Franzen: Great American Novelist," *Time,* August 12, 2010, http://content.time.com/time/magazine/article/0,9171,2010185,00.html.
38. Jacques Derrida, "The Word Processor," in *Paper Machine,* trans. Rachel Bowlby (Stanford, CA: Stanford University Press, 1995), 29.
39. Rushdie, quoted in *Conversations with Salman Rushdie,* ed. Michael Reder (Oxford: University Press of Mississippi, 2000), 172–173.
40. For one of the many press accounts of Rushdie's computers at Emory, see Patricia Cohen, "Fending Off Digital Decay, Bit by Bit," *New York Times,* March 16, 2010, http://www.nytimes.com/2010/03/16/books/16archive.html. For a professional description of the conservation work by the archivists who were involved, see Laura Carroll et al., "A Comprehensive Approach to Born-Digital Archives," *Archivaria* 72 (Fall 2011): 61–92.
41. See Jay David Bolter, *Writing Space: The Computer, Hypertext, and the History of Writing* (Hillsdale, NJ: Erlbaum, 1991).
42. The key historical document here is Von Neumann's *First Draft Report on the EDVAC,* Contract No. W-670-ORD-4926 between the United States Army Ordinance Department and the University of Pennsylvania Moore School of Electrical Engineering, June 30, 1945.
43. See Rowan Wilken's excellent essay "Peter Carey's Laptop," *Cultural Studies Review* 20, no. 1 (2014): 100–120.
44. For more on the working relationship between Mailer and McNally, see Gabriela Redwine, "Tracking Harlot's Ghost across Media," *Rough Cuts: Media and Design in Process,* June 4, 2012, http://mediacommons.futureofthebook.org/tne/pieces/tracking-harlots-ghost-across-media. For Michael Joyce's computers, see Kirschenbaum, *Mechanisms: New Media and the Forensic Imagination* (Cambridge, MA: MIT Press, 2008), chap. 4.
45. According to Larry Rohter, "*García Márquez:* Words into Film," *New York Times,* August 13, 1989, http://www.nytimes.com/1989/08/13/movies/garcia-marquez-words-into-film.html.

46. Jennifer Schuessler, "Gabriel García Márquez Archives Goes to University of Texas," *New York Times,* November 24, 2014, http://www.nytimes.com/2014/11/24/books/gabriel-garca-mrquezs-archive-goes-to-university-of-texas.html.
47. Charles McGrath, "The Afterlife of Stieg Larsson," *New York Times,* May 23, 2010, http://www.nytimes.com/2010/05/23/magazine/23Larsson-t.html. See also Elaine Sciolino, "A Word from Stieg Larsson's Partner, and Would-Be Collaborator," *New York Times,* February 17, 2011, http://artsbeat.blogs.nytimes.com/2011/02/17/a-word-from-stieg-larssons-partner-and-would-be-collaborator/.
48. Rachel P. Maines and James J. Glynn, "Numinous Objects," *Public Historian* 15, no. 1 (Winter 1993): 9–25.
49. Barry Longyear, email to the author, December 17, 2014.
50. Jerry Pournelle, "The User's Column," *Byte,* March 1983.
51. Ibid.
52. Ibid.
53. Ibid.
54. Ibid.
55. See "Information Age: People, Information and Technology," Smithsonian.com, accessed August 19, 2015, http://www.si.edu/Exhibitions/Details/Information-Age-People-Information-and-Technology-4069.
56. Letter from Jon Eklund to Pournelle, August 24, 1990, "The Information Age: People, Information and Technology" exhibit, Smithsonian National Museum of American History, Washington DC, May 9, 1990.
57. Jerry Pournelle, email to the author, January 11, 2012.
58. See Kamau Brathwaite, *The Zea Mexican Diary: 7 Sept 1926–7 Sept 1986* (Madison: University of Wisconsin Press, 1993), 33, 73.
59. Kamau Brathwaite, "Dream Chad," in *Dream Stories* (New York: Longman, 1994), 50; emphasis in original. That the hurricane destroyed Brathwaite's personal archives in the damage it inflicted on his home in Jamaica is lore that has been repeated by some critics and scholars, but the text of the 1993 preface to "Dream Chad" suggests otherwise: "I found that somehow that part of the houm that had housed the Eagle & the archives etc had somehow miraculously survived the storm" (49). He goes on to describe the increasing urgency he felt to make arrangements to safeguard his archives, "as Yale (1991) had so generously invited me to do" (50). I myself do not know what contingencies may exist for Brathwaite's archives.
60. See Ignacio Infante, *After Translation: The Transfer and Circulation of Modern Poetics across the Atlantic* (New York: Fordham University Press,

2013), 163. Or as Carrie Noland succinctly puts it, "a soft mother and a hard drive": Noland, "Remediation and Diaspora: Kamau Brathwaite's Video Style," in *Diasporic Avant-Gardes: Experimental Poetics and Cultural Displacement,* ed. Carrie Noland and Barrett Watten (New York: Palgrave, 2009), 87.

61. Infante, *After Translation,* 169. Infante refers to this archival capability as a "virtual temporality in which any point in time can be retraced and accessed instantaneously" (170). Not incidentally—though the SE/30 itself long predates it—Apple's current backup system for its computers is named Time Machine.
62. The key document from Brathwaite in regard to these events is entitled "Cultural Lynching or how to dis. man. tle the artist"; it was transmitted in an email message dated September 23, 2011, to the British poet and artist Tom Raworth (see http://tomraworth.com/SCPCL.pdf.)
63. See AirBorne, "Kamau Brathwaite Disgraced Abroad and at Home, Where Is the Justice? Literary Icon of Barbados Reports Continuous Theft of Memoirs and Souvenirs—NY Police Ignore Claims," *Bajan Reporter,* March 16, 2010, http://www.bajanreporter.com/2010/03/kamau-brathwaite-disgraced-abroad-and-at-home-where-is-justice-literary-icon-of-barbados-reports-continuous-theft-of-memoirs-and-souvenirs-%E2%80%93-ny-police-ignore-claims/.
64. Brathwaite, "Cultural Lynching"; emphasis in original.
65. Timothy Reiss, email to the author, August 28, 2015.
66. Amy Tan (@AmyTan), tweet to the author (@mkirschenbaum), October 15, 2014.
67. See Charles R. Acland, ed., *Residual Media* (Minneapolis: University of Minnesota Press, 2007). Debates over whether to replace the floppy icon are a staple on tech blogs and discussion forums. Connor (Tomas) O'Brien considers the issue from the standpoint of graphic design and notes that the floppy's distinctive shape, with one beveled corner, contributes to the icon's being uniquely recognizable: see O'Brien, "In Defence of the Floppy Disk Save Symbol," Connortomas.com, April 2013, http://connortomas.com/2013/04/in-defence-of-the-floppy-disk-save-symbol/.
68. See Jeff Jarvis, "Goodbye CTRL-S," Medium.com, May 20, 2014, https://medium.com/change-objects/goodbye-ctrl-s-8f424e463dbe.
69. Jason Scott, "Floppy Disks: It's Too Late," *ASCII* (blog), July 12, 2011, http://ascii.textfiles.com/archives/3191.
70. The diskettes can also be viewed online: "Exhibition of Abe Kobo at Tyohu City," posted by Kato Koiti, *Horagai* (blog), January 22, 1998, http://www.horagai.com/www/abe/xtadu.htm.

71. These details are from Guzzardi's editor's note to *The Salmon of Doubt: Hitchhiking the Galaxy One Last Time* (New York: Del Rey Books, 2002), Kindle loc. 57. At least one of Adams's many Macs turned up in the wild after his death, reportedly auctioned on eBay (the computer was apparently used predominantly by his spouse). The provenance had been unknown at the time of the sale, but as is often the case the hard drive had not been completely wiped. The buyer noticed Adams's name associated with the registration of some of its software, and eventually found—and is presently keeping to himself—a unique draft of a TV sketch entitled "Brief Re-encounter." See "Douglas Adams' Mac IIfx," VintageMacWorld.com, last modified October 14, 2008, http://www.vintagemacworld.com/iifx.html.
72. Guzzardi, "Editor's Note," Kindle loc. 65. Guzzardi thus elected to produce what textual scholars would call an eclectic edition, eschewing fidelity to any one single historical instance of the work in favor of his considered estimation of the author's intentions.
73. David Foster Wallace, quoted in D. T. Max, *Every Love Story Is a Ghost Story: A Life of David Foster Wallace* (New York: Viking, 2012), 319n29.
74. According to D. T. Max, "The Unfinished," *New Yorker*, March 9, 2009, http://www.newyorker.com/magazine/2009/03/09/the-unfinished.
75. David Foster Wallace, *The Pale King* (New York: Little, Brown, 2011), vi.
76. Michael Pietsch, interview with the author, April 19, 2012.
77. Adam Bradley, interview with the author, June 5, 2012.
78. See Ralph Ellison, *Three Days before the Shooting . . .* , ed. John F. Callahan and Adam Bradley (New York: Modern Library, 2010). The editorial notes contain extensive commentary on the role of digital files in establishing the text and the unique editorial conundrums thus presented. One example: two separate files duplicate the same prose, some seven pages' worth, but the ending of the first is truncated midsentence and the beginning of the second has a different opening phrasing. "Likely this was an accident in formatting," Callahan and Bradley sensibly conclude. "It nonetheless presents a challenge to the editor wishing to present Ellison in his own words, even if those words are likely the product of an accident or oversight" (496). They thus elected to include the duplicate passage verbatim in their edition, further noting that in a subsequent printout of the two files Ellison made no indication they were redundant. According to Bradley (in conversation), the duplicate passage was almost stricken from the final volume by a copy editor.
79. See "Conspiracy Theories," *Dune 7 Blog, Way Back Machine* (online archive), December 16, 2005, http://web.archive.org/web/20071012125808/http://dunenovels.com/dune7blog/page21.html.

80. Ibid.
81. Adam Begley, *Updike* (New York: HarperCollins, 2014), Kindle loc. 99.
82. Chris Hall, "J. G. Ballard: Relics of a Red-Hot Mind," *The Guardian*, August 4, 2011, http://www.theguardian.com/books/2011/aug/04/j-g-ballard-relics-red-hot-mind.
83. I personally became aware of the "other" Updike archive shortly after returning the page proofs for my essay "Operating Systems of the Mind: Bibliography after Word Processing (The Example of Updike)," *Papers of the Bibliographical Society of America* 108, no. 4 (December 2014): 381–412, in which I discuss the importance of the then still "missing" Wang disks.
84. See *The Other John Updike Archive*, http://johnupdikearchive.com/.
85. Adrienne LaFrance, "The Man Who Made Off with John Updike's Trash," *The Atlantic*, August 28, 2014, http://www.theatlantic.com/features/archive/2014/08/the-man-who-made-off-with-john-updikes-trash/379213/.
86. Ibid.
87. As of this writing, Moran has not made the diskettes available for an attempt at data recovery.
88. For an introduction to relevant practices and debates, see Matthew Kirschenbaum, Richard Ovenden, and Gabriela Redwine, *Digital Forensics and Born-Digital Content in Cultural Heritage Collections* (Washington, DC: Council on Library and Information Resources, 2010), http://www.clir.org/pubs/reports/reports/pub149/pub149.pdf.
89. "Screening Process Is Paper Chase," *New York Times*, March 20, 1981, http://www.nytimes.com/1981/03/20/opinion/topics-screening-processis-paper-chase-paper-chase.html.
90. See Michael Hancher's perceptive essay, "*Littera Scripta Manet:* Blackstone and Electronic Text," *Studies in Bibliography* 54 (2001): 115–132; see also Kirschenbaum, *Mechanisms*, 57–58.
91. Daniel Chandler, "Are We Ready for Word-Processors?," *English in Australia* 79 (1987): 11–17.
92. See "Philip Roth Talks to David L. Ulin," *LA Times*, October 1, 2010, http://latimesblogs.latimes.com/jacketcopy/2010/10/philip-roth-talks-to-david-l-ulin.html; also William Landay, "How Writers Write: Philip Roth," *William Landay* (blog), http://www.williamlanday.com/2009/06/27/how-writers-write-philip-roth/.
93. See Donadio, "Literary Letters."
94. See T. S. Eliot, *The Waste Land: A Facsimile and Transcript of the Original Drafts* (New York: Harcourt, Brace, 1971); the definitive textual history of the poem has been reconstructed by Lawrence Rainey, *Revis-*

iting The Waste Land (New Haven, CT: Yale University Press, 2005); finally, see Hannah Sullivan's reading of the revisions, *The Work of Revision* (Cambridge, MA: Harvard University Press, 2013), 120–146.

95. One notable example is *The Mongoliad,* an ongoing historical fantasy saga cowritten by Neal Stephenson and six other authors. The group used Microsoft Word and Track Changes to share drafts and comments. See Cesar Torres, "How Swords, Track Changes, and Amazon Led to *The Mongoliad: Book Two,*" *Ars Technica,* October 14, 2012, http://arstechnica.com/gaming/2012/10/14/how-swords-track-changes-and-amazon-led-to-the-mongoliad-book-two/.
96. Umberto Eco and Jean-Philippe de Tonnac, *This Is Not the End of the Book,* trans. Polly McLean (London: Harvill Secker, 2011), 116–117.
97. Jerry Pournelle, "WORMS and Friends: Words for the Ages and Pictures for the Millions," *InfoWorld,* March 7, 1988, 41.
98. Ibid.
99. I am grateful to Naomi Nelson for furnishing me with the specifics of this example.
100. According to the author's note in *Machine Man* (New York: Vintage, 2011); Subversion according to a tweet from @MaxBarry, June 3, 2015.
101. See Max Barry, "Nuts and Bolts," *Max Barry* (blog), October 5, 2011, http://maxbarry.com/2011/10/05/news.html.
102. Vikram Chandra, *Geek Sublime: The Beauty of Code, the Code of Beauty* (Minneapolis: Graywolf Press, 2014), 134.
103. Ibid., 135.
104. Details in these two paragraphs are from the author's interview with William Loizeaux, April 23, 2012. The book is *Anna: A Daughter's Life* (New York: Arcade, 1993).
105. Dave Eggers, *The Circle* (New York: Knopf, 2013), 328.
106. Benjamin Moser, "In the Sontag Archives," *New Yorker,* January 30, 2014, http://www.newyorker.com/books/page-turner/in-the-sontag-archives.
107. See Thorsten Ries, "Hard Drive Philology: Analyzing the Writing Process on Thomas Kling's Archived Laptops," *Digital Humanities 2014* (Lausanne, Switzerland), http://dharchive.org/paper/DH2014/Paper-786.xml.

After Word Processing

1. Friedrich Kittler, *Gramophone, Film, Typewriter,* trans. Geoffrey Winthrop-Young (Stanford, CA: Stanford University Press, 1999), 214.
2. Kittler has an essay dating from this period whose title translates as "Literature and Literary Studies as Word Processing": "Literatur

und Literaturwissenschaft als Word Processing," in *Germanistik-Forschungsstand und Perspektiven,* ed. Georg Stötzel (Berlin: Walter de Gruyter, 1985), 410–419. Stefanie Harris reads it as symptomatic of Kittler's by-then general abandonment of hermeneutics: "Literature is just a particular means of data processing through the medium of print and the materiality of the letter." See Harris, *Mediating Modernity: German Literature and the "New" Media, 1895–1930* (University Park: Penn State University Press, 2009), 8. According to information provided to the author by Kittler's widow, Susanne Holl, he does not appear to have begun using a computer for regular word processing himself until perhaps as late as 1989 (email to the author May 1, 2015).

3. See Umberto Eco, "The Holy War: Mac vs. DOS," *Espresso,* September 30, 1994, http://jowett.web.cern.ch/jowett/EcoMACDOS.htm.
4. Thomas J. Bergin, "The Proliferation and Consolidation of Word Processing Software: 1985–1995," *IEEE Annals of the History of Computing* 28, no. 4 (October–December 2006): 60.
5. See http://www.williamlanday.com/2009/06/27/how-writers-write-philip-roth/ and http://latimesblogs.latimes.com/jacketcopy/2010/10/philip-roth-talks-to-david-l-ulin.html.
6. Michael Crichton customized a stylesheet for writing screenplays and sent a description of how he did it to Microsoft. (An undated copy is present in the Microsoft Archives in Bellevue, Washington.)
7. Peter Rinearson, interview with the author, May 4, 2012.
8. Charles Simonyi, interview with the author, March 13, 2012.
9. See Matthew Fuller, "It Looks Like You're Writing a Letter," in *Behind the Blip: Essays on the Culture of Software* (Brooklyn: Autonomedia, 2003), 137–165.
10. See Edward Mendelson, "Escape from Microsoft Word," *New York Review of Books,* October 21, 2014, http://www.nybooks.com/blogs/nyrblog/2014/oct/21/escape-microsoft-word/.
11. See Charlie Stross, "Why Microsoft Word Must Die," *Charlie's Diary* (blog), October 12, 2013, http://www.antipope.org/charlie/blog-static/2013/10/why-microsoft-word-must-die.html.
12. Ibid.
13. My research in the marketing studies in Microsoft's archives confirmed this.
14. Len Deighton, quoted in Ray Hammond, *The Writer and the Word Processor* (London: Coronet, 1984), 221.
15. A claim I am basing on my interviews with various contemporary writers.
16. And yet, it is true that Word remains the de facto standard for submitting manuscripts throughout the publishing industry. Inasmuch as a writer now

faces a plethora of options when first creating a text, at some point before submission it will likely have to be converted to a .doc file. Stross on this point: "But somehow, the major publishers have been browbeaten into believing that Word is the sine qua non of document production systems. They have warped and corrupted their production workflow into using Microsoft Word .doc files as their raw substrate, even though this is a file format ill-suited for editorial or typesetting chores" ("Why Microsoft Word Must Die").

17. See Charlie Stross, "Writing a Novel in Scrivener: Lessons Learned," *Charlie's Diary* (blog), July 11, 2012, http://www.antipope.org/charlie/blog-static/2012/07/writing-a-novel-in-scrivener-e.html.
18. See Francesco Cordella's interview with Keith Blount, "Scrivener and Me," *L'avventura è l'avventura,* May 2013, http://www.avventuretestuali.com/interviste/keith-blount-english/.
19. See "Testimonials for Scrivener," Literature and Latte, https://www.literatureandlatte.com/testimonials.php.
20. Virginia Heffernan, "An Interface of One's Own," *New York Times Magazine,* January 6, 2008, http://www.nytimes.com/2008/01/06/magazine/06wwln-medium-t.html.
21. Ibid.
22. The politics and etiquette of social media and the expectations placed upon writers therein and thereon has been a subject of increasing conversation and controversy in the literary world. For just one entry point to the discussion, see Anne Trubek, "Only the Literary Elite Can Afford Not to Tweet," *SFGate,* October 25, 2013, http://m.sfgate.com/opinion/article/Only-the-literary-elite-can-afford-not-to-tweet-4926874.php.
23. In this it may be inspired by a movie scene Fuller also references, from *Blown Away* (1994), the otherwise forgettable Jeff Bridges and Tommy Lee Jones vehicle in which a bomb is set to explode if a typist's pace on her word processor falls beneath a certain threshold.
24. Leo Benedictus, "Write or Die: The Software That Offers Struggling Authors a Simple Choice," *The Guardian,* October 7, 2014, http://www.theguardian.com/technology/shortcuts/2014/oct/07/write-or-die-software-for-struggling-authors-david-nicholls.
25. See Kevin Lipe, "Markdown is the New Word 5.1," *512 Pixels* (blog), May 16, 2011, http://www.512pixels.net/blog/2011/05/markdown-new-word51.
26. One of the best known of such experiments, Jennifer Egan's "Black Box," was serialized over Twitter for a two-week period in May 2012; it was then published in its entirety in the *New Yorker*'s June 4, 2012, issue. Egan initially composed it not on Twitter, however, but using a Japanese

notebook whose pages were divided into cells approximating the length of a tweet. See "Coming Soon: Jennifer Egan's 'Black Box,'" *New Yorker,* May 23, 2012, http://www.newyorker.com/books/page-turner/coming-soon-jennifer-egans-black-box.

27. See Norimitsu Onishi, "Thumbs Race as Japan's Best Sellers Go Cellular," *New York Times,* January 20, 2008, http://www.nytimes.com/2008/01/20/world/asia/20japan.html.
28. For one recent example of industry analysis, see Kurt Mackie, "Google Apps Making Inroads against Microsoft Office, Gartner Says," *Redmond Channel Partner,* April 23, 2013, http://rcpmag.com/articles/2013/04/23/google-apps-vs-microsoft-office.aspx.
29. Since 2010 Microsoft has offered similar functionality through its Office Online product.
30. See Silvia Hartmann's "The Naked Writing Project," last modified November 11, 2012, http://silviahartmann.com/live/. The novel that resulted was called *Dragon Lords:* http://dragonrising.com/store/dragon_lords/.
31. See Tom Hanks's op-ed, "I Am TOM. I Like to TYPE. Hear That?," *New York Times,* August 4, 2013, http://www.nytimes.com/2013/08/04/opinion/sunday/i-am-tom-i-like-to-type-hear-that.html.
32. See Joanna Stern, "Handwriting Isn't Dead—Smart Pens and Styluses Are Saving It," *Wall Street Journal* (February 10, 2015), http://www.wsj.com/articles/handwriting-isnt-deadsmart-pens-and-styluses-are-saving-it-1423594704.
33. See John Brownlee, "Microsoft Research Invents a Stylus That Can Read Your Mind," *Design, Fast Company,* October 10, 2014, http://www.fastcodesign.com/3036931/microsoft-research-invents-a-stylus-that-can-read-your-mind?partner=rss.
34. See "Hemingwrite—A Distraction Free Smart Typewriter," Kickstarter.com, accessed August 19, 2015, https://www.kickstarter.com/projects/adamleeb/hemingwrite-a-distraction-free-digital-typewriter.
35. Ibid.
36. For example, Peter Swirski, *From Literature to Biterature* (Montreal: McGill-Queen's University Press, 2013).
37. For this last, see Shelley Poldony, "If an Algorithm Wrote This, How Would You Even Know?," *New York Times,* March 7, 2015, http://www.nytimes.com/2015/03/08/opinion/sunday/if-an-algorithm-wrote-this-how-would-you-even-know.html. This story became a minor sensation online for its quiz asking readers to guess whether sample passages were written by computers or humans.
38. See Steven Johnson, "Tool for Thought," *New York Times,* January 30, 2005, http://www.nytimes.com/2005/01/30/books/review/tool-for-thought.html.

39. Richard Polt documents what he characterizes as a contemporary "typewriter insurgency": "Our movement isn't nostalgic, much as we love our objects from a bygone world. We insurgents are pointing a way forward to greater focus, creativity and independence. Typewriters are one way to keep the digital in perspective and keep ourselves free." See Polt, *The Typewriter Revolution: A Typist's Companion for the 21st Century* (New York: Countryman Press, 2015).
40. In "developing" countries the technology of choice is not dial-up but cellular.
41. There is no evidence to suggest Hersey wrote his next novel, *The Walnut Door* (1977), on a computer system.
42. On screen essentialism, see Nick Montfort, "Continuous Paper: The Early Materiality and Workings of Electronic Literature" (2005), Nickm.com, accessed August 19, 2015, http://nickm.com/writing/essays/continuous_paper_mla.html.
43. Personal correspondence from Joyce to Howard Becker, January 7, 1982, The Michael Joyce Papers, Harry Ransom Center, University of Texas at Austin.
44. Please do so at http://trackchangesbook.info; see the Author's Note.

CREDITS

1. MT/ST prototype, 1957. Image courtesy of Leon Cooper. Reprint courtesy of International Business Machines Corporation, © (1957) International Business Machines Corporation.

2. Len Deighton at work on *Bomber.* Photograph by Adrian Flowers. Used with permission.

3. Ellenor Handley at work on the MT/ST. Photograph by Adrian Flowers. Used with permission.

4. "Think Tape." Reprint courtesy of International Business Machines Corporation, © (1964) International Business Machines Corporation.

5. "Death of the Dead-End Secretary" advertisement. Redactron Corporation. *New York* introduces *Ms. The New Magazine for Women* (December 20, 1971), p. 21.

6. Note from Nilo Lindgren to Larry Tesler. Image courtesy of Larry Tesler. Used with permission.

7. John Hersey's *My Petition for More Space.* Image courtesy of Brook Hersey and the Beinecke Rare Book and Manuscript Library, Yale University. Used with permission.

8. John Hersey's notebook. Image courtesy of Brook Hersey and the Beinecke Rare Book and Manuscript Library, Yale University. Used with permission.

9. The 1979 debut issue of *onComputing* magazine Vol. 1, No. 1, Summer 1979 (Peterborough, NH: onComputing, Inc.). Cover illustration by Robert Tinney.

10. Illustration from a 1982 Perfect Writer software manual. Perfect Software, Inc.

11. Stanley Elkin at the "Bubble Machine." Photograph by Lauren Chapin. Image courtesy of Joan Elkin. Used with permission.

12. Caricature drawing by Jack Gaughan from "Basic Genesis" by Barry B. Longyear and Jerry Pournelle in *Isaac Asimov's Science Fiction Magazine* Vol. 5, No. 8, August 3, 1981 (New York: Davis Publications, Inc.), p. 43.

13. *The McWilliams II Word Processor Instruction Manual* by Peter A. McWilliams (Los Angeles: Prelude Press, 1983), cover.

14. Stephen King with his Wang System 5 Model 3 word processor, from "I Sleep with the Lights On," directed by Henry Nevison, University of Maine at Orono, Public Information Office, in association with WABI-TV, Bangor, 1982. Image courtesy of Henry Nevison Productions. Used with permission.

15. "The Smartest Way to Write," advertisement for Tandy Radio Shack's TRS-80. *BYTE* Vol. 8, No. 1, January 1983 (Peterborough, NH: *BYTE* Publications, Inc., McGraw-Hill Publications Company).

16. John Updike in 1987. Copyright © Nancy Crampton. Used with permission.

17. John Updike, "INVALID.KEYSTROKE," draft. The John Updike Papers (2527), Houghton Library, Harvard University. Copyright © 1984 by John Updike, used by permission of the Wylie Agency LLC.

18. Ralph Ellison in 1986. Copyright © Nancy Crampton. Used with permission.

19. Eve Kosofsky Sedgwick at work. Photographer unknown. Image courtesy of Hal Sedgwick.

20. R. Crumb, "Old writer . . . leers into computer." Copyright © Robert Crumb, 1995. Used with permission.

21. One of John Updike's "lost" Wang computer diskettes. Image courtesy of Paul Moran, Curator, The Other John Updike Archive. Used with permission.

22. Ellenor Handley, 2013. Courtesy of Edward Milward-Oliver. Used with permission.

ACKNOWLEDGMENTS

Track Changes was an unusual book to write in that it has been in the public eye almost from its inception. Very early on in my research, December 2011 to be exact, I was invited to give a lecture at the New York Public Library. With some glee I titled it "Stephen King's Wang," a line I got from William Gibson (but King himself was the first to make all the jokes). A reporter from the *New York Times* was in the audience, and afterward a story appeared in the paper about an English professor's interest in the literary history of word processing. There was a publicity cycle and soon my inbox was flooded with tips, anecdotes, contacts, and suggestions that I never would have come by any other way. In retrospect, even though many who read that article undoubtedly feel like they have been waiting for the book for a very long time, it could not have been written without that initial high-profile burst of visibility. I am therefore enormously indebted to Doug Reside for the invitation to speak at the NYPL on that occasion, and to Jennifer Schuessler at the *Times*.

My initial research and writing was supported by a 2011 Fellowship from the John Simon Guggenheim Memorial Foundation. I hope what is in these pages fulfills the Foundation's trust in me. My work has also been generously supported by the University of Maryland's Department of English. I am grateful to Don Fehr and the Trident Media Group for my representation.

John Kulka at Harvard University Press has been unflaggingly enthusiastic about this project from its inception. He has also been patient, present, and genuinely critical when called for, not least in the course of his rigorous editing of the entire manuscript—a rare gift for any author to receive. Jacques Plante at the University of Maryland brought her keen eye for detail to the crucial tasks of proofing and formatting the manuscript and its citations, and Wendy Nelson proved an expert copy editor. Jonah Furman assiduously compiled the index. Joy

Deng, Kate Brick, and others at Harvard University Press, as well as Kim Giambattisto at Westchester Publishing Services, have also been part of an exceptional team to work with, and have my thanks. All of these unseen hands have improved the book on every page.

Several individuals extended themselves to facilitate access to archives: Donald Brinkman, Lee Dirks (RIP), and Amy Stevenson at Microsoft; Peggy Kidwell at the Smithsonian; Leslie Morris and Melanie Wisner at the Houghton; and Joel Minor at Washington University in St. Louis. Others helped furnish me with primary source documents essential to my research: for this I thank Larry Bond, Leon Cooper, Joan Elkin, Lori Emerson, Thomas Haigh, Till Heilmann, Sam Kalow, Lawrence Krakauer, Edward Milward-Oliver, Paul Moran, Stephen Olsen, Gabriela Redwine, Jason Scott, Hal Sedgwick, Steve Soboroff, and Larry Tesler. Eric Cartier and his staff in the University of Maryland Libraries' Digital Conversion and Media Reformatting department assisted with image scanning and preparation.

For interviews, conversations, personal emails, or other personal communications I am deeply indebted to Patricia Freed Ackerman, Roger Angell, Mary Jo Bang, Adam Begley, Evelyn Berezin, Tim Bergin, Edwin Black, Sarah Blake, Adam Bradley, David Brin, Maud Casey, Michael Chabon, Leon Cooper, Kathryn Cramer, John F. Cunningham, Andy van Dam, Marsha DeFilippo, Len Deighton, Michael Dirda, Joan Elkin, James Fallows, Sheila Finch, Robert Foothorap, William Gibson, Kenneth Goldsmith, Eileen Gunn, Ellenor Handley, David Hartwell, Susanne Holl, William Loizeaux, Barry Longyear, Bonnie MacBird, Bradford Morgan, Ted Nelson, Larry Niven, Michael Ondaatje, Michael Pietsch, Jerry Pournelle, Richard Powers, Timothy Reiss, Peter Rinearson, Seymour Rubinstein, Jeanne Sheldon, Charles Simonyi, Steve Soboroff, Bruce Sterling, Peter Straub, Don Swaim, Amy Tan, Larry Tesler, Miranda Updike, Peter Weiner, and Stuart Woods. Mention of their names here should imply absolutely no endorsement or prior review of my work: I merely wish to publicly thank some very busy people for being generous (often very generous) with their time and recollections.

Jane Donaworth, Lori Emerson, Neil Fraistat, Alan Galey, Lisa Gitelman, Thomas Haigh, Lee Konstantinou, Kari Kraus, Elizabeth Losh, and Clive Thompson all read and commented on various chapters, for which they have my deepest gratitude. Edward Milward-Oliver and Darren Wershler each deserve special mention: for reading chapters, but also for their interest and generous assistance throughout the entirety of my research. The readers for Harvard University Press likewise offered valuable critique and comments, which improved the whole book significantly. Pat Harrigan selflessly shouldered more than his share of another project to help me finish this one.

In addition—for advice, encouragement, information, and kindnesses of all kinds—I am grateful to colleagues and friends near and often far: Crystal Alberts,

Mark Amerika, James Ascher, Matthew Battles, Paul Benzon, Jed Birmingham, Brett Bobley, Alison Booth, Suzanne Bost, Gerry Canavan, Tim Carmody, Eric Cartier, Paul Ceruzzi, Daniel Chandler, Alex Chassanoff, Tita Chico, Tanya Clement, Bill Cohen, Matt Cohen, Linda Coleman, Michael Collier, Ryan Cordell, Florian Cramer, Brian Croxall, Tyler Curtain, Gabrielle Dean, Seth Denbo, Johanna Drucker, Morris and Georgia Eaves, Jason Farman, Andrew Ferguson, Kevin Ferguson, Ben Fino-Radin, Kathleen Fitzpatrick, Leo Flores, Amanda French, Matthew Fuller, Chris Funkhouser, Mike Furlough, Patricia Galloway, Oliver Gaycken, Alex Gil, Matt Gold, Ben Goldman, David Greetham, Stephen Gregg, Ray Guins, Barbara Heritage, Jennifer Howard, Allison Hughes, Jon Ippolito, Jeremy Leighton John, Andrew Johnston, Steven E. Jones, Nathan Kelber, Cal Lee, Bob Levine, Amanda Licastro, Tan Lin, Alan Liu, Henry Lowood, Jerome McDonough, Jerome McGann, Lev Manovich, Mark Marino, Mark Matienzo, Shannon Mattern, John Maxwell, Edward Mendelson, Trevor Muñoz, James Neal, Naomi Nelson, Bethany Nowviskie, Porter Olsen, Trevor Owens, Jussi Parikka, Carla Peterson, Richard Polt, Jessica Pressman, Leah Price, Rita Raley, Steve Ramsay, Sangeeta Ray, Gabriela Redwine, Jason Rhody, Lisa Rhody, Brian Richardson, Ben Robertson, Kellie Robertson, Mark Sample, Jentery Sayers, Matt Schneider, Nathan Schneider, Jason Scott, Stéfan Sinclair, Dag Spicer, Jonathan Sterne, Ted Striphas, Michael Suarez, Erin Templeton, Melissa Terras, Ted Underwood, John Unsworth, Siva Vaidhyanathan, Joe Viscomi, Amanda Visconti, Christina Walter, Orrin Wang, Ken Wark, Ethan Wattral, Joshua Weiner, Sarah Werner, Roger Whitson, George Williams, Kam Woods, and Glen Worthey. (It's a long list, but it's been a long time coming.) And while everyone else here was helping me get to 135K (words), Jonathan Hill got me to my first 5K. Members of my loving family asked when the book was going to be done only slightly more often than any other constituency—but had a larger and longer role than most any other, as I hope my dedication helps acknowledge. Finally, always and above all: my love and thanks again to my life companion, Kari Kraus.

In addition to the New York Public Library, material from this book has been presented in talks and lectures at the University of Toronto, the University of Western Ontario, Yale University, the University of Colorado Boulder, Ghent University, Loyola University Chicago, Northeastern University, the University of Texas, McGill University, the CUNY Graduate Center, Harvard University, the Bibliographical Society of America, Georgetown University, Penn State University, York College, the University of North Dakota, the University of Virginia, Washington University in St. Louis, the University of Coimbra, the National University of Ireland Galway, and the Archives Education and Research Institute. Every one of these occasions furnished me with additional feedback and further information, and I am grateful to the audiences, to the organizers, and for the honor of the invitations.

I first wrote about Len Deighton and Ellenor Handley's story in *Slate* magazine as "The Book-Writing Machine," March 1, 2013. A much more detailed and technical discussion of John Updike's word processing (which I draw upon in Chapters 4 and 10), can be found in my "Operating Systems of the Mind: Bibliography after Word Processing (the Example of Updike)," *Papers of the Bibliographical Association of America* 108, no. 4 (December 2014): 380–412. Portions of Chapter 10 are also informed by my essay "The .txtual Condition," *Digital Humanities Quarterly* 7, no. 1 (2013), abridged and reprinted in *Comparative Textual Media: Transforming the Humanities in the Postprint Era,* ed. N. Katherine Hayles and Jessica Pressman (Minneapolis: University of Minnesota Press, 2013). Likewise, I expand on ideas presented in "Tracking the Changes: Textual Scholarship and the Challenge of the Born Digital," co-authored with Doug Reside for *The Cambridge Companion to Textual Scholarship,* ed. Neil Fraistat and Julia Flanders (Cambridge: Cambridge University Press, 2013). A few sentences (and only a few) from one or two other things I have written have found their way in as well, as is the way with word processing. My epigraph from *A Star Shines over Mt. Morris Park* appears by kind permission of the Henry Roth Literary Trust.

For many of us who came of age with personal computers, word processing was the first real "killer app," the first hands-on experience that made owning a computer seem as indispensable as it was inevitable. I've heard hundreds of those stories over the last few years, and their patterns inform my work here even if the particulars are not always narrated. So last, I am grateful to everyone who has taken the time to tell me about their first word processor.

INDEX